Cinemas of the Black Diaspora

Contemporary Film and Television Series

A complete listing of the books in this series can be found at the back of this volume.

Cinemas of the Black Diaspora

Diversity, Dependence, and Oppositionality

WAYNE STATE UNIVERSITY PRESS • DETROIT

 Manufactured in the United States of America.
99 98 5 4 3 2

Library of Congress Cataloging-in-Publication Data

Cinemas of the Black diaspora : diversity, dependence, and oppositionality /
edited by Michael T. Martin.
p. cm.—(Contemporary film and television series)
Includes bibliographical references and index.
ISBN 0-8143-2587-4 (alk. paper).—ISBN 0-8143-2588-2 (pbk. : alk.
paper)
1. Blacks in the motion picture industry. 2. Blacks in motion pictures.
3. Afro-Americans in the motion picture industry. 4. Afro-Americans
in motion pictures. 5. Motion pictures—Africa. I. Martin, Michael T.
II. Series.
PN1995.9.N4C58 1995
284′.8′08996—dc20 95-31383

Text Designer: Elizabeth Pilon
Jacket and Cover Designer: Arthur Chartow

To my daughter

PILAR

[T]he history of the world is the history, not of individuals, but of groups, not of nations, but of races, and he who ignores or seeks to override the race idea in human history ignores and overrides the central thought of all history.

W.E.B. DuBois

"The Conservation of the Races," 1897

As I write these lines, modern aviation is transforming the world, bringing about a complete revolution (of which we have not yet chosen to take cognizance). . . . The conclusion to be drawn from it is this: that everything is becoming—indeed, has already become—interdependent.

Le Corbusier

The Modulor, 1951

In our time the photographic image of underdevelopment represents a worldwide phenomenon: it exists in the visual and historical consciousness of the world. It is what we dream. . . . It is the expression of our fears and desires. . . . And it is also what people imagine they could be when they have ideals.

Edmundo Desnoes

"The Photographic Image of Underdevelopment," 1967

Contents

Preface

This volume provides a global survey of the plurality of cinematic traditions and film practices in the black diaspora. It revisits and elaborates several of the themes addressed in the anthology, *Black Frames: Critical Perspectives on Black Independent Cinema,* edited by Mbye B. Cham and Claire Andrade-Watkins. Taking up their call for a theorized and alternative reading of filmmaking practices, the essays in the present book are devoted to the study of the politics of filmmaking in the black diaspora, and its aesthetic concerns and interrogations of colonialism and postcoloniality, culture and identity, exile and displacement.

The category "black" is used in this book largely to identify people of African descent, and not as an inclusive political term for a constellation of "minority" groups of color who perceive shared interests.

Together, the essays constitute a dialogue and debate among film scholars, film critics, and filmmakers from Europe, North America, and the Third World. The subjects under study and how they are conceptualized by the contributors foreground the conjunction of film practices and ideological presuppositions in black filmmaking. They also invite larger cultural debates about "authenticity" and national identity, and stand in counterpoint to the assumptions of colonialist and ethnocentrist discourses about Third World, Hollywood, and European cinemas.

This book has two principal goals. The first is to serve as a basic reference and research tool for the study of world cinema in general and black diasporic cinemas in particular. While the volume of research and the frequency of publications on Third World, including black, cinema has increased in recent years, it is largely dispersed among obscure film, cultural, and ethnic studies journals unavailable in the collections of many university libraries. Having a single volume of essential writings, readers can avail themselves of a useful resource for the study of filmmaking in the black diaspora in its global context; and works of enduring value are rendered more accessible to faculty and students in film studies courses.

The second goal of this book is to account for the diversity, innovativeness, and fecundity of black filmmaking in the diaspora, explicate its historical importance as a cultural form and political practice, and insinuate readers in a reflexive dialog with black filmmakers.

Finally, I should say that the perspectives represented in this collection hardly close the book on this topic. As black independent (and commercial) filmmaking emerges in other geographic sites and matures where its traditions and practices are established so, too, will our considerations of it deepen.

M.T.M.

Acknowledgments

During the development of this project it was my good fortune to have benefitted from individuals and organizations to whom I am indebted for encouragement, assistance and support. They include: E. Hope Anderson, Filippe Sawadogo, Director of FESPACO Festival, Claude Tarpilga, documentalist of FEPACI, Françoise Pfaff, Hamid Naficy, Lamont Yeakey, Claire Andrade-Watkins, Robert Crusz, editor of *Framework,* Dan Georgakas, editor of *Cineaste,* Lorna Johnson and Ada Gay Griffin of *Third World Newsreel,* Mbye Cham, Jackie Jones, former editor of *Black Film Review.* A special thanks to Chuck Kleinhans, co-editor of *Jump Cut,* whose review of the proposed project and suggestions about its reorganization were most helpful, and to Robert Burgoyne for his counsel and assistance in the review process.

To Adrienne Gregg and Claudia Willis of the Department of Africana Studies at Wayne State University I am grateful for their patience in expediting my numerous communications with authors in the preparation of the manuscript. Louise Jefferson for her translation, on short notice, of sections of the *Niamey Manifesto.* Michael Goldfield and Kathryne Lindberg whose editorial comments were especially helpful in the final draft of the introductory essay.

Arthur Evans, Lynn H. Trease, Miriam Jones, and Alice Nigoghosian of Wayne State University Press were consummate publisher, editor, copyeditor, and production design manager, respectively. Together, they reduced the stress I endured in preparing the manuscript and rendered the experience instructive.

For financial support to complete this project, I am grateful to Garrett Heberlein, Vice President for Research and Dean of the Graduate School, and to the Office of Research and Sponsored Programs at Wayne State University.

Framing the "Black" in Black Diasporic Cinemas

Michael T. Martin

One cannot understand black diasporic cinemas without understanding them as a part of Third World and minoritarian cinema. What constitutes Third World cinema and the Third World, frames and historicizes the subject of this volume on black filmmaking as an oppositional and cultural practice. The reasons for this should be clear. The culturally diverse populations of the black diaspora are largely concentrated in the Third World, primarily in Africa and the Americas, and they share a common history of colonialism with other Third World population groups. Thus, I begin with a discussion of the ambiguous categories "Third World cinema" and "Third World." The meaning of the two terms and their application to temporal, spatial and cultural processes is the subject of study and debate among both scholars and filmmakers. The categories denote historical and contemporary relationships, ideological and cultural presuppositions, and strategies and processes of development.

Further, the thematic concerns about culture, identity, history and social change, and the strategies and practices of black filmmaking in the diaspora, reflect and embody the aspirations and practices of Third World cinema, of which it is a part, especially the politically-oriented militant "Third Cinema." More generally though, black filmmaking in the diaspora bridges the North and South, as well as, the West and East. Practiced in the First and Third worlds, it draws on both Western and non-Western sources of culture for its stylistic and aesthetic forms.

Notwithstanding its innovativeness, cinematic traditions, and historical importance, Third World cinema was largely ignored by film scholars and film critics in Europe and North America until the 1960s. This is the case despite the fact that Argentina, India and Brazil, among other countries, had prolific and culturally distinct film movements during the first half of this century.[1]

In recent years, narrative and documentary productions by Third World filmmakers have attained recognition among Western audiences and circulation through film festivals, art houses, occasional broadcast on cable networks, and screenings in film courses. Concomitantly, scholars no longer able to ignore the cultural and political significance of Third World cinema, have generated a substantial, critical corpus of research that is increasingly featured in film and cultural studies journals, anthologies and monographs.[2]

I

Distinguished by national, cultural, linguistic, racial and ethnic categories, Third World cinema is informed by several fundamental and dislocating factors. First, the colonial process and the continuing neocolonial status of the countries in which Third World filmmaking was developed; second, the enduring impact of Western culture; and third, the determinacy of capitalist production and distribution practices.[3] Ella Shohat and Robert Stam, in a recent and important study of "multicultural media," delineate four tentative designations for Third World cinema:

> 1. A core circle of "Third Worldist" films produced by and for Third World people (no matter where those people happen to be) and adhering to the principles of "Third Cinema";
>
> 2. a wider circle of the cinematic productions of Third World peoples . . . , whether or not the films adhere to the principles of Third Cinema and irrespective of the period of their making;
>
> 3. another circle consisting of films made by First or Second World people in support of Third World peoples and adhering to the principles of Third Cinema;
>
> 4. a final circle, somewhat anomalous in status, at once "inside" and "outside," comprising recent diasporic hybrid films . . . , which both build on and interrogate the conventions of "Third Cinema."[4]

In the first, third and fourth denotations, Third World cinema is predicated on a relational concept of the Third World as political and historical categories, while in the second, the location of film production in the Third World is the defining category. Drawing on the humanistic impulse in the writings of Fidel Castro, about the role of art in the revolutionary process, Clyde Taylor has defined Third World cinema as a reflexive political and transnational project.

> Films become Third World, in short, by their function, once made, "by their usefulness for the people," as Fidel Castro said, "by what

> they contribute to man [and woman], by what they contribute to the liberation and happiness of man [and woman]."
>
> Yet even though the Third World is a mental state for which no one holds an official passport, it would be wrong to emphasize Third World cinema's local and national preoccupations at the expense of its resolute internationalism. The making of *O Povo Organizado* in Mozambique by Bob Van Lierop, an African American, or of *Sambizanga* (1972), about Angola, by Guadaloupian Sara Maldoror, or the Ethiopian Haile Gerima's *Bush Mama* (1975), set in Los Angeles, or Gillo Pontecorvo's *The Battle of Algiers* (1965), or the several Latin American and African films created by Cubans, or the many Third World films made by Europeans and white Americans—all suggest the cross-fertilization of an embryonic transnational Third World cinema movement.[5]

In theorizing the concept of black cinema, Tommy Lott has drawn on the political notion of Third World cinema. He argues that "What makes Third Cinema third (i.e., a viable alternative to Western cinema) is not exclusively the racial make-up of a filmmaker, a film's aesthetic character, or a film's intended audience, but rather a film's political orientation within the hegemonic structures of post-colonialism."[6]

For the purposes of this project and its concern with the politics and repertoire of black film practices, and in consideration of the classificatory demarcations of Shohat and Stam, and Taylor's and Lott's complementary emphases on the political orientation and transnational features of Third World cinema, I delineate (not define) black diasporic—black African, black British, Afro-Caribbean/Latin, black American—cinemas in broad terms:[7] First, as a filmmaking process—independent and commercial—concerned with black themes in which people of African ancestry participate as screen writers, directors and/or producers. And second, as a political orientation and oppositional cultural practice.

Correspondingly, in consideration of the second problematic term, Stam among others has defined the "Third World" as "the colonized, neo-colonized or de-colonized nations of the world whose economic and political structures have been shaped and deformed within the colonial process" by European imperial nations and the United States.[8] "Structural domination" is the key operative concept of the colonial relation and central to the definition of the Third World. In their recent work, Shohat and Stam have appended the category of "minorities" to their definition of Third World to account for the communities of Third World peoples who reside in the First World.[9] By doing so, the concept of the Third World is rendered mobile although contingent on specific historical conditions. Further, the authors contend that cultural, racial and geographical categories are imprecise approximations of the Third World situation. While these categories are ambiguous, especially as a *sine qua non* of a colonial or neocolonial situation, at the concrete level, the application of "Third World" must be against instances of historical specificity that include racial and cultural, and to a lesser extent geographical, distinctions.

II

The black diaspora exists over time and space, and is a historical formation of the capitalist world-system.[10] It is dynamic, plastic and transnational, intersecting First and Third worlds, across Africa, the Americas, and Asia to the metropolises of western Europe and North America. Migration and displacement, social oppression and resistance are among its key features. In defining the black diaspora, Ruth Hamilton contends that it represents:

> a type of social grouping characterized by a historical patterning of particular social relationships and experiences. As a social formation, it is conceptualized as a global aggregate of actors and subpopulations, differentiated in social and geographical space, yet exhibiting a commonality based on historical factors, conditioned by and within the world ordering system.[11]

Along with other diasporic population groups, whose descendants are from the Third World, the global dispersion of people of African descent is rooted in the colonial process.[12] However, unlike other colonized groups, Africans were enslaved for nearly four centuries and on a magnitude unprecedented in human history. This largely forced scattering of African descended peoples in other geographical sites, linked by a common social condition under colonialism and slavery, is the defining historical antecedent of the contemporary black diaspora.[13]

However tentative and problematic the category, race is also a defining, though historically contingent, feature of the black diaspora that distinguishes it from other, mostly cultural, religious or national diasporic populations. I do not intend to essentialize race, but rather to locate it in historical processes in which it is a determining social fact. Neither am I suggesting that "a set of essential physical and cultural traits, which emerged at a distant point in the past, have been preserved—unchanged in form, substance or meaning—by people of African descent wherever they may be found."[14]

European Expansion and Racism

While race is not a fixed feature of colonialism, it is a strategy of great consequence in the colonial experiences of African peoples, and it informs contemporary race relations and the social experience of black life in the diaspora. Any discussion then of race demands specificity because of the category's ambiguous definition and usage.

Racial antagonism developed in the modern world with the rise of capitalism and nationalism in 15th-century Europe, and later in North America.[15] Capitalist development has depended on the large-scale and international migration of people and population groups in the world-system. Significant international migrations that occurred during the period of capitalism's formation and early development include those from England to the Caribbean in the seventeenth century and to Australia and South Africa in the nineteenth century; from Africa to the Caribbean in the eighteenth century; and from India and China to the Caribbean and South Africa in the nineteenth century.[16] In each case, the migrant and indigenous populations were differentially incorporated into the colonial system under varying forms of capitalist relations, resulting in a division of labor in which racism became a social relation of production in capitalist development. By racism I mean "an ideological element of signification by which to *select* and to *legitimate* the selection of, a particular population," whose labor is exploited in specific production relations.[17]

During the formative period of capitalism's development,[18] four interrelated events occurred: (1) the European oceanic exploration of the coast of west Africa and later, the Americas; (2) the rise and expansion of protocapitalist centers in northern Europe which displaced feudal production in that region; (3) the consolidation of the "Bourgeois" revolution in England and the Low Lands of western Europe;[19] and (4) the development of the plantation system in the Americas, especially in Brazil, and the corresponding expansion of the slave trade and use of forced labor in plantation production.

Plunder, colonialism and slavery followed European exploration. Under the imperatives of the capitalist mode of production, colonial enterprise was above all else about profit. This was the case, whether through trade, plantation agriculture, gold and silver mining, wage or forced labor. According to J. M. Blaut, colonialism (and slavery and the slave trade from the mid-sixteenth century forward) "initiated the transformations in economy and society [in Europe] but which could not have done so without the continuing impact of colonialism throughout the period 1492–1688."[20] Colonialism removed the constraints on capitalist development and freed the bourgeoisie to seize political power in western Europe.

Conquest, enslavement and colonization were the basis of the distinctly Eurocentrist conception and mythological reconstruction of race, culture and world history. Though not fully articulated in colonialist discourses until the second half of the nineteenth century, as the essential legitimation of imperial policies and practices, blacks, for example, were dispossessed of the virtues Europeans appropriated as uniquely their own. The corollary of this dehistoricized ideological deformation was that blacks were unable to evolve and "progress," from a traditional and unchanging world, in the absence of the European. This culturally anti-universalist doctrine, simplified here, was a central feature and *raison d'être* of the Euro-hegemony.[21]

This stage in the development of capitalism is without precedent in modern history. Cross-cultural and racial encounters occurred on a world-scale, in which the outcomes were a *sine qua non* for the formation and development of capitalism as a global system. Racial differences were not circumstantial. They were decisive to the formation of capitalism, precisely because racial and cultural distinctions were the bases upon which to establish the more specific social legitimization of colonialism and slavery. The following explication clarifies the global impact and scope of the European expansion:

> Certainly expansion and the creation of empire were not unusual in history. Nor was the looting of lands and the enslavement of peoples violently incorporated into these empires unprecedented. Yet some unique features, associated with the expansion of Western Europe, laid the foundation for the system of modern race relations. First, the area that Western Europe came to control was greater than any area previously conquered. The influence of the Western European empires was more extensive than that of any previous empires, including the Roman Empire. . . . Third, through either formal political control or informal economic dominance, the structural evolution of most of the non-European world was shaped by Western Europe for several hundred years. . . . Finally, and perhaps most significantly, the skin color of Western Europeans differed from that of most of the peoples they conquered. Thus, the world stratification system created by the new empires was affected by racial differences, for Western Europe was a white world expanding into a non-white world. . . . Although skin color distinctions have served in other parts of the world as the basis of social differentiation, they have not had the determinate impact on world history and the lives of so many people as the Western European expansion had between 1400 and 1900.[22]

The fateful first encounters in the late fifteenth century were the harbingers of what would, several centuries later, become the global antinomy between First and Third worlds. Further, they anticipated the enslavement of a "race" of peoples and the forcible migration and displacement of other population groups, the destruction of sovereign nations, the formation of new states under colonial rule, and the creation and propagation of false cultural claims and representations that persist to this day. Hamilton, summarizes the enduring impact of race on the collective identities of African-descended peoples:

> The dislocation of millions of Africans—their uprooting and transportation to various parts of the world—significantly changed their identity and the sense of peoplehood that had been theirs in various African societies. From the inception of the diaspora to the present, race became a central defining factor. For this reason, the primary experience of being defined in racial terms (and as an inferior race) is pertinent to the people-formation process. Racial definitions have constituted a fundamental reality imposed upon the African diaspora peoples, and have informed their fate within a racially divisive system. In response to such a system and in terms of their own efforts to survive and develop as a group, it is virtually impossible for peoples of the black diaspora to avoid being conscious of race. Therefore, the race dimension must be examined as a prime factor in the group's own social formation and in the development of their sense of peopleness. Thus, populations such as the black diaspora can become, under particular historically conditioned circumstances, a distinctive people.[23]

III

The present moment is distinguished by three significant and contingent developments that further contextualize the study of black diasporic cinemas. First, as capitalism expanded on a world-scale, during the late fifteenth to nineteenth centuries, its development, as noted earlier, has been associated with and dependent on large-scale international migrations of people from one geographical site to another in the world-system.

Late twentieth-century international migration is markedly different from earlier migrations in magnitude, composition, pervasiveness and, arguably, trajectory toward the First World. International migration has significantly grown as the global capitalist economy has expanded. The composition of the migratory populations is more differentiated than in earlier waves by type—labor, refugee and settlement. A case in point illustrative of the magnitude of international migration is the number of refugees in the world. Estimated at 49 million in 1994, nearly 23 million refugees have crossed international borders compared to 2.4 million in 1974.[24] One observer has characterized ours as the "century of the displaced person," while another has referred to Africa as a "continent of refugees."

Along with the refugees from ethnic conflicts, civil wars and repression, migration is linked to industrialization, changing labor markets, population growth, and the cumulative effects of the continuing economic deterioration of the Third World.[25] Declining economic conditions have put enormous pressure on some governments in the Third World to adopt policies that encourage the outward migration of populations to those areas of the world economy where there is demand for low-wage labor. For example, one-half of the Jamaican population migrated to the United States (and Britain) during the postwar period,[26] where they have established permanent communities.

The second factor is population growth. The world's population more than doubled from 2.5 billion in 1950 to 5.7 billion in 1994. By the year 2025, the global population is projected to increase to 8.5 billion, and then further climb to 11 billion in 2100, before stabilizing at 11.6 billion between 2150 and 2200.[27] Ninety-five percent of the population growth will be in the Third World, largely in Asia and Africa. And by 2025, 84 percent of the world's population will reside in the Third World. Population growth has been associated with the massive and rapid rise in Third World urbanization. In the next 20 years three billion people will live in cities in the Third World, and by the year 2015, more people will be living in urban than in rural areas in the Third World. Rapid urbanization, on the scale we are today witnessing, is creating mega-cities in which traditional cultures and ethnic affiliations are increasingly stressed, reconstituted and displaced by poverty, unemployment and the growing inequality between social classes.

The third condition is a result of the historical and contemporary processes described above in the ubiquitous, though uneven, development of capitalism as a global system. The colonial process and slavery, of which international migration and settlement are a part, and population growth and urbanization, especially but not only in the Third World, during the second half of this century, have forged distinct multi-racial, multi-ethnic and multi-cultural formations in the world. The development of these "mixed" formations, where race, gender, ethnicity, and class intersect, pose interesting and important intellectual and practical questions about *"cultural"* and *"national"* and *"transnational"* identities. What populations constitute a nation and what are nations' (extra-national) boundaries, and how diasporan communities reshape "national" cultures, reconfigure and create "new" identities, while maintaining "bi-national citizenship," are the subjects of renewed study and redeployment that challenge our notions and the normative assumptions and definitions about these categories. Diasporan and cultural studies scholars, filmmakers and activists concerned with marginalized communities in, but not only in, First World metropolises have increasingly engaged these subjects; along with the cultural-technological forms that communicate and maintain solidarity among dispersed social groups in a mediated and interconnected world.

"Mixed" formations have preceded colonialism, but they were expanded and reconfigured in the colonial process. These formations are most evident in Third World cities, where diverse population groups are more likely to converge, cohabitate and interact, or in nations like Brazil, perhaps the most mixed and "hybrid" of countries in the Third World. Mixed formations are also manifest in many First World nations, where Third World people are largely defined as "minorities" though they may be among the numerical "majorities" in the Third World. In Europe, for example, the racial and ethnic formations that have evolved were forged largely by Third World immigrants who have historically served as a source of cheap labor in the post-war period. Distinguishing racial and cultural features of the immigrants that set them apart from and "fixed" their relationship to the peoples of the host nations followed labor from the peripheries of the Third World. Britain is a case in point:

> [Britain] was virtually a homogeneous society until the 1950's. Its primary experience with nonwhites had been as their colonial master in the vast British empire. But in the 1950's, a postwar labor shortage, a damaged economy and a weakening grip on its territories forced Britain to encourage people from the Caribbean, the Indian subcontinent and Africa to come here to fill manual jobs.[28]

More generally, in ethnic enclaves and on the streets of any European metropole are the living artifacts of a colonial past, peoples who selfconsciously parade down the *Damstraats* of Amsterdam, Paris, Berlin and London, forever in a binary world that tolerates them as wards of the former imperial state. Culturally and physically distinct, "they all carry their passports on their faces," as one observer has crudely but unerringly put it.[29]

While some cultural and physical traits of social groups are clearly discernable, the categories that they presumably signify are immutable in racialist discourse. In describing the features of "polycentric multiculturalism," Shohat and Stam have presented a reconstituted and dynamic framework for understanding "identity" in its variegated, multilayered and transnational dimensions.

> It thinks and images "from the margins," seeing minoritarian communities not as "interest groups" to be "added on" to a preexisting nucleus but rather as active, generative participants at the very core of a shared, conflictual history. . . . [It] rejects a unified, fixed, and essentialist concept of identities (or communities) as consolidated sets of practices, meanings, and experiences. Rather, it sees identities as multiple, unstable, historically situated, the products of ongoing differentiation and polymorphous identification. . . . [It] goes beyond narrow definitions of identity politics, opening the

> way for informed affiliation on the basis of shared social desires and identifications. . . . [It] is reciprocal, dialogical; it sees all acts of verbal or cultural exchange as taking place not between discrete bounded individuals or cultures but rather between permeable, changing individuals and communities.[30]

The above authors' critical reworking and redeployment of identity, within the "polycentric multiculturalism" model they have proposed, is especially troublesome to Eurocentrist and black nationalist discourses about racial and cultural identities.[31] More importantly, the model is relevant to the study of identity and culture in the black diaspora, and to the cinematic concerns and practices of black filmmakers because of its central focus on human agency, relationality and power relations.

The model's organizing themes also intersect and frame, within broader historical and cultural categories, the more recent writings of Stuart Hall and Paul Gilroy about black diasporic culture(s). In discussing the sources of black popular culture, Hall observes that:

> The point of underlying overdetermination—black cultural repertoires constituted from two directions at once—is perhaps more subversive than you think. It is to insist that in black popular culture . . . , there are no pure forms at all. Always these forms are the product of partial synchronization, of engagement across cultural boundaries, of the confluence of more than one cultural tradition, of the negotiations of dominant and subordinate positions, of the subterranean strategies of recoding and transcoding, of critical signification, of signifying. Always these forms are impure, to some degree hybridized from a vernacular base. Thus, they must always be heard, not simply as the recovery of a lost dialogue bearing clues for the production of new musics . . . , but as what they are—adaptions, molded to the mixed, contradictory, hybrid spaces of popular culture. They are not the recovery of something pure that we can, at last, live by.[32]

And in contradistinction to the "idea of blacks as a 'national' or proto-national group with its own hermetically enclosed culture . . . ," and the construction of "canons which seems to be proceeding on an exclusively *national* basis—Afro-American, Anglo-phone Caribbean and so on," Gilroy, in challenging the "essentialist" and "pluralist" views of culture, contends that black "Atlantic" culture is at once African, Caribbean, American and European.[33] Formed through the African dispersion and the experience of slavery, and crossing national, cultural and racial divides, the black diaspora, argues Gilroy, is a transnational and intercultural formation,[34] and historical process of renewal, innovation and change.

IV

This anthology of manifestoes, interviews and critical essays by film critics, theorists, historians, and prominent black filmmakers, constitutes a dialogue and engagement about the cultural politics of filmmaking in the black diaspora. Cinematic traditions have not emerged autonomously, spontaneously or evenly in the black diaspora as the essays and documents in the anthology substantiate. Derived from and shaped by the historical and contemporary processes I have discussed, filmmaking in the black diaspora is not a unified and monolithic project. Although many black diasporan filmmakers have developed comparable representational strategies and practices, share similar thematic concerns and aesthetic values, their films (video and broadcast productions) are distinguished by national, cultural, gender, class and ideological orientations. Social differences among black filmmakers are most evident in the cinematic representations of culture and identity. And they are most apparent in the productions of First World "minoritarian" filmmakers, who reside and work in the metropolises where the postcolonial situation and the problems associated with settlement, exile and displacement, appear more pronounced and conditional than in the Third World.

The anthology is organized into three parts. Part One maps the study of black cinemas, along cultural and political lines, and theorizes the constituents of an aesthetically and radically alternative cinema in the diaspora. It positions black cinema as part of larger issues of cultural politics and struggle. Roy Armes, for example, examines the duplicitous role of Third World elites, as a class, in the foreign domination of the postcolonial state, and the problematic and ambivalent location of Third World intellectuals in the postcolonial situation. By means of a reading of Amilcar Cabral, Armes takes up the important question of the oppositional nature and trajectory of popular culture in the national liberation struggle, and by reference to another Third World theorist, Frantz Fanon, he discusses the question of the location of culture in the colonial/anti-colonial process. Both of these themes resonate in the recent writings of Hall and Gilroy, among others, about the politically inflected stance of popular culture. Moreover, Armes discusses the hazards of idealizing the "traditional" in traditional culture, and the imperative of language in the formation and/or recovery of national identity.

In an important essay that rejects essentialist and "aesthetic" based perspectives, Tommy Lott calls for a reworking of the concept of black cinema in political rather than racial terms. In doing so, Lott theorizes a concept of black cinema that is at once oppositional, historically contingent and transitional.

Mark Reid's essay follows with a discussion of the "black womanist film." He argues that in contradistinction to a "raceless feminism" and "phallocentric pan-Africanism," black womanist discourse constitutes a form of affirmation and resistance. Analyzing several black-directed independent films, Reid theorizes a black feminist transnational perspective which "rejects the closures of nation, race, gender, and class exclusivity." Reid's essay is linked to the larger issues of identity, patriarchy and racism within the context of global capitalism and neocolonialism.

Part One concludes with the pioneering and controversial essay by Teshome Gabriel. Criticized for homogenizing Third Cinema, avoiding the "national question,"[35] and attributing fixed meanings to western film practices,[36] Gabriel, nevertheless, elaborates a nuanced and novel schema of Third World cinema. He is concerned with the central question of what are the "essential qualities" of Third World films and how are they distinguished from Euro-American cinema? Within a Fanonian framework, he outlines a three "phase" model of development of Third World film culture and then formulates, in the third phase, a theory of Third World cinema, based on its common thematic, industrial, stylistic traits and filmic conventions, that are in contrast to Western film practices.[37]

Within a comparative framework, Part Two, is organized geographically, rather than by national or linguistic categories, and is differentiated by region in the black diaspora. The regions under study are Africa, the Caribbean/South America, Europe and North America. Criticism of the comparative method aside, it facilitates rather than inhibits identifying commonalities (and differences) between the myriad of filmmaking practices in the black diaspora. Focusing on each geographical site, essays address the history and development of black cinema, its innovative and conventional practices, and location in larger cultural and political processes. The African and U.S. sections are more fully represented for two reasons: first, the cinematic traditions in Africa and the United States are, generally, more developed than in other regions of the diaspora, problems of control of financing, distribution, exhibition notwithstanding. And second, because of the differing colonial experiences and development trajectories of African states, and the myriad of cultural, religious, and linguistic distinctions between and within African states, the cinematic traditions that have developed on the African continent are more diverse than in Europe or in other diasporic sites.

Most of the essays in this first section address sub-Saharan and southern Africa.[38] The lead essay by Manthia Diawara, surveys and critiques the development, beginning in 1969, of the Fédération Panafricaine des Cinéastes (FESPACI); its historical and strategic importance in contributing to the formation of national film centers especially in Francophone

Africa,[39] the establishment of the inter-African distribution agency, the Consortium Interafricain de Distribution Cinématographique (CIDC); the production center, Centre Interafricain de Production de Films (CIPRO-FILM); and to the promotion of African cinema through the creation of the Ouagadougou festival. Diawara's study is especially important for locating African cinema in the context of other cinematic movements in the Third World.

Complementing the essay by Diawara, the Tunisian filmmaker, Férid Boughedir, outlines the history of African cinema, as a parallel and contingent movement, in the postindependence era; its early achievements in the 1960s and 1970s at international film festivals, and concomitant struggles against foreign domination and "total" nationalization by some African governments during the 1970s.

The next essay by Françoise Pfaff explores the concept and presence of the "griot" in several of the works of the celebrated and pioneering Senegalese filmmaker, Ousmane Sembene. Pfaff's perceptive and nuanced study, describes the griot's or storyteller's social role in traditional African societies, showing how "identical structures and intentions" exist between Sembene's films and African storytelling. More importantly, his films, through "visual and aural elements" that resemble the griot's performance, are rendered accessible to popular audiences.

Kenyan Tomaselli's essay follows the development of independent cinema (and video) in apartheid South Africa. He shows why, under historical and present conditions, a radical or oppositional film movement has not been forged in South Africa (except for the productions sponsored by the African National Congress), unlike in other regions of Africa and Latin America during the 1960s and 1970s. Then Claire Andrade-Watkins, in her review of the 1991 FESPACO festival, in Ouagadougou, Burkina Faso, takes up the important question of why the "invisibility of African women behind the camera. . . ." Cognizant of the central and strong roles accorded women characters, especially in films from sub-Saharan Africa, Andrade-Watkins discusses the obstacles to filmmaking in the region that have more adversely effected African women filmmakers than their male counterparts. In recounting her observations at the festival, Andrade-Watkins also discusses the issues of identity that surfaced among women participants at a workshop devoted to "Women, Cinema, TV, and Video in Africa."

In his essay, Roy Armes revisits the historical and contemporary cultural and political factors that have prevented the development of an industrial infrastructure for film production and, consequently, the failure to develop a black African film industry in the 1980s. In a detailed study, Armes surveys film production during this period, and suggests that, under the conditions at the time, "In many ways African films are defined more by the source of their production finance than in terms of a national identity."[40]

The final two essays by Andrade-Watkins and N. Frank Ukadike as well as the interview by Ukadike follow the Armes essay into the current "state" of black filmmaking in sub-Saharan and southern Africa. Ukadike's interview with Nigerian Chief and independent filmmaker, Eddie Ugbomah, while largely confined to the Nigerian context, foregrounds the economic obstacles to film production in Africa, and the thematic concerns and representational strategies of African filmmakers in the 1990s. Andrade-Watkins's critical study follows the three phases of the development of lusophone African cinema, from the anticolonial period to 1993. And in a lengthy essay that concludes this section, Ukadike reviews and critiques the practices of the "new breed" of African filmmakers, their renewed interest in and creative reworking of traditional culture, as well as, the misappropriation of this central category by some of these filmmakers.

The second section of Part Two studies the Caribbean and South America. It features three essays and an interview with the Afro-Cuban filmmaker, Sergio Giral. In the first essay, Cham surveys, on a country by country basis, the state of filmmaking in the Caribbean by Caribbeans from its nascent formation in the 1970s to 1990. He critically assesses the ambiguous categories of race, nation and geographical location, along with the thematic interests of Caribbean filmmakers against the question of what constitutes a "Caribbean film practice" and identity? And in his analysis of the complex interplay of historical, economic and cultural factors that have shaped film practices in the region, Cham concludes that, except for Cuban cinema, "Caribbean cinema is in its infancy at the moment, the most recent 'arrivant' . . . in the domain of Black world film practice, hence its designation as 'un cinéma au rez-dechaussée des nègres,' ('a cinema at the basement of cinema by Blacks')."[41]

Cham's comprehensive overview of black filmmaking in the Caribbean complements Keith Q. Warner's essay. Warner examines the specific implications of the failure to adopt Caribbean literature to film. He addresses the absence of an "indigenous" cinema in the West Indies, in order to query the cultural relevance of this film practice to and reception by Caribbean audiences. In his analysis of Euzhan Palcy's adaption of Joseph Zobel's novel, *La Rue Cases-Nègrès* (Sugar Cane Alley), Warner discusses the director's fidelity to the novel, the innovative cinematic strategies in her work, and the implications of the success of her film to developing an indigenous film practice concerned with the recovery of history and identity in the Caribbean.

In consideration of the film practices in the Spanish speaking Caribbean, and the adaption of literature to film addressed in the previous essay by Warner, I have included an interview with the Afro-Cuban filmmaker, Sergio Giral, by Ana M. Lopez and Nicholas Peter Humy. Perhaps best known in the English speaking world for his important film on Cuban slavery, *El Otro Franscisco* (The Other Francisco), Giral makes the pointed distinction that "black issues" are not part of the "social reality" in ICAIC and in post revolutionary Cuban society, although race, as a historical category, is an important theme of Cuban history.

Robert Stam's essay on the Afro-Brazilian cultural contribution to Brazilian cinema concludes the section. Stam examines black performance—music and religion—in cinema within the context of African cultural traditions and the historical evolution of race relations in Brazilian society. Critical of Cinema Novo's reductive portrayal of blacks, Stam asserts that, ironically, unlike its later development in the 1980s, many films, during the movement's formative period, "evoke political and social conflict through the manipulation of cultural expression, and Afro-Brazilian music is crucial in this regard."[42] Stam's insightful historical overview and analyses of important films of the Cinema Novo movement, provide a framework useful for studying the complex interplay of race and culture in the development of Brazilian cinema.

Identity, in its variegated and multi-layered dimensions—racial, cultural, gender, national and transnational—is a problematic and contested category, hotly and seemingly endlessly, debated by cultural critics, filmmakers and scholars, especially, but not only, in the First World, as I have shown. Because of the complexity of modern life under "late" capitalism and the enduring legacy of the colonial process, the intersection and inscription of class, racial, cultural, gender, and so forth, differences on social groups have rendered identities more fluid and tentative in metropolitan sites than in the peripheries, although as I have noted earlier, in urbanized areas of the Third World and among elite and intellectual strata, the effects of these processes on traditional culture and group affiliations are also increasingly manifest.

Equally important, though less often discussed, are the effects of dislocation on the sensibilities and cultural practices of Third World filmmakers living in exile in First World metropolises. Moreover, how does exile alter the conception of a "national" cinema when filmmakers are dispersed abroad by choice or political necessity? Are Third World filmmakers' cultural productions altered when produced or co-produced and financed abroad?[43] These and other issues are addressed in this third section on black filmmaking in Europe.

In Coco Fusco's essay, the state of black filmmaking in Britain is reviewed against the development of a new generation of politically engaged filmmakers associated with the Sankofa Film and Video Collective and the Black Audio Film Collective. In counterpoint to "radical" film theory's perfunctory treatment of race issues, Fusco explicates how these black media collectives have positioned race at the center of their cultural practice, especially with regard to identity within contemporary black British communities. Fusco analyzes several of the more important film and video productions of the two workshops, and discusses their efforts to develop an aesthetic "from diasporic experiences common to Black peoples." José Arroyo's essay follows in which he argues that the films of Isaac Julien, in challenging colonialist and racialist discourses, allows for the differentiation and affiliation of black identity along sexual, class, gender lines, among others.

The essay by Mauritanian filmmaker, Med Hondo, concerns exile. The location of the Third World filmmaker in the host country: his or her relation to fellow-exiles and nationals; how to communicate with them and the people of the host country; how to finance your films, enter into co-productions, and struggle to retain your vision and control of your work. Hondo takes up these questions in this short and important essay. Mark Reid's interview with film editor, Andrée Daventure, concludes the section. Conducted between 1985 and 1986, during the period when Rightist movements in France fueled anti-Semitic and racist sentiments, the interview addresses cross-cultural issues in film editing and the tenuous relationship of the French government to filmmakers in francophone Africa.

In the fourth and final section of Part Two, contemporary black film practices in the United States are surveyed largely in the context of the development of black independent cinema—an alternative cinema to Hollywood that is at once oppositional and affirmative. Essays focus on a range of important subjects from broad political and transnational to race based definitions of black film by Thomas Cripps (see also Tommy Lott essay); the historical development of black film production and the contemporary challenges to black independent cinema by James A. Snead; the cultural and financial obstacles to producing *Daughters of the Dust* by Julie Dash; the career, in the context of documentary film practice, of the politically conscious and innovative filmmaker, William Greaves, by Charles Musser and Adam Knee; the cinematic strategies, aesthetics of and "new realism" in black independent cinema by Manthia Diawara; and from an exchange between film scholars and critics, Clyde Taylor and David Nicholson, and filmmaker, Zeinabu Irene Davis, on the "state" and future of the black independent film movement in the United States.

Part Three concludes the anthology with four important documents on the political and economic concerns and counter-hegemonic institutional organizing efforts—regional and transcontinental—of black and Third World filmmakers from the 1970s to early 1990s. The manifestoes included are the *Resolutions of the Third World Film-makers Meeting*, held in Algiers, 1973; the document, "Seminar: The Role of the African Film-maker in Rousing an Awareness of Black Civilization," the summaries and resolutions of an international seminar organized by the Society of African Culture and sponsored by the Government of Burkina Faso (formerly the Upper Volta) in 1974; the *Niamey Manifesto*, resolutions of an international colloquium held in Niamey, Niger, 1982; and the *FeCAViP Manifesto*, resolutions of the second colloquium of the "Images Caraïbes Festival," held in Fort-de-France, 1990.

In counterpoint to bourgeois aesthetic theory that divorces art from politics, the essays in this anthology affirm the centrality of politics to black film practice, and open new possibilities of agency. In the sense that as we theorize black cinema, we mediate the world by reconstituted images.

Notes

1. For example, see the studies by Erik Barnouw and S. Krishnaswamy, *Indian Film* (New York: Oxford University Press, 1980); Jacob M. Landau, *Studies in the Arab Theater and Cinema* (Philadelphia, NJ.: University of Pennsylvania Press, 1958); Allen L. Woll, *The Latin Image in American Film* (Los Angeles, CA.: UCLA Latin American Center Publications, 1977), 76–83; *Brazilian Cinema*, ed. Randal Johnson and Robert Stam (East Brunswick, NJ.: Associated University Presses, 1982), 17–30.

2. See select bibliography of this anthology for readings about film production and politics in the Third World.

3. For a detailed explanation of these three factors, see Roy Armes's important study, *Third World Film Making and the West* (Berkeley: University of California Press, 1987), 9–49.

4. Ella Shohat and Robert Stam, *Unthinking Eurocentrism: Multiculturalism and the Media* (NY.: Routledge, 1994), 28.

5. Clyde Taylor, "Third World Cinema: One Struggle, Many Forms," *Jump Cut: Hollywood, Politics and Counter Cinema*, ed. Peter Steven (New York: Praeger, 1985), 332–33. For a discussion of the constituents of a "Third Cinema," see Teshome Gabriel's pioneering essay, "Towards a critical theory of Third World film," in this collection.

6. Tommy L. Lott, "A No-Theory Theory of Contemporary Black Cinema," *Black American Literature Forum* 25, no. 2 (1991): 11.

7. There is considerable controversy over what constitutes black cinema in national, transnational and political terms. For varying perspectives, see the Tommy Lott and Thomas Cripps essays included in this anthology; for a definition of black independent and black commercial films, see Mark A. Reid, *Redefining Black Film* (Los Angeles: University of California Press, 1993), 2, 4.

8. Robert Stam and Louise Spence, "Colonialism, Racism, and Representation: An Introduction," *Movies and Methods*, Volume 2, ed. Bill Nichols (Berkeley: University of California Press, 1985), 635. For a detailed consideration of the origin and usage of the term "Third Word," see Leslie Wolf-Phillips, "Why 'Third World'?: Origin, Definition and Usage," *Third World*

Quarterly 9, no. 4 (1987): 1311–27; for a broader definition, see L. S. Stavrianos, *Global Rift: The Third World Comes of Age* (New York: William Morrow, 1981), 33, 39–40; for an explication and differentiation of the term by economic criteria, see Timothy Shaw, "The South in the 'New World (Dis) Order': Towards a Political Economy of Third World Foreign Policy in the 1990s," *Third World Quarterly* 15, no. 1 (1994): 18–19; and for a critical reassessment of the term, in consideration of recent world developments, see Mark Berger, "The End of the 'Third World'?," *Third World Quarterly*, 15, no. 2 (1994): 257–75.

9. *Unthinking Eurocentrism*, 25. In an earlier work, Shohat discusses the peculiar circumstance of Israel's dual status of being at once a First and Third world nation, see Ella Shohat, *Israeli Cinema: East/West and the Politics of Representation* (Austin, TX.: University of Texas Press, 1989), 4–5.

10. For a comprehensive and critical survey of the black (African) diaspora, see *Global Dimensions of the African Diaspora*, ed. Joseph E. Harris (Washington, D.C.: Howard University Press, 1982).

11. Ruth Simms Hamilton, "Toward a Paradigm for African Diaspora Studies," *Creating a Paradigm and Research Agenda for Comparative Studies of the Worldwide Dispersion of African Peoples*, Monograph No.1, ed. Ruth Simms Hamilton (East Lansing, MI.: African Diaspora Research Project, Michigan State University, 1988), 18.

12. For the purposes of this project, the terms black and African descent are used interchangeably.

13. The presence of Africans in Europe and Asia occurred in antiquity, preceded Islam and was then globalized by the slave trade. For further study, see Frank M. Snowden, *Before Color Prejudice, An Ancient View of Blacks* (Cambridge: Harvard University Press, 1983); Hans Werner Debrunner, *Presence and Prestige: Africans in Europe* (Basel: Basler Afrika Bibliographien, 1979); and St. Clair Drake, *Black Folk Here and There*, 2 vols. (Los Angeles, CA.: Center for Afro-American Studies, UCLA, 1987; 1990).

14. While referring to African Americans, this statement, by Drexel G. Woodson, is relevant because it rejects essentialist notions of race and culture as pure and unchanging. (memorandum) "Reflections on the Diasporan Issues Group Meeting of 15 January 1993," National African American Museum Project, January 20, 1993, p. 3.

15. See Oliver C. Cox, *Caste, Class, and Race: A Study in Social Dynamics* (New York: Monthly Review Press, 1970).

16. Of course, large-scale migrations occurred from Europe to North America during this period, especially in the nineteenth century. For a discussion of the relationship of international migration to the development and extension of capitalism, see Robert Miles, *Capitalism and Unfree Labour: Anomaly or Necessity?* (London: Tavistock, 1987).

17. Ibid., 188. Cultural and psychological explanations of racism are inadequate because they fail to account for how racism is a social relation of the capitalist mode of production. For example, the marginalization of Third World immigrants in Europe's metropoles is related more to the economic performance

of national economies, technological developments in the work place, and globalization than to cultural differences between social groups, or the xenophobic concerns of insular and homogeneous communities. To be sure, the cultural histories and colonial encounters of European states particularize race relations, but they do not adequately account for their origins or trajectory.

18. There is considerable disagreement about the period in which the transition occurs. For example, see Oliver C. Cox, *Capitalism as a System* (New York: Monthly Review, 1964) and Immanuel Wallerstein, *The Modern World-System: Capitalist Agriculture and the Origins of the European World Economy in the Sixteenth Century* (New York: Academic Press, 1974).

19. J. M. Blaut, "Colonialism and the Rise of Capitalism," *Science & Society* 53, no.3 (1990): 280.

20. Ibid., 280.

21. For a systematic and critical discussion of Eurocentrism, see Samir Amin, *Eurocentrism* (New York: Monthly Review, 1989). By "hegemony," I mean political, and, as developed by Antonio Gramsci, and later elaborated upon by Raymond Williams, a system that informs the whole of social life, and that pervades daily existence as a "culture" of internalized dominance. A hegemonic changes in response to pressures from within its sphere of practice, as well as from without. Mutable yet defended, it controls, transforms or incorporates alternative and counter-hegemonic formations. As Williams asserts, a dominant hegemony must be "transformational." See the works of Williams, *Problems in Materialism and Culture* (London: NLB, 1980), 37–40; and *Marxism and Literature* (New York: Oxford University Press, 1977), 110.

22. William Barclay, Krishna Kumar and Ruth P. Simms, eds., *Racial Conflict, Discrimination, & Power: Historical & Contemporary Studies* (New York: AMS Press, 1976), 80.

23. Creating a Paradigm," 20. During this moment of "late capitalism," race correlates with *structural violence*. The worldwide system of racial stratification that developed between the fifteenth and nineteenth centuries is now largely reproduced without the violence of colonial conquest and rule. Its manifestations are malnutrition, high infant mortality, mass poverty and unemployment, and environmental degradation ("environmental racism"). These conditions define the lives of most people in the Third World, and in the First World where blacks and other "minority" populations reside. Observers have often obscured this racial distinction with systemic concepts of the "North/South," "developed/underdeveloped," and, until the collapse of the socialist bloc, the "East/West."

24. See John Darnton, "U.N. Faces Refugee Crisis That Never Ends," *New York Times*, 8 August 1994, sec. A1, A5.

25. Other important factors not discussed include the prolonged economic recession in the First World, "globalization" and "restructuration" of enterprises, advances in new technologies and their impact on the workplace, environmental degradation, natural disasters (e.g. droughts) affecting food production and so forth.

26. See Hilbourne A. Watson, "Surplus Labor, Unequal Exchange, and Merchant Capital: Rethinking Caribbean Migration Theory," in Michael T. Martin

and Terry R. Kandal, eds., *Studies of Development and Change in the Modern World* (New York: Oxford University Press, 1989), 261.

27. See Williams Stevens, "Feeding a Booming Population Without Destroying the Planet," *New York Times*, 5 April 1994, sec. B5–6.

28. A. Sivanandan, "Editorial," *Race & Class*, 32, no. 3 (1991): vi.

29. Ibid., 30, no. 4 (1989): v.

30. *Unthinking Eurocentrism*, 48–49.

31. Ironically, the more extreme versions of the Eurocentrist and black nationalist perspectives converge, advancing quasi-biological claims for racial and cultural superiority.

32. Stuart Hall, "What is This "Black" in Black Popular Culture?," *Social Justice*, 20, nos. 1–2 (1993): 110.

33. Paul Gilroy, *Third Text*, no. 13 (1990/91): 6.

34. See Paul Gilroy, *The Black Atlantic: Modernity and Double Consciousness* (Cambridge: Harvard University Press, 1993).

35. For example, see the review and critique of Third Cinema by Paul Willemen, "The Third Cinema Question: Notes and Reflections," *Questions of Third Cinema*, ed. Jim Pines and Paul Willemen (London: BFI, 1989), 14–20.

36. See Felix Thompson's recent study, "Metaphors of space: polarization, dualism and Third World cinema," *Screen* 34, no. 1 (1993): 44–53.

37. Gabriel's views on the distinction between Western and non-Western film conventions are also discussed in an earlier essay, "Teaching Third World Cinema," *Screen* 24, no. 2 (1983): 60–64.

38. North Africa is considered by some film scholars as, arguably, part of the Arab world, and therefore the filmmaking traditions in that region are associated with variants of Arab cinema. For example, in the case of Egyptian cinema, see Lizbeth Malkmus and Roy Armes, *Arab & African Film Making* (London: Zed, 1991), 28–32.

39. The noted African film scholar, Francoise Pfaff, suggests that Diawara has overemphasized the impact of *FEPACI* in the development of African cinema, see Françoise Pfaff, "African Cinema Inside," *Black Film Review* 7, no. 3 (1993), 24.

40. Roy Armes, "Black African Cinema in the Eighties," *Screen* 26, nos. 3–4 (1985): 69.

41. Mybe Cham, "Shape and Shaping of Caribbean Cinema," *Ex-ILES: ESSAYS ON CARIBBEAN CINEMA*, ed. Mbye Cham (Trenton, NJ.: Africa World Press, 1992), 1.

42. Robert Stam, "Samba, Candomble, Quilombo: Black Performance and Brazilian Cinema," *The Journal of Ethnic Studies* 13, no. 3 (1985): 65.

43. A growing corpus of published work in these areas are available, for example, see Zuzana M. Pick, *The New Latin American Cinema: A Continental Project* (Austin, TX.: University of Texas, 1993), 157–161; Kathleen Newman, "National Cinema After Globalization: Fernando Solanas's *Sur* and the Exiled Nation," *Mediating Two Worlds*, ed. John King, Ana M. Lopez, and Manuel Alvarado (London: BFI, 1993), 242–57; Hamid Naficy, *The Making of Exile Cultures: Iranian Television in Los Angeles* (Minneapolis, MN.: University of Minneapolis, 1993); Andrew Higson, "The Concept of National Cinema," *Screen*

30, no. 4 (1989): 36–46; and on the formation of "hybrid" cinemas, see the recent studies by Laura U. Marks, "A Deleuzian Politics of Hybrid Cinema," *Screen* 35, no. 3 (1994): 244–64 and Hamid Naficy, "Phobic Spaces and Liminal Panics: Independent Transnational Film Genre," *East-West Film Journal* 8, no. 2 (1994): 1–30.

I
Mapping the Terrain

Culture and National Identity

Roy Armes

> A national culture is the whole body of efforts made by a people in the sphere of thought to describe, justify, and praise the action through which that people has created itself and keeps itself in existence. A national culture in underdeveloped countries should therefore take its place at the very heart of the struggle for freedom which these countries are carrying on.
>
> *Frantz Fanon*

The elites that emerged under colonialism at virtually all the key points of interaction between traditional societies and the West—in Alexandria, Buenos Aires, and Madras, as much as in Manila, Calcutta, and Dakar—are historically unique in terms of their westernization. They owe their power neither to initial wealth (though they may indeed *become* wealthy) nor to direct involvement in production or even ownership of the means of production (though the state they control is likely to become the nation's major economic force). They form a group which has the role of mediating the conflicting interests of both landowning classes and an emerging business class, but which does not owe its position directly to either (though family ties will probably link individuals to one or both of these classes). Their particular position vis-à-vis the West has been characterized by Cedric Robinson as constituting the "conduit through which the technical, political and commercial relations of the technologically sophisticated northern hemispheric people articulate with that vast majority of mankind which supports them materially and economically." These elites are also the stratum through which "the ideological, moral and philosophical traditions of Western civilisation have been transferred, at least superficially, to non-Western societies."[1]

Nationalism and Development

The style of nationalism adopted by the elite was conceived before independence as an opposition to colonialism, and it took Western political forms. It was in fact a search for political independence within a framework of foreign economic and ideological dominance. The concept of the nation was defined in the territorial terms of the colonial state, whose boundaries took no account of social or cultural groupings. The nationalists did not seek to revive a traditional form of society or to mobilize mass support for the independence movement in terms of ethnic identity—denigrated as "tribalism" by the colonizers. Instead, their ambition was to create a modern state, using concepts of democracy, elections, and political parties borrowed from the West. Even the underlying democratic definition of "one man, one vote" conceals a concept of Western origin: individualism.

This modern nationalism has shown the same dynamism as other social and cultural forms derived from the West, but it has done nothing to remedy the elite's inevitable limitations as a force for development and progress. Firstly, since the colonial state is founded on an unbridgeable gulf between rulers and ruled, by taking over the mechanisms of the colonial state, the elite separates itself from the mass of the people whom its nationalist impulse aims to serve. Such a state is designed not to develop the people but to hold them in check. Any move to give real power or representation to the people or to allow the emergence of political opposition parties is a weakening of the structures to which the elite owes its position. Secondly, in moving into the position of the colonial administration, the elite takes on the same subservient role vis-à-vis the Western metropolitan centers, for ultimate power in this structure of government rests not within the colonial state itself, but in the metropolis. The new elite that comes to control the postcolonial state can operate in a wide context only under foreign economic dominance, of which it now becomes the local agent.

This situation is reinforced by the effects of foreign aid. Since far more is extracted annually by the West from the Third World than is returned in the form of aid, the latter can be little more than a cosmetic device. As Pierre Jalée aptly puts it, "It is not the imperialist countries which aid the Third World, but the Third World which aids imperialism."[2] Moreover, even if there were a balance between aid and exploitation, the effect would still be disastrous, since it implies taking money from areas of Third World productivity, such as agriculture and mining, and returning it in the form of "aid" to be used by the westernized elite on prestige projects of "modernization," which are often largely irrelevant to the country's economic development and always carry the invitation to corruption. This is a theme spelled out by Gérard Chaliand, a participant in and observer of Africa's struggles for independence: "The political and social uses of foreign aid in tropical Africa make it an instrument of corruption, a basic neo-colonialist trait. Here it can be seen that aid aims at reinforcing a leadership group linked by self-interest to Western capitalism."[3] Even less can be said in favor of the military aid dispensed with such largesse by the United States to its client dictators in Latin America and elsewhere.

The political complicity of the elite in the economic system underpinning the imperialism from which the elite has claimed or received independence is strikingly illustrated by comparison with another distinctive group created by the colonial system, namely, the settler communities. Once the momentum of decolonization was under way, the old empires were dissolved with quite remarkable speed and with few if any disadvantages to the former colonizing powers. European economic imperialism could come to terms with the political demands of the emergent nationalist elites in most parts of Africa and Asia and still maintain the kind of economic dominance that first Great Britain and then the United States had exercised over Latin America for a hundred and fifty years. With few exceptions, Western-educated elites could move smoothly into place, without disturbing the world system—indeed, major confrontations came only where, as in Nigeria, colonialism had fostered particularly uneven development. By contrast, settler communities have always proved awkward for metropolitan states, since they refuse to accept a subordinate role in the world system, expecting and struggling, rather, to achieve metropolitan status. The process began with the independence of settlers in the United States, who have since been joined in the Western group of "core" nations by settler communities in Canada, Australia, and New Zealand. In the period of decolonization, the Belgians found themselves confronted by their settler communities in the Congo, the British by theirs in Rhodesia, and the French by the *pieds noirs* in Algeria. In each case the Western governments repudiated such communities, sometimes even to the extent of waging war on them.

But insistence that independence be granted only to a native majority was not an act of altruism on the part of the Western governments; rather, it was an example of sound economic sense. From the standpoint of international capitalism, settler communities are, in Arghiri Emmanuel's terms, "a dead weight—if not a parasitic and harmful element" and a "competitive and anarchic sector." The logic of the situation whereby the French first contained the National Liberation Front (FLN) and then fought those who claimed to want a "French Algeria" is clear in Emmanuel's terms: "If the partisans of 'French Algeria' had won, Algeria would have been much less French than she still is today, in spite of the profound breaches made by the revolution and the war. It was so that Algeria might remain as French as possible that [De Gaulle] fought the OAS [Organisation Armée Secrète—the movement within the French army opposing Algerian independence]."[4]

Here we see the central paradox of the postcolonial situation: the settlers, who, aware of their "historic ties," see themselves as part of Europe, are repudiated by Western capitalist interests, while the Western-educated elites, who define their identity in terms of independence from colonialism, are in fact incorporated into the capitalist world system. This progression, worked out in Africa over the past thirty years or so, merely echoes that which has evolved over a much longer period in Latin America. In all Third World countries caught in this way, the state is the crucial element: it is the only potential force for real social change, but at the same time it is "frozen" in an exploitative form, since its present situation is the source of profit and power for those who command it.

National Culture

Within the network of contradictions that characterizes the Third World, there is probably no more problematic area than that of culture. The artistic intelligentsia of Africa, Asia, and Latin America displays the tensions between the forces of westernization and tradition to a particularly intense degree. Even if they are driven to oppose the political and social policies of the ruling elite, they cannot cease to be a part of it, through either shared origins or achieved social status. In respect to the gulf between rulers and ruled, those concerned with culture, whether as politicians and intellectuals, organizers and administrators, or as writers, artists, and film makers, are inevitably closer to the rulers. Yet ultimately in postcolonial society, culture can only be valid if it is the product of individuals able and willing to work against their own narrow class interests.

The position of Third World artists is a unique one. Though often at odds with the ruling members of the elite to which they belong by virtue of their education, they are equally cut off from the mass of the people by the literary forms and language that they choose. Their position could hardly be more different from that of the traditional storyteller or craftsman, whose identification with his audience or clientele was direct and immediate. As Ousmane Sembene noted at the First Festival of Negro Arts at Dakar in 1965, "All of us who are writers are also people who have to some extent lost their roots."[5] A film maker, a radio or television producer, will be using a Western technology; a writer will often be employing the former colonizer's language, in which only a tiny minority of the population is literate. In both cases he or she will almost certainly be using formal structures derived from a foreign source, for all the basic forms of print literature, as well as those of the filmic narrative or television documentary report, are imported.

Perhaps the most valid criteria for the evaluation of work produced under these conditions are to be derived from the national liberation struggles in the Third World over the last thirty years or so. Africa is particularly fortunate in that two of its principal theorists of the national liberation struggle, Frantz Fanon and Amilcar Cabral, have both attributed to culture a key role and offered valuable insights into its priorities. As a starting point, Cabral in his lecture "National Liberation and Culture," delivered in New York in 1970, makes clear that the basis of culture—as of national liberation as a whole—must lie outside the narrow confines of the elite: "Without minimizing the positive contribution which privileged classes may bring to the struggle, the liberation movement must, on the cultural level just as on the political level, base its action in popular culture, whatever may be the diversity of levels of culture in the country."[6] Cabral's definition uses basically Marxist concepts and terminology, and he is at pains to stress the fact that culture "has as its material base the level of the productive forces and the mode of production," whose contradictions, manifested through the class struggle, are "the principal factor in the history of any human group" and "the true and driving power of history."[7] We have seen from Albert Memmi's account of colonization that one of its effects is the elimination of the colonized's culture. As a corollary of this, Cabral points out that "it is generally within the culture that we find the seed of opposition, which leads to the structuring and development of the liberation movement." In fact, for Cabral, "national liberation is necessarily an act of *culture,*" since

> a people who free themselves from foreign domination will be free culturally only if, without complexes and without underestimating the importance of positive accretions from the oppressor and other cultures, they return to the upward paths of their own culture, which is nourished by the living reality of its environment, and which negates both harmful influences and any kind of subjection to foreign culture.[8]

A fascinating assessment of what this process involves is to be found in Frantz Fanon's celebrated essay "On National Culture," contained in *The Wretched of the Earth.* Here Fanon singles out three separate stages: a concern primarily with the culture of the colonizing power; a return to an idealized view of the traditional culture; and a third stage of uniting with the people's struggle in an authentic way. Fanon's definitions are worth quoting at some length, beginning with his initial stage, when the power of attraction of the colonizer's culture is at its strongest:

> In the first stage, the native intellectual gives proof that he has assimilated the culture of the occupying power. His writings correspond point by point with those of his opposite numbers in the mother country. His inspiration is European and we can easily link up these works with definite trends in the literature of the mother country. This is the period of unqualified Assimilation.[9]

Fanon's second stage is one in which the native artist or intellectual attempts to reestablish his links with an abandoned traditional culture:

> In the second phase we find the native disturbed; he decides to remember what he is. . . . But since the native is not a part of his people, since he has only external relations with his people, he is content to recall their life only. Past happenings of the bygone days of his childhood will be brought up out of the depths of his memory; old legends will be reinterpreted in the light of a borrowed aestheticism and of a concept of the world which was discovered under other skies.

In Fanon's third phase, which he calls "the fighting phase,"

> the native, after having tried to lose himself in the people and with the people, will on the contrary shake the people. Instead of according the people's lethargy an honoured place in his esteem, he turns himself into an awakener of the people; hence come a fighting literature, a revolutionary literature and a national literature."[10]

The great relevance of Fanon's analysis is not in doubt, though of course any generalization that in this way divides a complex cultural process into just three clearly differentiated stages is open to criticism. The three focal points Fanon picks out are key areas of discussion, even if they cannot be taken to represent the chronologically successive stages of a single development.

What Fanon's formulation perhaps underplays is the indivisibility of the three stages, that is, the necessity of both an interaction with Europe and a concern with local tradition. For many Third World countries, continuing interplay with Europe provides a crucial and by no means wholly negative thread. The example of Brazil can stand as a particularly clear example of trends present to some extent throughout Latin America and continuing across the divide of political independence. The Brazilian literary historian Afrânio Coutinho offers a striking definition of the dynamic force that can stem from the relationship between Europe and Latin America and the effort to create a local tradition:

> Starting with an original stock of notions that had been transplanted with the first impulses of colonization, Brazilian cultural life went its way in relation to European culture, sometimes repeating its forms, sometimes reinterpreting them and adapting them to individual and social native conditions, sometimes trying to turn its back on them in its anxiety to create a peculiarly Brazilian tradition. But the persistence and strength of European intellectual influences, the habit of looking to Europe in search of inspiration, became the fulcrum of Brazilian mental life, which was marked by an extraordinary capacity for the assimilation of foreign values. The very idea of Europe was the center of our concentric waves of culture.[11]

If one had to pick one key aspect of this interplay, it would be that of drawing the Third World culture into the contemporary world, bridging the time gap that so often exists to separate metropolis and satellite, the giver and the receiver of cultural influence. In Brazil, for example, the birth of an authentic national literature is usually dated from the 1920s, which was precisely a time of strong European influence, typified by the Modern Art Week held in February 1922 in São Paolo. Likewise, it was the impact of surrealism that liberated the Caribbean and African poets of Négritude from the constraints of a borrowed language and enabled them to forge a new and wholly contemporary black poetry.

This sense of entering the present thanks to the assimilated influence of Europe contrasts strongly with the retreat into the past entailed in the confrontation with local tradition, which forms Fanon's second stage. The attempt at a reinsertion into tradition can be extremely difficult for an individual, and one of the central themes of the West African novels published over the last thirty years has been the destructive impact of westernization on an educated person who subsequently finds himself caught between two cultures.[12] Even when this crisis is mastered and becomes the subject of a successfully completed novel, a certain retreat into the past still inevitably seems involved. As Dorothy S. Blair writes in her study of the francophone African novel:

> It is rare to find traces of contemporary European innovations in the structure and style of this African novel in French: no suggestion

> of a Proust, of a James Joyce, not to mention a Robbe-Grillet or a Claude Simon. By setting out to be the faithful mirror of the African condition of the first half of the twentieth century, the French-African novel remains obstinately based on the French novel of the nineteenth century.[13]

Once more the West African experience can be seen as analogous with a slightly earlier tendency in Latin American literature, where a similar return to the past and to rural life had been attempted by a number of regionalist writers in the latter part of the nineteenth century and the first three decades of the twentieth. Like the francophone African novel, this is essentially a literature *about* tradition *for* a westernized audience. Its self-evident roots in European realist styles allow the customary lines of argument to be effortlessly stood on their head by Jorge Luis Borges in his essay "The Argentine Writer and Tradition": "The idea that a literature must define itself in terms of its national traits is a relatively new concept; also new and arbitrary is the idea that writers must seek themes from their own countries. . . . The Argentine cult of local colour is a recent European cult which the nationalists ought to reject as foreign."[14]

The caution is salutary, for any ostensible rejection of westernized culture in favor of a return to superficially understood traditional roots is hazardous after a period of colonization. As Terence Ranger has demonstrated, those who adopt this strategy "face the ironic danger of embracing another set of colonial inventions instead."[15] Certainly in Africa, much of what passes for "tradition" is

> the result of a conscious determination on the part of the colonial authorities to "re-establish" order and security and a sense of community by means of defining and enforcing "tradition." . . . The most far-reaching inventions of tradition in colonial Africa took place when the Europeans believed themselves to be respecting age-old African custom. What were called customary law, customary land-rights, customary political structure and so on, were in fact *all* invented by colonial codification.[16]

This reified tradition, which has little historical validity, since it was the product of colonial boundaries, laws, and classifications, continues to color much of the thinking about precolonial Africa. A sense of tradition may be a vital component of a national culture, but it is by no means easily attained, particularly by one whose thought processes have been shaped by a Western education. The kind of intellectual task that the reconstruction of the past after a period of colonization entails is spelled out clearly by the Filipino historian Renato Constantino, in his "Notes on Historical Writing for the Third World":

> A people's history must rediscover the past in order to make it reusable. . . . The past should not be the object of mere contemplation if the present is to be meaningful. For if the past were viewed as a "frozen reality" it would either dominate and immobilize the present or be discarded as irrelevant to today's concerns. The past as a concrete historical reality must be viewed as an integral part of the process of unfolding total reality.[17]

Since not only the "new" notions of nationalism, political independence, and the nation-state but also the very definitions of tradition itself are shaped by Western concepts, Fanon's plea for a third stage of political commitment might seem self-evident. Yet even here there are traps for the unwary. At least as far as African leaders are concerned, seeking an alternative in Marxism also implies westernization, since, as Ali A. Mazrui points out, it is for the time being "a socio-linguistic impossibility" for an African to be a sophisticated Marxist without at the same time being substantially westernized. The basic Marxist texts are unavailable in African languages, and "no Africans are ever admitted to Soviet universities directly from some village compound."[18] Political leaders and theorists like Cabral and Fanon acquired their radical views abroad, and considerable care is needed in applying these views to Third World contexts. But such modernizing influences are vitally necessary, particularly in situations where national liberation takes the form of armed struggle. The dual need to draw on the local past while at the same time participating actively in the contemporary world is very clear, but its achievement remains elusive. Albert Memmi, stressing like Fanon the inseparability of cultural development from political advance, asserts that "the most urgent claim of a group about to revive is certainly the liberation and restoration of its language."[19] Certainly many of the paradoxes with which we are confronted in this study are to be found within language itself.

Language, Literature, and Theater

As we have seen, the concept of the nation-state is one that has been inherited from Europe, and a part of this inheritance is the particular role attributed to language. Richard M. Morse has pointed out the inextricable link between definitions of language and national identity: "In nineteenth-century Europe the linguistic concerns of the intelligentsia gave predominance to the languages of that continent and how they revealed the 'genius' of nations considered to be world-historical. The identification of 'language' and 'nation' became so strong as to shape the self-image of peoples, ambitions of leaders, and political demarcations of the continent."[20]

Such an identification could not be other than problematic—though to a varying extent—for all colonized countries, beginning with those Latin American states that achieved their political independence in the first quarter of the nineteenth century. Here the crux of the problem was the necessity to express the distinctive identity of the new nation in the language that had been inherited from the colonizer—Spanish or Portuguese. The search for an authentic national language capable of supporting a national literature was enormously long, but the fact that the new economically dominant powers—Britain for most of the nineteenth century and the United States since World War I—used a different language eventually allowed the Spanish and Portuguese languages to become a source and focus of independence and identity. Even so, most critics see the literatures of Spanish America and Brazil as emerging from provinciality only long after political independence, with the Hispanic American modernism of the 1880s and the Brazilian modernism of 1920. Significantly, it was not until the turn of the century that, as George Pendle tells us, reform "had begun to acquire a more radical meaning. Equality in law, free elections, and universal education would not be enough. There was a growing demand for the redistribution of land, and then for the fixing of minimum wages, and for social insurance. Reformers were becoming increasingly insistent, too, that the hold of the foreigner on the national economy must be broken."[21]

It is instructive to compare the delayed growth of this elite culture with the emergence of a national literature in the United States, where political independence was matched by economic independence. The title of Larzer Ziff's book on the generation of writers who, in the fifteen years before the outbreak of the Civil War in 1861, established American literature as "a distinctive way of imagining the world" is itself highly significant: *Literary Democracy: The Declaration of Cultural Independence in America.* In his analysis of Poe, Emerson, Hawthorne, Thoreau, Whitman, and Melville, Ziff shows the close connections between the emergence of the literature they produced and the development of society as a whole. The shift to the new idea that literature was "the voice of the latent forces of the land," which these writers' work illustrates, had to be preceded by "a redefinition of American society," and this redefinition needed a significant economic development to accompany the larger political one: "In the 1820s and 1830s the growth of the factory system produced two new social classes, the industrial capitalists and the factory workers. Strikingly, the major writers came from the economic groups that were the most threatened by the new classes." The result was not a simple opposition but a complex interaction. "The same commercial and industrial developments that dispossessed the classes from which the writers came also provided them with improved means of printing and circulating what they wrote, as well as a widening number of countrymen who had both the knowledge and leisure to read what they wrote."[22]

This complex interplay between an emerging generation of writers and a context of wider economic and political developments is usually lacking in the Third World. While British rule cannot be held responsible for the complexity of India's linguistic problems (some 845 separate languages or dialects were reported in the 1951 census), elsewhere in Asia the colonial era has bequeathed many potentially explosive entanglements of language, religion, education, and status. In Africa it is only in the Muslim north that we find a single unifying language—Arabic—which despite its considerable variations in dialect can to some degree constitute a focus of independence and a rich and viable alternative to the inherited language of the colonizer. South of the Sahara we find an unbridgeable gulf between European and traditional languages. English and French may serve a certain unifying function—as, say, the languages in which a conference on the linguistic problems of Africa may be held—but they remain administrative languages, and their application outside official life remains limited. They are alien to the mass of the people, and, as Abiola Irele puts it in the best discussion of the subject, none of them is able "to carry fully with it the reality of African experience as it exists today. . . . The new literature that is being expressed in them, for all its value and significance, must be seen for this reason, from the African point of view, to be placed in a most ambiguous, not to say precarious situation."[23]

The amount of skill, energy, and persistence that has gone into shaping an African literature in European languages is undeniable, but such work is by definition inaccessible to the masses, and it has the added effect of relegating traditional literatures to a marginal position. African writers working in European languages are not in fact bringing modernizing influences and new sources of knowledge to their fellow countrymen, since their readership is, by definition, limited to precisely those whose linguistic abilities allow them immediate access to this knowledge. In truth, as Irele points out, they are merely "carrying over into the European language a whole stock of symbols derived from the African environment,"[24] and so the extent to which they are thereby truly enriching these European languages remains, in many cases, doubtful.

In terms of literary expression, there is no single, universal way in which these contradictions of language can be overcome in the foreseeable future. One personal response that may have wider application, however, is that of the Kenyan novelist Ngugi wa Thiong'o. He began his writing career as "James Ngugi," writing in English, but after Africanizing his name in 1970 he began, toward the end of the decade, to write drama and fiction in Gikuyu. By translating his work into English and Kiswahili and encouraging translations into other Kenyan languages, Ngugi is today working toward the creation of a national culture in which indigenous languages play a full part. The political significance of this use of African languages is clear from the reaction of the Kenyan government to this new and vital interaction between an intellectual and the peasant masses. Though his work in Gikuyu is no more overtly political in content than his earlier work in English was, Ngugi was detained without trial for a year in 1977–78, has been prevented from resuming his post as professor of English at Nairobi University, and at present lives in exile.[25]

Even more crucial for future film making than the establishment of the novel was that of a European-style theater. By the beginning of the twentieth century, many local traditions of popular drama had already declined as a result of social changes. In India, for example, "theatre and dance had lost their standing and become a domain of the degraded castes, the occupation of prostitutes. So strong had become the association between the performing arts and the prostitute that measures to combat prostitution seemed likely, for a time, to eradicate what was left of Indian drama, dance and music."[26]

Other popular traditions of dramatic performance did not lend themselves to modern adaptation, however. In black Africa it was well into the twentieth century before theater separated itself from religious ritual, as had occurred in the European tradition as early as Aeschylus. Thus Ola Balogun, himself a dramatist and film maker, observes that

> to speak of the "performing arts" is in a sense a misnomer, since this term is understood in a very different manner in other societies. Perhaps it would be more appropriate to speak of ritual or folk performances or of communicative arts in the African setting, in view of the fact that one does not have in mind theatrical or other entertaining displays performed for the benefit of a passive public, but ceremonies of a social or religious nature into which dramatic elements are incorporated.[27]

The introduction of European-influenced drama was, moreover, initially limited to the urban elites. In the Arab world, in Arlette Roth's view, the appearance of the need for dramatic expression "probably has to be linked to the irruption of Western civilisation," since "theatre was unknown in classical Arabic literature. Though in certain literary genres such as the 'maqāma,' there are dramatic elements, these have never detached themselves or developed to form a distinct dramatic genre."[28] Theater in the modern sense, then, first appeared in the Arab world in Syria during the 1850s—with a translation of Molière into Arabic—and during the latter part of the nineteenth century Syrian émigrés introduced theater into Egypt. But there it remained an imported form, quite distinct from any popular folk tradition and largely derivative of European models, and so could offer pioneer Egyptian film makers no roots for a viable dramatic form that would relate directly to a popular audience.

Western theater was introduced into China even later. There, the first experiments were heavily influenced by Japanese borrowings from the West, and indeed the beginnings of a modern theater movement can be dated to 1907, when a group of Chinese students in Tokyo (where Dr. Sun Yat-sen then had his headquarters) staged a version of Alexandre Dumas's *La Dame aux camélias*. This production was imitated in Shanghai the same year, but it was not until after the overthrow of the Manchus in 1911 that the new ideas of drama began, in a limited way, to establish themselves:

> The new theatre commanded attention as a political and social propaganda medium in these early years, but no significant Chinese playwright appeared to develop it further, and production methods were rudimentary. A play would be prepared, often borrowing freely from other literary sources, and arranged in a number of scenes. It was then presented to the actors, who were relied upon to pull the story together on the stage. Although the substance of the dialogue was decided on beforehand, there was a good deal of improvisation.[29]

This style of performance by students and amateurs for a Western-educated urban audience was common in India, too. But there, as S. Theodore Baskaran has shown, the influence was gradually diffused by traveling groups of actors who shaped a popular form of commercial drama in which mythological subjects were treated with songs and rough comedy. Baskaran traces a continuity between such groups and the Tamil-language sound cinema of South India, which took over the traveling groups' performers and song writers, turned the primitive rural theaters into cinema halls, and prolonged the current of nationalist sentiment that had made itself felt initially within the crude popular drama of the itinerant performers.[30]

Early Chinese cinema shared the confusion of styles to be found in early modern drama in China, for example, in the continuing practice of assigning male actors to play female roles in otherwise realistically intended scenes. But after the upheaval of the May Fourth Movement of 1919 there was an increasing identity between stage and screen, as the new dramatists who emerged in the 1920s and 1930s became heavily involved in the most exciting period of Chinese cinema, between 1931 and 1937, when numerous important left-wing film productions were made in Shanghai.

Though modern theater was slower to establish itself in black Africa, there too it developed political importance as well as popular appeal. The interaction of local traditions, colonial influences, and political aspirations is well illustrated by the career of the Nigerian Hubert Ogunde, a number of whose plays were filmed in the late 1970s and early 1980s. Ogunde began producing as an amateur in a missionary context, as a student teacher and choir master, and his work shows a succession of Western influences. But as a child Ogunde had performed in the Alarinjo Theater (the classical Yoruba theater, in which the players are masked), the influence of which can be seen in his lifelong devotion to researching Yoruba traditions of music and dance. Moreover, he first achieved widespread fame in 1945 with an anticolonial piece, *Strike and Hunger.* His concern to promote a national culture is such that a recent Nigerian critic of this work has claimed that "through his theatre Ogunde was a leading figure in the nationalist movement."[31]

Notes

1. Cedric Robinson, "Domination and Imitation: *Xala* and the Emergence of the Black Bourgeoisie," *Race and Class* 22 (Autumn 1980): 147–48.
2. Pierre Jalée, *The Third World in World Economy* (New York: Monthly Review Press, 1969), p. 117.
3. Gérard Chaliand, *Revolution in the Third World* (Harmondsworth: Penguin, 1978), p. 27.
4. Arghiri Emmanuel, "White Settler Colonialism and the Myth of Investment Imperialism," *New Left Review,* no. 73 (May–June 1972): 48.
5. Ousmane Sembene, in *African Writers on African Writing,* ed. G. D. Killam (London: Heinemann, 1973), p. 150.
6. Amilcar Cabral, *Return to the Source* (New York: Monthly Review Press/ Africa Information Service, 1973), p. 47.
7. Ibid., p. 42.
8. Ibid., p. 43.
9. Frantz Fanon, *The Wretched of the Earth* (Harmondsworth: Penguin, 1967), pp. 178–79.
10. Ibid., p. 179.
11. Afrânio Coutinho, *An Introduction to Literature in Brazil* (New York: Columbia University Press, 1969), p. 45.
12. See William Lawson, *The Western Scar* (Athens: Ohio University Press, 1982).
13. Dorothy S. Blair, *African Literature in French* (Cambridge: Cambridge University Press, 1976), p. 182.
14. Jorge Luis Borges, "The Argentine Writer and Tradition," in *Labyrinths* (Harmondsworth: Penguin, 1970), pp. 214–15.
15. Terence Ranger, "The Invention of Tradition in Colonial Africa," in *The Invention of Tradition,* ed. Eric Hobsbawm and Terence Ranger (Cambridge: Cambridge University Press, 1983), p. 262.

16. Ibid., pp. 249–50.

17. Renato Constantino, "Notes on Historical Writing for the Third World," *Journal of Contemporary Asia* 10 (1980): 234.

18. Ali M. Mazrui, *The African Condition: A Political Diagnosis* (London: Heineman, 1980), p. 63.

19. Albert Memmi, *The Colonizer and the Colonized* (London: Souvenir Press, 1974), p. 110.

20. Richard M. Morse, "The Americanization of Languages in the New World," *Cultures* 3 (1979): 28.

21. George Pendle, *A History of Latin America* (Harmondsworth: Penguin, 1963), pp. 187–88.

22. Larzer Ziff, *Literary Democracy: The Declaration of Cultural Independence in America* (Harmondsworth: Penguin, 1982), p. 301.

23. Abiola Irele, *The African Experience in Literature and Ideology* (London: Heineman, 1981), p. 44.

24. Ibid., p. 54.

25. See "Kenya: The Politics of Repression," special issue, *Race and Class* 24 (Winter 1983).

26. Erik Barnouw and S. Krishnaswamy, *Indian Film* (New York: Oxford University Press, 1980), p. 13.

27. Ola Balogun, "Traditional Arts and Cultural Development in Africa," *Cultures 2* (1975): 159.

28. Arlette Roth, *Le théâtre algérien* (Paris: François Maspéro, 1967), p. 13.

29. A. C. Scott, *Literature and the Arts in Twentieth Century China* (London: George Allen and Unwin, 1965), pp. 36–37.

30. S. Theodore Baskaran, *The Message Bearers: Anglo-Indian Nationalist Politics and the Entertainment Media in South India, 1880–1945* (Madras: Cre-A, 1981), p. 41.

31. Ebun Clark, *Hubert Ogunde: The Making of Nigerian Theatre* (London: Oxford University Press, 1979), p. xi.

A No-Theory Theory of Contemporary Black Cinema

Tommy L. Lott

When film scholars are asked to decide which are best among a body of films they identify as "black," what is at stake is something more than merely the aesthetic question of what counts as a good black film. Indeed, they must consider a more fundamental definitional question regarding the nature of black cinema, a question which raises deeper issues concerning both the concept of black identity and the concept of cinema itself. I suspect that film criticism has not offered much assistance in clarifying the concept *black cinema* because there exist no uncontested criteria to which an ultimate appeal can be made to resolve these underlying issues. This scholarly morass must be understood in terms of the inherently political context in which the concept of black cinema has been introduced.

In his book *Black Film as Genre,* Thomas Cripps demonstrates how difficult it is to provide an adequate definition of black cinema. He employs a notion of black cinema that refers almost exclusively to theater films about the black experience that are produced, written, directed, and performed by black people for a primarily black audience (3–12). But this leaves us to wonder what to do with a well-known group of films about black people by white filmmakers. Although Cripps displays a rather tenuous allegiance to his initial statement of an essentialist paradigm, he has nonetheless presented an idea which lends credence to those who would exclude films such as King Vidor's *Hallelujah,* Shirley Clarke's

African American Review (formerly *Black American Literature Forum*), vol. 25, no. 2, 1991, pp. 221–36.

The Cool World, Michael Roemer's *Nothing But A Man,* Charlie Ahearn's *Wild Styles,* and John Sayles's *The Brother From Another Planet* from the newly emerging black canon. On strictly aesthetic grounds, however, these films may strike some critics as being better than many others which would more adequately satisfy Cripps's essentialist criteria.

Some black film cataloguers have sought to avoid the essentialist problem of being overly restrictive by opting for all-inclusive criteria. Klotman's *Frame by Frame: A Black Filmography* and Parish and Hill's *Black Action Films,* for instance, seem to identify films as black if they meet *any* of Cripps's several criteria. As might be expected, some critics have complained that not all of the films they list ought to count as black cinema.[1]

Missing from both the narrowness of essentialist criteria and the broadness of non-essentialist criteria are criteria that would account for the political dimensions of black filmmaking practices. Although audience reactions may vary from film to film, black people have a deep-seated concern with their history of being stereotyped in Hollywood films, a concern which provides an important reason to be skeptical of any concept of black cinema that would include works which demean blacks. Some would seek to abate this concern by specifying a set of wholly aesthetic criteria by which to criticize bad films about black people by both black and white filmmakers. Unfortunately, this approach contains undesirable implications for black filmmaking practices. We need only consider the fact that low-budget productions (e.g., *Bush Mama, Bless Their Little Hearts,* and *Killer of Sheep*) frequently suffer in the marketplace, as well as in the eyes of critics, when they fail to be aesthetically pleasing, or the fact that a film's success will sometimes be due largely to its aesthetic appeal, despite its problematic political orientation (e.g., *Roots* or *Shaka Zulu*).

For this reason, such commentators as Teshome Gabriel and Kobena Mercer ("Diaspora") have urged the need for film criticism to address the politics of black filmmaking practice, with an awareness that what is often referred to as "aesthetics" is linked with important issues pertaining to the control of film production and distribution. Incorporating aesthetics into a more politicized account of black filmmaking practices would seem to allow critics to evade the narrowness of the essentialist view, but there is some reason to wonder whether this move toward aesthetics would allow the accommodation of a strictly cultural criterion for the definition of black cinema without invoking a notion of "black aesthetics," upon which some reconstituted version of biological essentialism may again be reinstated.

The political aspects of the notion of aesthetics in film theory is sometimes shielded by the latent connection between biological essentialism and issues of control in film practice. We can, for example, see a tendency to racialize the political concern with control of the black film image in August Wilson's recent demand for a black director for the movie version of his play *Fences*: "Let's make a rule. Blacks don't direct Italian films. Italians don't direct Jewish films. Jews don't direct black American films. That might account for about 3 percent of the films that are made in this country" (71). Although Wilson's claim might be taken to commit him to accepting any director who is biologically black, he clearly would not want a black director who lacked the cultural sensibility required for a faithful rendering of his play. But if even a black director could prove unsatisfactory for aesthetic reasons, how do we make political sense of Wilson's demand for a black director, given that there could be some white director who might be more suitable from a cultural standpoint?

I want to advance a theory of contemporary black cinema that accords with the fact that biological criteria are neither necessary nor sufficient for the application of the concept of black cinema. I refer to this theory as a no-theory, because I want to avoid any commitment to an essentialized notion by not giving a definition of black cinema. Rather, the theoretical concern of my no-theory is primarily with the complexity of meanings we presently associate with the political aspirations of black people. Hence, it is a theory that is designed to be discarded when those meanings are no longer applicable.

The Aesthetic Critique of Blaxploitation

The history of black cinema can be roughly divided into four periods: Early Silent Films (1890–1920), Early Soundies and Race Films (1920–1945), Post-War Problem Films (1945–1960), and Contemporary Films.[2] With regard to the history of black cinema, the so-called "blaxploitation" period is a relatively recent, and short-lived, phenomenon. Although there has been a siphoning off of black audiences since the early days of race films, nothing approximating the Hollywood onslaught of the early seventies has occurred at any other time.

The term *blaxploitation* has been used to refer to those black-oriented films produced in Hollywood beginning in 1970 and continuing mainly until 1975, but in various ways persisting until the present (Miller; Pines, "Blaxploitation"; Ward). However, in addition to its being an historical index, the term is a way of labeling a film that fails in certain ways to represent the aesthetic values of black culture properly.[3] Mark Reid, for instance, expresses this view in his account of the shortcomings of blaxploitation era films:

> Having established the fact that there was a young black audience receptive to thoughts about violence, it should have been possible

> to create black action films that appealed to this audience while satisfying a black aesthetic. The commercial black action films of the 1970s, however, never reached this ideal because they were not independent productions or because black independent producers relied on major distributors. (25)

Although, as I shall indicate shortly, Reid's criticism rests on a misleading dichotomy between independent and non-independent films, his remarks inherently acknowledge the role that production and distribution play in shaping the aesthetic characteristics of a film. At a time of financial exigency, some Hollywood studios discovered that there was a large black audience starving for black images on the screen. This situation provided an immediate inducement for them to exploit the box office formula of the black hero (first male, later female) which, subsequently, became the earmark of the blaxploitation flick.[4]

Although there are many issues raised by blaxploitation era filmmaking that deserve greater attention, I want to focus on the problem of commercialism in order to highlight the influence of the market on certain aesthetic characteristics of black movies.[5] First of all, it needs to be stated, and clarified, that not all blaxploitation era films conformed to the box office formula. Some were not commercially oriented, while others were very worthwhile from a social and political standpoint.[6] To reduce them all to the hero formula, provided somewhat inadvertently by Van Peebles's *Sweet Sweetback's Baadasssss Song,* is to overlook their many differences in style, audience orientation, and political content.

Secondly, given the history of black cinema, there is a certain logic to the development of the box office formula. The idea of depicting black men as willing to engage in violent acts toward whites was virtually taboo in Hollywood films all the way through the sixties. But once the news footage of the sixties rebellions, along with the media construction of the Black Panther Party, began to appear, mainstream films such as *In the Heat of the Night* made an effort to acknowledge (albeit to contain) this "New Negro" (Ryan and Kellner 121–29). Even within these limits, however, what had made Malcolm X appear so radical to mainstream television audiences at that time was the fact that he had *publicly* advocated self-defense.

When *Sweetback* was shown in 1971, it was an immediate success with black audiences because it captured an image of self-defense that gave on-screen legitimation to violent retaliation against racist police brutality. Black heroic violence against white villains rapidly became a Hollywood commodity, and literally dozens of films were produced for black audiences that capitalized on this new formula. It is worth noting here that it was, in many respects, a Hollywood-induced taboo that created a need for such images in the black audience, a need which was then fulfilled by Hollywood. The ultimate irony is that, once these films began to proliferate, there was an organized effort in the black community to demand their cessation (see Miller 149).

It is also worth noting that *Sweetback* provoked a critical response that varied among different political factions within the black community, as well as among film critics. Community-based activists who opposed the film's image of black people ranged from cultural nationalists, who wanted a more culturally educational film, to middle-class black protesters, who wanted a film that projected a more positive image of the race (Reid 29). As Mark Reid has noted, the film's political orientation, quite interestingly, received both "high praise" from Huey Newton and the Black Panther Party newspaper in Oakland and "denunciation" from a Kuumba Workshop nationalist publication in Chicago (30). Although Newton was not alone in giving the film critical praise, his allegorical interpretation of the film's sexual imagery was not widely shared among critics, especially feminists concerned with its portrayal of women (Bowser 51).

The critical controversy around *Sweetback's* image of black people is not amenable to resolution on strictly aesthetic grounds, for *Sweetback* clearly represents some version of the black aesthetic. A political debate seems to have transpired between the film's supporters and detractors in an attempt to make the case for either accepting or rejecting the film. Indeed, some critics have argued that *Sweetback* lacks a politicized image (Reid 26), while others have argued that it politicized the image of black people to the point of lapsing into propaganda (Pines, "Blaxploitation" 123).

With regard to the role of aesthetics, blaxploitation era films pose a rather peculiar problem for a theory of contemporary black cinema. Can a film count as black cinema when it merely presents a blackface version of white films, or when it merely reproduces stereotypical images of black people?

Commentators have maintained quite different views in answer to this question. James Snead has argued for a very sophisticated notion of recoding that requires of black cinema what he calls the "syntagmatic" revision of stereotyped images through the selective use of editing and montage ("Recoding Blackness" 2). According to Snead, the syntax of traditional Hollywood cinema must be reworked to recode the black image effectively. However, against the backdrop of Hollywood's pre-blaxploitation era stereotype of black men as sexually castrated buffoons, what rules out the less sophisticated blaxploitation practice of substituting a highly sexualized black male hero who exercises power over white villains as an attempt to recode the Hollywood image of black men? Mark Reid asserts, with little hesitation, that "blacks who would find psychological satisfaction in films featuring black heroes have just as much right to have their tastes satisfied as do whites who find pleasure in white heroes such as those in Clint Eastwood and Charles Bronson films" (30). If the creation of black heroic images through role reversals can be considered a recoding technique utilized by blaxploitation filmmakers, then how does this practice compare with other, more avant-garde recoding practices of black independent filmmakers?

It can be quite troublesome for a theory of black cinema that relies too strongly on aesthetics to give an account of the influence of blaxploitation films on subsequent black independent films (see Taylor, "We Don't Need" 84–85). Given that aesthetic-based theories, such as Snead's, want to contrast black independent films with Hollywood-produced films about black people, where do blaxploitation films fit into such a juxtaposition? How do we make sense of the charge, brought by a black independent filmmaker, of a fellow black independent filmmaker's having irresponsibly produced a blaxploitation film?[7] The fact that the charge was made suggests that black independent filmmaking is not immune from the aesthetic pitfalls of blaxploitation cinema.

For present purposes, I am less interested in deciding the question of what films to count as blaxploitation than I am interested in the implication the appeal to aesthetics, inherent in the accusation, seems to carry for our understanding of the place of aesthetics in a theory of black cinema. To denounce a film, such as *Sweetback,* as exploitative is to suggest that aesthetic criteria provide the highest ground of appeal for deciding definitional questions regarding black cinematic representation, for the charge presupposes that there is some sense in which to produce a blaxploitation film is to have compromised black aesthetic values. What must be explained, however, is how such films stand in relation to independently made films that were not constrained to violate black aesthetic values in this way. Apparently, the term *independent* does not always mean that a filmmaker has eschewed market concerns. When a blaxploitation film is independently made by a black filmmaker for a black audience, however, to whom has the film's aesthetic orientation been compromised and, further, to what extent do such compromises affect a film's status as a black cinematic work?

Recently there has been a major shift towards independently produced blaxploitation films. This practice makes clear that the biologically essentialist view of black cinema (those films about black people, produced by a black filmmaker, for a black audience) is much too simplistic. One important implication of the aesthetic critique of blaxploitation is that certain aesthetic qualities of a film can sometimes count as much against its being inducted into the canon of black films as the filmmaker's race or the film's intended audience. While the insights derived from the aesthetic critique of black filmmaking practices are undoubtedly healthy signs of sophistication in black film commentary, we must not overlook the fact that these critiques also give rise to many difficulties connected with the problem of how film criticism should relate to a plurality of standards by which black films are evaluated.[8]

One such difficulty that must be faced by aesthetic-based theories of black cinema arises from the fact that, since the mid-1980s, there has been a growing interest in black-oriented cinema, especially black comedy, by white audiences. The success of black television sitcoms, as well as Arsenio Hall's nightly talk show, provides some indication that white audiences are more willing to indulge not so completely assimilated black people than network executives had previously supposed. Spike Lee's humorous social commentary has opened the door for other, similarly inclined, black filmmakers and television producers. All of this comic relief in the television and movie industry has been spearheaded, of course, by the mass appeal of Richard Pryor, Eddie Murphy, and Bill Cosby. Given their influence on the present context for black filmmaking, it seems that a theory of contemporary black cinema cannot postulate the black audience as a necessary ingredient.

A related difficulty that carries greater significance for our understanding of the influence of the crossover audience on the aesthetics of certain films about black people arises from the manner in which Eddie Murphy's attempt to signify on black minstrelsy has simply replaced the old-fashioned minstrel show. Murphy's success in Hollywood was quickly followed by that of his "black pack" cohort Robert Townsend, whose humorous criticism of Hollywood in his very popular film was largely reduced to a shuffle with a critique of itself.[9] As though the hegemony of the Hollywood industry were not enough to contend with, more politically astute filmmakers working in the realm of comedy, such as Spike Lee, are now challenged with finding ways to distinguish themselves from such neo-minstrelsy. Indeed, some filmmakers formerly aligned with the counter-hegemonic practices of the post-sixties black independent movement seem to have allowed the white audience for black-oriented humor to so influence their filmmaking that we now have a new generation of blaxploitation cinema. This influence is displayed in the Hudlin brothers' film *House Party,* which seems rigorously to avoid dealing with certain very pressing issues raised in the film (e.g., police brutality) in order not to offend the potential white audience. Unlike Spike Lee's probing satire, which engages in a black-oriented humor that sometimes seems intended to offend white audiences, *House Party* is closer to mindless slapstick.[10]

Although some film commentators have attempted to acknowledge the disparity between the aesthetic values of black audiences and the aesthetic values of filmmakers and critics, film criticism generally tends to adhere to a top-down view of aesthetics, as though audiences have no role to play in the determination of aesthetic values. What the black audience appeal of blaxploitation films (old and new) indicates, against the wishes of many film critics, is that it is misguided to suppose that a filmic work of art, or entertainment, has black audience appeal simply because it aims for a black audience by promoting certain black aesthetic values. In the case of the black independent cinema movement of the seventies in America, as well as the eighties black workshop movement in Britain, the attempt to reclaim and reconstruct a black film aesthetic that would somehow counteract the influence of Hollywood's blaxploitation filmmaking has, by and large, not been well-received by black audiences, although many of these films have been frequently presented at international festivals, in art museums, and in college courses devoted to film study.[11] How can we best understand the fact that films which aim to present a more authentic black aesthetic are largely ignored by and unknown to black audiences, while being extremely well-received in elite white film circles? Despite their admirable political orientation, such films seem to have achieved the status of art-for-art's sake, with mainly an all-white audience appeal.

This lack-of-a-black-audience problem shows the need to resist the tendency of aesthetic-based theories of black cinema to position the aesthetic values of the black artist above those of the black audience. In order for black film commentary to acknowledge more pluralistic criteria by which to assess the artistic value of cinematic works, some weight must be given to the viewpoint of black audiences, inasmuch as it is imprudent at best continually to posit a black aesthetic which very few black audiences share.

Some of these considerations regarding the audience crossover phenomenon in contemporary black cinema argue against the cultural essentialist attempt to define black cinema in terms of aesthetics. As the divisions between independent and Hollywood-produced films about black people begin to dissolve, as a result of the mainstream market for both, it has become extremely difficult to maintain that either a black filmmaker or a black audience is required for a film with a black orientation. To see this, we need only consider the fact that, in addition to his crossover status in the record industry, Prince is virtually neck-and-neck with Spike Lee as a filmmaker, each having four major releases. There is no reason to suppose that, despite a preference among commentators for Spike Lee's version of the black aesthetic, the aesthetic in Prince's movies has any less box office appeal to much of the same audience.[12]

The need for an essentialist theory diminishes, along with the idea of a monolithic black film aesthetic, once we realize that there is no monolithic black audience. There certainly are black-oriented films, some of which are much better than others, but not all of those approved by critics manage to touch base with black audiences (e.g., *To Sleep with Anger*) and many of those condemned by critics have become black audience classics (e.g., *Superfly*). These facts may be difficult to accept, but to advocate a "better" cinema which is significantly different requires a political argument. I will now turn to consider the argument I think is presently most viable in a politically confused era dominated by neo-conservative ideology.

Black Identity and Black Cinema

Before I take up the question of how politics and aesthetics can be situated into a theory of black cinema, I would like to insert a word of clarification regarding the prevailing use of the term *cinema* to refer to films as such; i.e., movies that were made to be shown in theaters. I believe that this restrictive usage is unfortunate, since some fairly good films about black people have been made for television.[13] The misconception that underlies this narrow focus on box office movies is exacerbated by the fact that some of the most innovative black filmmaking is presently occurring in music videos. Indeed, the dominant influence of television on black popular culture has some rather interesting implications for black film practices which no theory of black cinema can afford to overlook. Because black urban youth culture has been visually promulgated primarily through television, this segment of the black movie audience has been heavily influenced by black images presented on television. Added to this television orientation is a large black youth market for blaxploitation films on video cassettes. These influences are displayed quite regularly in what Nelson George refers to as "blaxploitation rap"; i.e., rap lyrics that have been heavily influenced by blaxploitation films.[14] It is, to say the least, perilous for filmmakers interested in reaching black youth to ignore the single most important medium of visually representing their cultural values.

In Britain, black filmmaking and television are much more structurally connected, since the workshops produce their films for Channel 4 (see Fountain; Pines, "Cultural"). Undoubtedly, this structural relation between filmmaking and television will eventually obtain in America once high-definition television is introduced since, with the advent of this new technology, movies, as such, seem certain to be superceded by television. For all these reasons, I think it wise at this point to expand the concept of cinema to include television.

With regard to politics, there is a very good reason that the biological version of the essentialist definition of black cinema will invariably fall short. Any definition which requires films to be made by black filmmakers in order to be included in the category of black cinema will simply not match the ambivalence engendered by having to place biological over cultural criteria in deciding questions of black identity. This does not mean that, generally speaking, most of us have no idea of what to count as a black film. Indeed, the definition of black cinema is a problem by virtue of the fact that, whether it is based on biological or cultural criteria, its viability can easily be called into question.

The Du Boisian worry about the adequacy of biological criteria as the ultimate ground of appeal when faced with questions of black identity poses the greatest difficulty for the essentialist notion of black cinema. For Du Bois, the problem stems from the fact that there is no agreement about how best to define a black person, although there is some sense in which we all operate with some ideas about what constitutes black identity.[15] We need only consider the manner in which we must still grapple with the age-old problem of the "non-black" black person; i.e., the person who, though biologically black, does not identify with black culture. Although there can be little doubt that, in the context of the American system of apartheid, the question of whether a particular person counts as black is most often decided by skin color and physical appearance, there are numerous instances in which this honor is withheld strictly on cultural grounds. It is far too common for black people to feel the yoke of oppression at the hands of a white-identified black person. Consequently, as someone perceived to be disloyal to the group, an overly assimilated (Eurocentric) black person can sometimes lose his or her standing in the eyes of other black people. In such cases we can notice how the tension between biological and cultural criteria of black identity is resolved in terms of a political definition of black people.[16] It is for some reason such as this that I am motivated to develop the concept of black cinema within the context of a political theory.

The Concept of Third Cinema Revisited

Without any pretense that I can offer a replacement for the essentialist definition of black cinema, I want to suggest why I think the Third Cinema movement of the sixties seems to have been on the right track, although in America certain mainstream cooptational factors have basically derailed it. According to various conflicting reports, the advocates of Third Cinema have come under heavy criticism lately for being, of all things, overly nationalistic.[17] Unfortunately, in an attempt to address this worry, some commentators tend needlessly to equate nationalism with the essentialist view.[18] But the concept of Third Cinema should not be saddled with the myopic vision of essentialists who are constrained by an overemphasis on biological criteria for resolving questions of national identity.[19] What makes Third Cinema third (i.e., a viable alternative to Western cinema) is not exclusively the racial make-up of a filmmaker, a film's aesthetic character, or a film's intended audience, but rather a film's political orientation within the hegemonic structures of post-colonialism. When a film contributes ideologically to the advancement of black people, within a context of systematic denial, the achievement of this political objective ought to count as a criterion of evaluation on a par with any essentialist criterion.

The best way to meet the criticism that the concept of Third Cinema is too vague because it allows under its rubric many diverse cultural groups is to recognize that this objection misleadingly imputes an uncontested essentialist paradigm.[20] The Third Cinema movement represents a break with, and resistance to, the cultural imperialism fostered by the global expansion of the Hollywood industry. There is an important sense in which it aims to do what Hollywood has done—namely, to reach beyond national boundaries. There is no reason to deny that cultural diversity is a problem among the many ethnically distinct black people living together in America, much less a problem among various Third World people from widely different backgrounds in faraway places. But clearly if Europeans, who for centuries have waged war against each other, and are still caught up in their own ethnic rivalries, can construct a concept of themselves as a globally dominant white group, how can it be so much more objectionable for non-white people to construct a global counter-concept by which to defend themselves? The white cultural nationalism of Hollywood's Eurocentric empire requires something like a Third Cinema movement to help non-white people survive the oppressive and self-destructive consciousness that empire seeks to perpetuate.

With regard to black filmmaking practices, the concept of Third Cinema provides the rudiments of a theory of black cinema that is most conducive to this political function. As a primarily oppositional practice engaged in resistance and affirmation, black cinema need not be presently defined apart from its political function (see Espinosa). I call this a no-theory theory because I see no need to resolve, on aesthetic grounds, the dispute over what counts as blaxploitation. Neither do I see a need to choose between realist and avant-garde film techniques.[21] I am more interested in understanding how any aesthetic strategy can be employed to challenge, disrupt, and redirect the pervasive influence of Hollywood's master narrative. To accomplish this decidedly political objective, black filmmaking practices must continue to be fundamentally concerned with the issues that presently define the political struggle of black people. Hence, I want to advance a theory of black cinema that is in keeping with those filmmaking practices that aim to foster social change, rather than participate in a process of formulating a definition of black cinema which allows certain films to be canonized on aesthetic grounds so as to occupy a place in the history of cinema. The theory we need now is a political theory of black cinema that incorporates a plurality of aesthetic values which are consistent with the fate and destiny of black people as a group engaged in a protracted struggle for social equality.

Notes

1. Phyllis Klotman's black filmography prompted Gladstone L. Yearwood to complain that "to identify a black film as any film with black faces is to trivialize or nullify a definition of black film" ("Towards" 68–69). In his review of *Black Action Films,* Roland Jefferson takes Parish and Hill to task for including *Rocky I–IV* as black films (22).

2. Various periodizations have been offered by other commentators. See, for instance, Snead, "Images"; Taylor, "L.A."

3. In this sense, Fred Williamson's independently produced films (which have been continuous since his participation in the early phase of Hollywood blaxploitation) would count as a perpetuation of this style. Eddie Murphy's recent *Harlem Nights* is a throwback to Williamson's *Black Caesar* and *Hell Up in Harlem.* I would also include the recent spate of hip hop movies as a neo-blaxploitation genre, although, within this genre, we must again distinguish "positive-image" films such as Harry Belafonte's *Beat Street* from more violence-laden films such as Run DMC's *Tougher Than Leather.*

4. For a discussion of the cooptational use of black heroic characters to legitimate oppression, see Gladstone Yearwood's "The Hero." David E. James provides a wise bit of cautionary reflection on the bildungsroman narrative of *Sweetback* as a self-defeating contributor to the film's commodification into blaxploitation (135–37). And Clyde Taylor ("We Don't Need") takes issue with the master narrative in all of its various guises.

5. For instance, there is a need to examine more fully the transition period in the late sixties as a precursor to blaxploitation era filmmaking, especially with regard to independently produced films about black people. Some attention should also be given to the carryover effect of blaxploitation era films on the image of black people in mass-audience films and to the intertextual influences of blaxploitation on television programming.

6. Several factors helped shape the movie industry's multifarious output of blaxploitation era films, but the most outstanding was the market orientation of each film. Given that some of the larger budgeted productions (e.g., *100 Rifles, The Great White Hope,* and *The Learning Tree*) were intended for a mainstream audience, whereas low-budget productions (e.g., *Superfly, Blacula,* and *Coffey Brown*) were limited to box office showings in black communities, it would be a serious oversight to ignore the guerrilla tactics employed by Melvin Van Peebles to produce *Sweetback* and Bill Gunn to film *Ganja and Hess.* The avant-garde styles mastered by Van Peebles and Gunn owe much to the clandestine context in which their projects were pursued and contrast sharply with the more standard approaches displayed in mainstream productions such as *Claudine, A Hero Ain't Nothing But A Sandwich, The River Niger, Brothers, The Spook Who Sat by the Door,* and *Gordon's War.* These latter films, nonetheless, were a far cry from the more typical black action movies of the period.

7. See the text of the panel discussion on *Sweetback* with Melvin Van Peebles, St. Clair Bourne, Haile Gerima, and Pearl Bowser held at a conference on black independent filmmaking at Ohio State University in 1981 (Yearwood, "*Sweet*").

8. This point was brought to my attention in a fall 1987 lecture at Northeastern University, during which Clyde Taylor presented an analysis of the class differences in audience reactions to *Sweetback* as a methodological device by which to interpret the film. See the very interesting discussion of this issue by Mercer in his "Recoding Narratives," but also see Mercer and Julien, and Willemen.

9. In a similar vein, Keenan Ivory Wayans's *I'm Gonna Git You Sucka* and his television show *In Living Color* present a black-oriented variety of post-Eddie Murphy humor that relies heavily on the ridicule of white stereotypes of black people.

10. The social taboo against public statements regarding the media hype of Larry Bird as the greatest basketball player ever was quite deliberately violated by Lee in *She's Gotta Have It.* Lee seemed to rail against the de facto censorship of what many black people believe about Larry Bird by having his character Mars Blackmon flaunt anti-Bird jokes in the face of the audience which had witnessed Isaiah Thomas's being coerced on national television to demeaningly recant his truthful comments regarding the racist commentary in sports broadcasting.

11. For a pointed discussion of this dilemma facing black independent filmmakers, see Taylor, "Black Films"; Gilroy. See also Larry Rohter's discussion of Charles Burnett's *To Sleep with Anger.*

12. Prince's films remain closer to Hollywood's assimilationist paradigm of "crossover" black cinema, whereas Spike Lee entered the crossover market with black-oriented films that are closer to the black independent tradition.

13. I have in mind here, specifically, films such as *The Killing Floor, Minstrel Man, The Autobiography of Miss Jane Pittman, Go Tell It on the Mountain,* and *The Women of Brewster Place.*

14. Some rap artists have characterized their cultural practice as "black America's TV station" (see Leland 48).

15. Du Bois argued for a socially constructed concept of black identity that black people should "invent" to advance themselves (483–92).

16. Within the unspoken norms of African-American culture, black people with a caucasian appearance generally bear the burden of proving their loyalty, given that they have an option to "pass," despite their known biological heritage.

17. See the alternative accounts of Kobena Mercer ("Third Cinema"), David Will, Paul Willemen, and Clyde Taylor ("Eurocentrics").

18. Willemen displays this tendency when he attempts to utilize Bakhtin's thesis regarding socio-historical specificity to reconstruct Gabriel's internationalist account of Third Cinema practice (see 23ff.).

19. While it would be inaccurate to attribute this view to Willemen, some of his declarations lend themselves to an interpretation along these lines. For instance, he maintains that "the question of the national cannot be divorced from the question of Third Cinema" (20) and that those engaged in black film practices must refuse "to homogenise every non-Euro-American culture into a

globalised other" (29). But surely Willemen does not mean to deny the possibility of new social formations, perhaps international in scope, which stand opposed to neo-colonial structures that are ultimately rationalized on biological notions of national identities.

20. Stuart Hall ("Cultural Identity" and "New Ethnicities") has advocated an extreme version of the non-essentialist view of cultural practice.

21. For a critical discussion of how black films that are modernist in style have gained a greater currency than those which are steeped in realism, see Willamson; Fusco. See also Valerie Smith's insightful commentary on what counts as an "experimental" black film.

List of Works Cited

Bowser, Pearl. "Sexual Imagery and the Black Woman in Cinema." Yearwood, *Black Cinema* 42–51.

Cham, Mbye B., and Claire Andrade-Watkins, eds. *BlackFrames: Critical Perspectives on Black Independent Cinema.* Cambridge: MIT P, 1988.

Cripps, Thomas. *Black Film as Genre.* Bloomington: Indiana UP, 1978.

Davis, Mike, et al., eds. *The Year Left 2: An American Socialist Yearbook.* London: Verso, 1987.

DuBois, W. E. B. "The Conservation of Races." *Negro Social and Political Thought 1850–1920.* Ed. Howard Brotz. New York: Basic, 1966. 483–92.

Espinosa, Julio Garcia. "For an Imperfect Cinema." 1969. *Liberation, Socialism.* Vol. 2 of *Communication and Class Struggle: An Anthology in 2 Volumes.* Ed. Armand Mattelart and Seth Siegelaub. New York: International General, 1979–83. 295–300.

Fountain, Alan. "Channel 4 and Black Independents." Mercer, *Black Film* 42–44.

Fusco, Coco. "The Other is In." Mercer, *Black Film* 37–39.

Gabriel, Teshome H. "Third Cinema as Guardian of Popular Memory: Towards a Third Aesthetics." Pines and Willemen 53–64.

Gilroy, Paul. "Nothing But Sweat Inside My Hand: Diaspora Aesthetics and Black Arts in Britain." Mercer, *Black Film* 44–46.

Hall, Stuart. "Cultural Identity and Cinematic Representation." *Framework* 36 (1989): 68–81.

———. "New Ethnicities." Mercer, *Black Film* 27–30.

James, David E. "Chained to Devilpictures: Cinema and Black Liberation in the Sixties." Davis 125–38.

Jefferson, Roland. Rev. of *Black Action Films,* ed. James R. Parish and George H. Hill. *Black Film Review* 5.4 (1989): 21–22.

Klotman, Phyllis, ed. *Frame by Frame: A Black Filmography.* Bloomington: Indiana UP, 1979.

Leland, John. "Armageddon in Effect." *Spin* 4.6 (1988): 46–49, 76.

Mercer, Kobena, ed. *Black Film/British Cinema.* London: Institute of Contemporary Arts, 1988.

———. "Diaspora Culture and the Dialogic Imagination: The Aesthetics of Black Independent Film in Britain." Cham and Andrade-Watkins 50–61.

———. "Recoding Narratives of Race and Nation." *Black Film* 4–14.

———. "Third Cinema at Edinburgh: Reflections on a Pioneering Event." *Screen* 27.6 (1986): 95–102.

Mercer, Kobena, and Isaac Julien. "De Margin and De Centre." *Screen* 29.4 (1988):2–10.

Miller, James A. "From Sweetback to Celie: Blacks on Film into the 80s." Davis 139–59.

Parish, James R., and George H. Hill, eds. *Black Action Films.* Jefferson: McFarland, 1989.

Pines, Jim. "Blaxploitation: 'Ethnic Adjustment' in Hollywood." *Blacks in Films: A Survey of Racial Themes and Images in the American Film.* London: Studio Vista, 1975. 118–27.

———. "The Cultural Context of Black British Cinema." Cham and Andrade-Watkins 26–36.

Pines, Jim, and Paul Willemen, eds. *Questions of Third Cinema.* London: British Film Institute, 1989.

Reid, Mark A. "The Black Action Film: The End of the Patiently Enduring Black Hero." *Film History* 2 (1988): 23–36.

Rohter, Larry. "An All-Black Film (Except for the Audience)." *New York Times* 20 Nov. 1990: B1.

Ryan, Michael, and Douglas Kellner. *Camera Politica.* Bloomington: Indiana UP, 1988.

Smith, Valerie. "Reconstituting the Image: The Emergent Black Woman Director." *Callaloo* 11 (1988): 709–19.

Snead, James. "Images of Blacks in Black Independent Films: A Brief Survey." Cham and Andrade-Watkins 16–25.

———. "Recoding Blackness: The Visual Rhetoric of Black Independent Film." *Circular for The New American Filmmakers Series* 23. New York: Whitney Museum of American Art, 1985. 1–2.

Taylor, Clyde. "Black Films in Search of a Home." *Freedomways* 23 (1983): 226–33.

———. "Eurocentrics Vs. New Thought at Edinburgh." *Framework* 34 (1987): 140–48.

———. "The L.A. Rebellion: A Turning Point in Black Cinema." *Circular for The New American Filmmakers Series* 26. New York: Whitney Museum of American Art, 1986. 1–2.

———. "We Don't Need Another Hero: Anti-Theses on Aesthetics." Cham and Andrade-Watkins 80–85.

Ward, Renée. "Black Films, White Profits." *Black Scholar* 7.8 (1976): 13–24.

Will, David. "Edinburgh Film Festival, 1986." *Framework* 32–33 (1986): 197–209.

Willemen, Paul. "The Third Cinema Question: Notes and Reflections." Pines and Willemen 1–29.

Williamson, Judith. "Two Kinds of Otherness." Mercer, *Black Film* 33–37.

Wilson, August. "I Don't Want to Hire Nobody Just 'Cause They're Black." *Spin* 6.7 (1990): 71.

Yearwood, Gladstone L., ed. *Black Cinema Aesthetics: Issues in Independent Black Filmmaking*. Athens: Ohio U Center for Afro-American Studies, 1982.

———. "The Hero in Black Film: An Analysis of the Film Industry and Problems in Black Cinema." *Wide Angle* 5.2 (1982): 42–51.

———. "*Sweet Sweetback's Baadasssss Song* and the Development of the Contemporary Black Film Movement." Yearwood, *Black Cinema* 53–66.

———. "Towards a Theory of a Black Cinema Aesthetic." Yearwood, *Black Cinema* 67–81.

Dialogic Modes of Representing Africa(s)

Womanist Film

Mark A. Reid

One has only to peruse the shelves of a university bookstore or page through "academic" journals to discover that members of the academy have indirectly reified feminism as a discourse by, for, and about white, heterosexual, Western women. Corrective energy is best directed toward the development of theories of representation which describe the voices that mainstream feminism ignores, for the employment of critical approaches which consider race, class, and sexual affiliation would cure academic feminism of its narcissistic gaze upon the white, middle-class female subject.

In an effort to assist in the building of a black feminist theory of film production and reception, this paper presents a "womanist" interpretation of three aspects of black independent film. First, I explore the term *black womanist film* in the context of the production and reception of any womanist film. Second, I discuss the politics of black womanist discourse as a form of resistance to a raceless feminism and a phallocentric pan-Africanism. Third, I analyze three spectatorial relationships that black womanist films elicit from two "interested" spectatorial groups—feminists and pan-Africans. Finally, informed by my description of black womanist film, its politics of blackness and womanness, and its three receptive modes, I interpret how certain narrative films present African and African diasporic women's subjectivity as a polyvalent idea.

I do not intend to prescribe conditions for black womanist filmmaking or its spectatorial readings. Nonetheless, I do consider how certain

African American Review (formerly *Black American Literature Forum*), vol. 25, no.2, 1991, pp. 375–88. Reprinted by permission of the publisher and author.

processes of representation and reception permit different readings of black womanist films. Finally, when I speak of African or black women, I am making reference to the many black African women who exist, survive, and struggle in Africas throughout the world. Therefore, I am explicating an international black womanist theory—a theory which rejects the closures of nation, race, gender, and class exclusivity. However, I must limit this present study to an analysis of black-directed independent films.

Black Womanist Film: A Working Definition

Throughout this essay, I will employ the term *black womanist film* to refer to narrative constructions as well as viewing positions which permit "womanish," as opposed to "girlish," processes of black female subjectivity. Films belonging to this category dramatize the shared experiences of black women. I have borrowed the term *womanist* from Alice Walker, who provides, as one of its principal definitions:

> A woman who loves other women, sexually and/or nonsexually. Appreciates and prefers women's culture, women's emotional flexibility . . . and women's strength. Sometimes loves individual men, sexually/or nonsexually. Committed to survival and wholeness of entire people, male *and* female. Not a separatist, except periodically, for health. Traditionally universalist, as in: "Mama, why are we brown, pink, and yellow, and our cousins are white, beige, and black?" (xi).

In relation to black film, the term *womanism* refers to certain black-oriented films (not their authors) and reading strategies whose narrative and receptive processes permit polyvalent female subjectivity.

The term *black womanist film* describes two levels of filmic production: (1) the narrative content which constructs black womanist subjectivity and (2) the various processes by which an audience might receive the narrative's construction of this subjectivity. The two levels conjoin and articulate an ideology of black womanism as a twofold process of construction and reception. *Black womanist film* does not describe *all* films made by African and African diasporic women. However, black womanist film results from imaginatively representing the socio-psychic and socio-economic experiences of African and African diasporic women. This film calls forth "black womanist spectators" and creates a spectatorial space for such an audience.

The concept *black womanist spectatorship* should *not* be taken "to refer directly to the [black] woman who buys her ticket and enters the movie theater as the member of an audience, sharing a social identity but retaining a unique psychical history. Frequently, [it] do[es] not even refer to the spectator as a social subject but, rather, as a psychical subject, as the effect of signifying structures."[1] Black womanist spectatorship, then, is a socio-psychical process, not a biological trait. It cannot totally exclude or include its audience based on race or gender.

Womanism and Post-Negritude: Theorizing Black Experience

Black womanism is a form of resistance to a raceless feminism and a phallocentric pan-Africanism. As a theoretical tool, it requires that one scrutinize how its closest supporter decodes black womanist film. This is especially true for feminism and pan-Africanism, because each of these movements has ignored the triple oppression of black women. Black female subjectivity, as woman and African, is historically positioned at the boundaries of gender, sexuality, and race. This subjectivity conjoins the two limit-texts of feminism and pan-Africanism, but has been heretofore on the margins of these discourses. As an effect of a womanist ideology, black womanist films resist dramatizing one-dimensional struggles which ignore the black woman's three-pronged oppression. Therefore, Safi Faye's *La passante (The Passerby,* Senegal, 1972) criticizes the sexist expectations of a white French man as well as a black African man; Sara Gomez's *De cierta manera (One Way or Another,* Cuba, 1974/1977) presents sexism in post-revolutionary Cuba; and Michelle Parkerson's *Storme: The Lady of the Jewel Box* (USA, 1987) constructs black gay history while documenting the life of Storme, a male impersonator. These black womanist films question patriarchal and heterosexist notions of black female subjectivity. Additionally, each film creates a spectatorial position which speaks to micro- as well as macro-struggles against *phallic* forms of knowledge and power within and without the black community.[2]

A theory and practice of black film, if based on a womanist ideology of the polyphony of blackness, initiates *possible* receptive processes. A womanist reception occurs only after an "interested" spectator actively participates in a systematic critique of the singularity of canonical (con)texts. Stressing the importance of an "interested" reader in the production of a plural text, Catherine Belsey writes,

> In the writable . . . wholly plural text all statements are of indeterminate origin, no single discourse is privileged, and no consistent and coherent plot constrains the free play of the discourses. The totally writable, plural text does not exist. At the opposite extreme, the readable text is barely plural. The readable text is merchandise to be consumed, while the plural text requires the production of meanings through the identification of its polyphony. (105)

Thus, black womanism, as a theory of reception and production, requires an "interested" spectator to decode the plurality of (con)texts, which include intra- as well as interracial forces that dehumanize the community. Equally, black womanist critical strategies deconstruct narrative systems and viewing positions which reduce racial, sexual, and class differences to one of gender. Black womanism, as Alice Walker suggests, represents universalist notions of blackness which conjoin African and African diasporic cultures. Consequently, the black womanist film project constructs a post-*Negritude* theory of reception and production.

Before defining *post-Negritude,* I would like to suggest the polyphonic quality of the movement and concept that preceded it and shares in its global concerns. Nigerian philosopher Abiola Irele writes that Negritude

> has acquired, in the way it has been used by different writers, a multiplicity of meanings covering so wide a range that it is often difficult to form a precise idea of its particular reference at any one time or in any one usage. The difficulty stems from the fact that, as a movement and as a concept, Negritude found its origin and received a development in a historical and sociological context whose implications for those whom it affected were indeed wide-ranging, and which ultimately provoked in them a multitude of responses [readings] that were often contradictory, though always significant. . . . The term has thus been used in a broad and general sense to denote the black world in its historical being, in opposition to the West, and in this way resumes the total consciousness of belonging to the black race, as well as an awareness of the objective historical and sociological implications of that fact. (67)

Similarly, *post-Negritude* refers to any moment when members of the black community, through their literature, art, and politics, recognize that black culture "is, concretely, an open-ended, creative dialogue of subcultures, of insiders and outsiders, of diverse factions." Correspondingly, these members share a language of black subjectivity that celebrates "the interplay and struggle of regional dialects, professional jargons, generic commonplaces, the speech of different age groups, individuals, and so forth" (Clifford 46). Contrary to Negritude, the recognition and shared productivity of the post-Negritude project result from the active participation of an *interested* audience that decodes the black subject using discourses which surround and construct representations of gender, race, class, sexuality, and nation. An "interested" audience scrutinizes the incorporation of these images as an internal regulatory agency of the representation as well as themselves. Through this shared post-Negritude recognition, respective black audiences will question their own (and, by implication, others') interpretations (Hutcheon 180).

Reception: Resistance, Accommodation, Assimilation

In the analysis of black womanist films, I propose three general modes of reception—resistance, accommodation, and assimilation. These three receptive modes represent the dialogic quality of any particular black womanist film. In the *resistance* mode, the source of the rejection of a particular black womanist film lies in its inability to mirror a given audience's partial vision of themselves. The desired image that a group holds of itself may be derivative of an authorized hegemonic image, a resistant radical image, or a combination of the two which produces a constant tension for both the viewer and artist in question. This tension combines race, sexuality, and gender subjectivity. The blurring of racial, sexual, and gender hierarchies permits free zones of discourse and makes the black womanist project a most dynamic movement. Theorizing this post-Negritude moment permits one to understand the reception of any black womanist film.

Since spectatorship is a psychical investiture, certain black womanist texts permit spectatorial positions that criticize sexism but maintain black communal solidarity. Contrarily, black womanist films which depict non-sexist men may threaten the psychical desires of certain feminists—for example, separatists who deplore any feminism which includes men. In addition, non-black feminist spectators who maintain a hierarchic allegiance to a raceless, classless, and nondescript sexuality may deplore this form of black feminism, because it criticizes racial, class, and sexual discrimination within the women's movement. Nonetheless, such feminists may still enjoy black womanist films. Enjoyment demands that black womanist films represent, and their viewers/listeners identify with, some *nexus of desire* which results in *accommodative* readings of these films. The same criterion is equally true for raceless feminists and for phallocentric pan-African spectators.

Certain pan-Africans might regard black womanist narratives as too hostile toward black patriarchal figures or might view black feminism as a threat to black communal solidarity because, according to them, the natural place for black women is within a nuclear family as the provider of moral support to her husband. Generally speaking, this point of view naturalizes the subordination of women; it reflects a phallic hierarchical system which processes false notions of black masculinity and femininity. Consequently, this phallic type of pan-African thought denies both male and female pan-Africans the ability to appreciate black womanist ideology as represented in film. Black psychologist Vickie M. Mays writes,

> The Afro-American woman has been denied power and privilege. She has been raised expecting to work, as she will need to assist in supporting her family. She will also be asked to do all she possibly can to advance the Black man and the Black race—at the cost of ignoring the oppression of sexism. Indeed, the Black woman has been taught from early childhood that one way to survive in this society is through marriage or in a male-female relationship. (75)

Pan-Africans who deny or devalue the oppression of black women are as demeaning as white feminists who disregard racial and class factors. Yet, on viewing black womanist films, these same pan-Africans might identify with the cinematic use of African and African diasporic elements, while resisting the elements that evoke identification among non-black feminists. Thus, like some non-black feminists, pan-Africans may assume an accommodative viewing relationship to a black womanist film.

Contrary to the two discussed accommodative positions, an *assimilative* reception tends to accept a film as a "realistic" vision, rather than an imaginative representation open to *revision* of black womanhood. Assimilative spectatorial positions, as organizing concepts, promote authoritarian conformity among creator, discursive apparatus, and consumer. Authoritarian discourses eliminate the tension between the desire to resist and the desire to accommodate. In the assimilative mode, audience members' consumptive habits are eased when their identity (their imagined subjectivity) is not threatened by what they feel are contradictory images of themselves. Singularity is also part of the spectatorial mode which produces resistance among certain feminists and pan-Africans when they view the more radical aspects of black womanist ideology as constructed in films.

The authoritarian discourse process is present in rigid systems as well as those contesting on the margins of discourse. Like any discursive apparatus, the black womanist film has an assimilative mode of construction and reception capable of denying "unauthorized" readings. If effective, this mode protects against uncritical acceptance and duplication of black womanist discursive strategies. Once black womanist theory moves from the periphery to the center, assimilative readings of black womanist discourses might retain their protective strategies and oppose their initial, heteroglossic conception of black womanism. The move from liberation to repression occurs because modernist projects require a closure of their *imagined* boundaries—a reflection of hegemony's need to contain polyvalent subjectivity by maintaining segregated sites of "resistance," such as women's liberation, gay liberation, and black liberation. Black womanist subjectivity, however, blurs these imagined boundaries of black selfhood. This post-Negritude blurring reflects the dialogic possibilities of blackness in womanist film.

Black Womanist Praxis

Some of the earliest work of black women filmmakers appeared in the 1970s and was filmed by women from the Americas and Africa. Madeline Anderson, Sarah Maldoror, Safi Faye, and Sara Gomez are representative of the first wave of black womanist filmmakers. The beginning of an international black feminist film practice is reflected in Anderson's *I Am Somebody* (USA, 1970), Maldoror's *Des Fusils pour Banta (Guns for Banta,* Guinea-Bissau, 1971), and Gomez's *De cierta manera (One Way or Another,* Cuba, 1974/1977). Admitting the historical importance of these filmmakers and their works, I confine my analysis to Alile Sharon Larkin's *A Different Image* (USA, 1981), Safi Faye's *La passante (The Passerby,* Senegal, 1972), and Kathleen Collins's *Losing Ground* (USA, 1982). I will conclude with some general statements on black womanist films and post-Negritude.

In 1981, Alile Sharon Larkin focused on sexism and African heritage in *A Different Image.* Interestingly, the film previewed at the Second International Women Filmmakers Symposium, which underlines the film's importance to this audience. In addition to being a woman-centered film. *A Different Image* is equally resilient in its celebration of a pan-African consciousness which is at odds with the sexism and racism of the West. Larkin says, "The film deals with assimilation . . . of Western sexism and how racism is inherently a part of that, . . . can't be separated from it. When I say that the racism and the sexism are inseparable, this also applies to definitions of manhood as well as womanhood" *("Your Children").* Larkin underscores the black womanist belief that racism is an inherent part of sexism, and to ignore this fact is counterproductive to the black womanist struggle for equality.

The film focuses on Alana, a young black woman, and her friend Vincent, an equally young black man. After a male acquaintance has ridiculed Vincent's platonic relationship with Alana, Vincent tries to assert his manhood by forcing himself on Alana, who has fallen asleep on his bedroom floor. After Alana accuses him of attempted rape and severs their friendship, Vincent, missing her companionship, reexamines the question of the appropriate kind of relationship between a man and a woman. Finally, he initiates a reconciliation by giving her earrings shaped like the Egyptian goddess Isis.

In the film, Larkin focuses on a touchy gender-related problem within the black community: Black women are expected to submit to the sexual desires of their black male friends. Larkin does not limit this problem to America. By introducing the film with a collage of photographs of black women, and by using an African goddess as the symbol of Vincent's effort of reconciliation, Larkin underscores the correspondence between the experiences of black women in America and Africa. She writes,

> If you look at Africa today, you see that people there face the same problem with sexism[; therefore,] I am saying that there were other cultures and other ways that men and women related to each other . . . before the African slave trade, before Islam, before the patriarchal religions came in. So, we have to study those ways before we [black people] took on these [patriarchal] values. (*"Your Children"*)

Larkin understands the need for solidarity among black filmmakers around a commonly shared ideology that reflects a post-Negritude openness to feminism, classism, and sexuality. She writes, "An ideology or belief which attempts to compartmentalize the nature and form of the oppression of African people solely into gender and/or class is ultimately destructive to our achieving genuine equality and liberation" ("Black Women" 158). Larkin refuses to permit authoritarian ideologies of opposition to determine or speak for a black womanist film practice. She recognizes the ease with which some groups have coopted the civil rights struggle to benefit their very white agendas:

> As a Black woman I experience all areas of oppression—economic, racial, and sexual. I cannot "pick and choose" a single area of struggle. I believe it is in this way that feminists and other progressive whites pursue their own interests at the expense of those of us subjected to racism. They do not have to deal with the *totality of oppression,* and instead may conform to, or accept the policies of, institutional racism. . . .
>
> Feminism succumbs to racism when it segregates Black women from Black men and dismisses our history. ("Black Women" 158)

To understand *A Different Image* as an articulation of feminist ideas, as opposed to black womanist ideology, would be aberrant reading and would instance the authoritarian mode of feminism. Larkin argues,

> The assumption that Black women and white women share . . . similar histories and experiences presents an important problem. . . . Both historically and currently, white women participate in and reap

> the benefits of white supremacy. Feminism must address these issues[;] otherwise its ahistorical approach towards Black women can and does maintain institutional racism. ("Black Women" 158–59)

A pan-African reading which recognizes the abuses black women experience will produce a different set of articulations and assumptions about the similar histories of women and men in the black community. For example, if the viewer focuses on the visual allusions to African women and Alana's colorful dress, and if this viewer listens to the musical score, which connotes an African rhythm, then this viewer/listener might perceive Alana as the female representative of pan-African subjectivity, which, for the woman of color, is more dynamic than a generalized female subjectivity. I am not proposing that the black womanist and the feminist readings are binary opposites. I merely want to show that the same text can activate certain *accommodative* readings for these two audiences.

A black womanist reading would combine the latter two spectatorial positions—feminist and pan-African—and articulate the importance of the *refigured* African-American male, Vincent. This dialogic reading of *A Different Image* would resist the processes of gender and racial closure. In discussing the problematics of feminist thinking, Hortense J. Spillers writes, "Sexuality as a term of power belongs to the empowered. Feminist thinking often appropriates the term in its own will to discursive power in a sweeping, patriarchal, symbolic gesture that reduces the human universe of women to its own image" (78). Equally, certain forms of black nationalism belong to the empowered, and such nationalist thinking often appropriates the term *pan-African culture* in its own will to discursive power in a sweeping, patriarchal, symbolic gesture that reduces black humanity to its own image. Both of these reductive forms of empowerment deny the micro-politics of struggle within their specific groups. They attempt to force the assimilation of radical members by creating a politics of binary opposition which discounts individual differences. Alile Sharon Larkin's film dramatizes the micro-politics of difference within the black community and portrays the shared phallic notions that some men and women act upon. The film denies a hypothetical sweeping, patriarchal gaze as biologically determined, since the patriarchal gaze is an "ideological construct" which is figured in men as well as women. The film privileges a text constructed on pan-African historical subjects who are embedded in the *maternal* voice of Alana"s mother and the embodied portraits of African women.

The black womanist reading/viewing position avoids the inherent sexual dualism which engenders both the "dominant, male-centered position" and its appropriation by a "female-phallic position." It acknowledges that the goals of black feminist theory are a revision of gender relations and an open-ended sexuality. Womanist goals do not attempt a simple appropriation of hierarchic systems to continue oppressive processes of subjectivity. Consequently, the black womanist film project proposes the collective refiguration of the pan-African community. Such a purpose requires an artist/critic/audience to re-vision "possible" black others in a post-Negritude world.

Alile Sharon Larkin's *A Different Image* pursues the refiguration of gender through deconstructing the illusion that all black men are phallocrats and all black women are oppressed by such men. The film presents sexism as an ideological construct rather than a biological trait of the male. Larkin's film portrays how phallic ideas call forth (hail) men, women, boys, and girls. According to the film, phallocentric thought is not limited to men; it is present in the baiting that Vincent receives from his male friend as well as the teasing that Alana receives from her female friend. Larkin's womanist film demystifies the totalizing effects of phallic ordering. The reconciliation of the Alana-Vincent friendship resists patriarchal closure. The idea that men are contentedly contained within a sexist system is revised, and the unquestioned fate of women as sex objects or reproductive apparati-in-waiting is denied.

Black filmmakers in Africa have represented black Africa through womanist discourse strategies in which the experiences of women are central to the film narrative. This is true of the first black-oriented African feature film, Ousmane Sembene's *La Noire de* (Senegal, 1968), as well as other African films directed by black men (Vieyra 162).

Concurrent with the development of black African cinema in the late '60s, Safi Faye, an African-born Senegalese, directed African films. Faye's first film, *La passante* (*The Passerby,* Senegal, 1972), depicts the Parisian experiences of an African woman who is preyed upon by African and French men. Interestingly, Faye was both filmmaker and the filmed subject, since she performed the role of the female who is the object of the gaze of both men. Faye does not agree with the closed status of the male gaze[3] but views the male gaze as a matter of who is behind, who is in front of, and who belongs to the audience that participates in this relationship. In *The Passerby*, a 10-minute narrative film, she disrupts the male gaze upon female body parts by giving the female object a transitive quality and, thereby, both reversing the source of the gaze and providing an "other" meaning. A series of shot-reverse shots and point-of-view shots decenters the authority of the male point of view, as well as its discourse, and exposes the polyvalent quality of the male gaze.

Thus, as early as 1972, Faye was constructing a black womanist gaze to resist dominant viewing relationships. This permitted spectatorial positions of resistance, accommodation, and assimilation. While I have not attempted an in-depth study of the possible receptions of this film,

I find Faye's *The Passerby* equal in receptive possibilities to Larkin's *A Different Image*. Each film criticizes a sexism that crosses racial boundaries. Yet Safi Faye rejects identification as a woman filmmaker:

> I never say that I am a woman filmmaker. I think I am a human being like the others. I never put someone under me but I never accept to be placed in a submissive position. The only difference between me and a man is my familial responsibilities. When I think about my daughter Zeiba, I say OK, I am a woman. But when I think of my job or a film project, I am like . . . African men. (Reid)

Faye's response raises two important issues. First, the product should not be confused with its author any more than authorial intent should describe the final product. Second, Faye refuses the label *woman filmmaker* because, according to her, her films are not solely about women. She views the struggle in a larger frame—a sort of womanist struggle to counter dehumanizing relationships which oppress men and women.

The representation of black African women should not be limited by a choice between motherhood and a career. Faye, for instance, acknowledges that African Serrer customs form her concept of womanhood—womanism. She feels a deep commitment to motherhood, but is equally committed to filmmaking:

> Since I am an African woman, I was educated in this way. I have remained close to the traditional role of mother, and this responsibility is not a burden to me. I try to keep my job near my eight-year-old daughter. Men do not have the responsibilities of caring for children, because they have wives to care after them. When I think about this, I realize that I am a woman. But I cannot be a filmmaker without being a mother. As an African woman filmmaker, the two are very much a part of me. (Reid)

Like Safi Faye, there are other black women filmmakers who see motherhood and filmmaking as inseparable. Kathleen Collins was, until her death in 1988, a mother, a writer, a university professor, and the first African-American woman to direct a feature-length film, *Losing Ground.*

Womanist Film and the Black Professional

In 1982, Kathleen Collins directed *Losing Ground,* in which she dramatized a critique of sexism. Here, however, she did not employ the pan-African sensibility of Larkin's film. Sara, her protagonist, is a black professor of philosophy who is married to an insensitive black artist. Perceiving her husband's interest in an attractive, extroverted Puerto Rican woman, the introverted Sara reevaluates her career, marriage, and life. Altering her inhibited beliefs about actions appropriate to women professors, Sara takes a role as Johnnie in a dramatization of the story in which Johnnie shoots and kills her boyfriend Frankie because he has proved unfaithful. At the end of the film, Sara, turning the blank pistol away from the man playing Frankie, symbolically points it at her own philandering husband. *Losing Ground* marks the first appearance of a black professional woman as protagonist in black independent film, and it is one of very few portrayals of developing consciousness in a black professional woman in any feature-length film.

It is beneficial neither to feminism nor to black womanist creativity to collapse the socio-economic differences of race and gender into a generalized feminist or women's film aesthetic of purity. Collins speaks of an "imperfect synthesis" of the African-American condition. She is "willing to recognize that being Black is without purity. That one cannot achieve [racial] purity in . . . [American] culture. That one can only achieve some kind of emotional truth. . . . That's what characterizes my work. . . . I am much more concerned with how people resolve their inner dilemma in the face of external reality" (Nicholson 12). But Collins is quick to affirm the cultural differences between black women filmmakers and their white colleagues, and acknowledges that such differences inform the aesthetic choices of these filmmakers:

> I would think that there is a Black aesthetic among Black women filmmakers. Black women are not white women by any means; we have different histories, different approaches to life, and different attitudes. Historically, we come out of different traditions; sociologically, our preoccupations are different. However, I have a lot of trouble with this question because I do not feel that there has been a long enough tradition. (Campbell 62)

Granted, the history of black filmmaking presents a discontinuous pattern of false starts and long production gaps. Yet there are films to document these attempts. The archaeology of the black woman's participation is no more than a twenty-year effort. Nevertheless, black womanist films have pioneered creative processes for the reception and production of racialized, sexualized, and engendered black subjectivities. Cultural studies and visual research must freely consult the sacred (legitimized) discourses on blackness as well as those deemed profane in order to make visible the histories of the ignored members of the black community, and to permit a dialogue within this community and across its *imagined* boundaries of race, gender, sexuality, class, and nation.

If the dialogic aspects of blackness(es) remain invisible and ignored, then we further the fragmentation of black communities. It is time to affirm and celebrate the sonorous polyphony of black voices. Let us not welcome them only to deny their voices and our attention. Why imitate the history that humbled and shamed us?

Notes

1. Mary Ann Doane writes, "It is important to specify precisely what is meant by the 'female spectator' or 'female spectatorship.' Clearly these terms are not meant to refer directly to the woman who buys her ticket and enters the movie theater as the member of an audience, sharing a social identity but retaining a unique psychical history. Frequently, they do not even refer to the spectator as a social subject but, rather, as a psychical subject, as the effect of signifying structures. . . . Women spectators oscillate or alternate between masculine and feminine positions, . . . and men are capable of this alternation as well" (8).

2. When referring to phallic forms of knowledge and power within the black community, I make reference to discourses which reflect a "tendency in every form to harden its generic skeleton and elevate the existing norms to a model that resists change. . . . Canonization is that process that blurs heteroglossia; that is, that facilitates a naïve, single-voiced reading" (Bakhtin 425). Phallic gender or race reasoning reflects any socio-psychical attempt to disavow the fact that the male (and white) subject, like the female (and black) subject, is surrounded by and constructed through discourses which he also incorporates as an internal regulatory agency (see Silverman 99).

3. Taking a different approach, Laura Mulvey argues that the "male gaze" is an inherent and uncontestable property of narrative cinema.

Works Cited

Bakhtin, M. M. *The Dialogic Imagination: Four Essays by M. M. Bahktin.* Ed. Michael Holquist. Trans. Caryl Emerson and Michael Holquist. Austin: U of Texas P, 1981.

Belsey, Catherine. *Critical Practice.* New York: Methuen, 1980.

Campbell, Loretta. "Reinventing Our Image: Eleven Black Women Filmmakers." *Heresies* 4.4 (1983): 58–62.

Clifford, James. *The Predicament of Culture: Twentieth-Century Ethnography, Literature, and Art.* Cambridge: Harvard UP, 1988.

Doane, Mary Ann. *The Desire to Desire: The Woman's Film of the 1940s.* Bloomington: Indiana UP, 1987.

Faye, Safi. Personal interview. 26 May 1986.

Hutcheon, Linda. *A Poetics of Postmodernism: History, Theory, Fiction.* New York: Routledge, 1988.

Irele, Abiola. *The African Experience in Literature and Ideology.* London: Heinemann, 1981.

Larkin, Alile Sharon. "Black Women Filmmakers Defining Ourselves: Feminism in Our Own Voice." Ed. E. Deidre Pribram. *Female Spectators: Looking at Film and Television.* London: Verso, 1988. 157–73.

———. "*Your Children Come Back To You* and *A Different Image.*" Debate "Lutte actuelle des noirs américains" (Debate on "The Afro-American Contemporary

Struggle"), 8eme Festival du Cinéma des Minorités Nationales (Eighth Festival of National Minority Cinema). Douarnenez, France, 28 Aug. 1985.

Mays, Vickie M. "I Hear voices But See No Faces." *Heresies* 3.4 (1981): 74–76.

Mulvey, Laura. "Visual Pleasure and Narrative Cinema." 1975. *Visual and Other Pleasures.* Bloomington: Indiana UP, 1989. 14–26.

Nicholson, David. "A Commitment to Writing: A Conversation with Kathleen Collins Prettyman." *Black Film Review* 5.1 (1988–89): 6–15.

Pfaff, Françoise. *Twenty-Five Black African Filmmakers.* New York: Greenwood, 1988.

Silverman, Kaja. *The Acoustic Mirror: The Female Voice in Psychoanalysis and Cinema.* Bloomington: Indiana UP, 1988.

Spillers, Hortense J. "Interstices: A Small Drama of Words." *Pleasure and Danger: Exploring Female Sexuality.* Ed. Carole S. Vance. Boston: Routledge, 1984. 73–100.

Vieyra, Paulin Soumanou. *Le Cinéma africain: Des origines à 1973.* Paris: Éditions Présence Africaine, 1975.

Walker, Alice. *In Search of Our Mothers' Gardens.* New York: Harcourt, 1983.

Filmography

Anderson, Madeline. *I Am Somebody.* USA, 1970. 30 min., 16mm. Documentary on the 1969 hospital workers' strike in Charleston, SC.

Collins, Kathleen. *Losing Ground.* USA, 1982. 86 min., 16mm.

Faye, Safi. *La passante (The Passerby).* Senegal, 1972. 10 min., 16mm.

Gomez, Sara. *De cierta manera (One Way or Another).* Cuba, 1974/1977. 97 min., 16mm. Semi-documentary depicting post-revolutionary sexism in an individual's relationship that affects his lover as well as the men with whom he works; ends with a political analysis of the protagonist's actions.

Larkin, Alile Sharon. *A Different Image.* USA, 1981. 51 min., 16mm.

Maldoror, Sarah. *Des Fusils pour Banta (Guns for Banta).* Guinea-Bissau, 1971. 105 min., 35mm. Portrays a Guinea-Bissau woman who develops a program of armed resistance against the Portuguese but dies before the actual struggle begins. According to Françoise Pfaff, this film "never went beyond the editing stages" due to a verbal confrontation Maldoror had with a high-ranking Algerian officer (207).

Parkerson, Michelle. *Storme: The Lady of the Jewel Box.* USA, 1987. 21 min., 16mm. Archival materials used to document the story of Storme DeLarverie, a male impersonator and former master of ceremonies at the Jewel Box Revue, considered to be America's first integrated female impersonation show.

Towards a Critical Theory of Third World Films

Teshome H. Gabriel

Wherever there is a film-maker prepared to stand up against commercialism, exploitation, pornography and the tyranny of technique, there is to be found the living spirit of *New Cinema.* Wherever there is a film-maker, of any age or background, ready to place his cinema and his profession at the service of the great causes of his time, there will be the living spirit of *New Cinema.* This is the correct definition which sets *New Cinema* apart from the commercial industry because the commitment of industrial cinema is to untruth and exploitation.

From The Aesthetics of Hunger, *Glauber Rocha [Brazil]*

Insert the work as an original fact in the process of liberation, place it first at the service of life itself, ahead of art; *dissolve aesthetics in the life of society:* only in this way, as [Frantz] Fanon said can decolonization become possible and culture, cinema, and beauty—at least, what is of greatest importance to us—become *our culture, our films, and our sense of beauty.*

From Towards a Third Cinema, *Fernando Solanas and Octavio Gettino [Argentina]*

Frantz Fanon, in his attempt to identify the revolutionary impulse in the peasant of the Third World, accepted that culture is an act of insemination upon history, whose product is liberation from oppression.[1] In my search for a methodological device for a critical inquiry into Third World films,

From Raana Gauhar, ed., *Third World Affairs 1985* (London: Third World Quarterly, 1985), pp. 355–69. Reprinted by permission of the publisher.

I have drawn upon the historical works of this ardent proponent of liberation, whose analysis of the steps of the genealogy of Third World culture can also be used as a critical framework for the study of Third World films. This essay is, therefore, divided into two parts and focuses on those essential qualities Third World films possess rather than those they may seem to lack. The first part lays the formulation for Third World film culture and filmic institutions based on a critical and theoretical matrix applicable to Third World needs. The second part is an attempt to give material substance to the analytic constructs discussed previously.

From precolonial times to the present, the struggle for freedom from oppression has been waged by the Third World masses, who in their maintenance of a deep cultural identity have made history come alive. Just as they have moved aggressively towards independence, so has the evolution of Third World film culture followed a path from "domination" to "liberation." This genealogy of Third World film culture moves from the First Phase in which foreign images are impressed in an alienating fashion on the audience, to the Second and Third Phases in which recognition of "consciousness of oneself" serves as the essential antecedent for national, and more significantly, international consciousness. There are, therefore, three phases in this methodological device.

Phases of Third World Films

Phase I—The unqualified assimilation

The industry: Identification with the Western/Hollywood film industry. The link is made as obvious as possible and even the names of the companies proclaim their origin. For instance, the Nigerian film company, Calpenny, whose name stands for California, Pennsylvania and New York, tries to hide behind an acronym, while the companies in India, Egypt and Hong Kong are not worried being typed the "Third World's Hollywood," "Hollywood-on-the-Nile," and "Hollywood of the Orient" respectively.

The theme: Hollywood thematic concerns of "entertainment" predominate. Most of the feature films of the Third World in this phase sensationalise adventure for its own sake and concern themselves with escapist themes of romance, musicals, comedies, etc. . . . The sole purpose of such industries is to turn out entertainment products which will generate profits. The scope and persistence of this kind of industry in the Third World lies in its ability to provide reinvestable funds and this quadruples their staying power. Therefore, in cases where a counter-cinematic movement has occurred the existing national industry has been able to ingest it. A good example is in the incorporation of the "cinema novo" movement in the Brazilian Embrafilme.

Style: The emphasis on formal properties of cinema, technical brilliance and visual wizardry, overrides subject matter. The aim here is simply to create a "spectacle." Aping Hollywood stylistically, more often than not, runs counter to Third World needs for a serious social art.

Phase II—The remembrance phase

The industry: Indigenisation and control of talents, production, exhibition and distribution. Many Third World film production companies are in this stage. The movement for a social institution of cinema in the Third World such as "cinema moudjahid" in Algeria, "new wave" in India and "*engagé* or committed cinema" in Senegal and Mozambique exemplify this phase.

The theme: Return of the exile to the Third World's source of strength, i.e., culture and history. The predominance of filmic themes such as the clash between rural and urban life, traditional versus modern value systems, folklore and mythology, identify this level. Sembene Ousmane's early film *Mandabi* about a humble traditional man outstripped by modern ways characterises this stage. *Barravento* (The Turning Wind), a poetic Brazilian film about a member of a fishermen's village who returns from exile in the city, is a folkloric study of mysticism. The film from Burkina Faso (Upper Volta), *Wend Kûuni* (God's Gift) attempts to preserve the spirit of the folklore in a brilliant recreation of an old tale of a woman who is declared a witch because of her conflicts with custom when she refused to marry after the disappearance of her husband. While the most positive aspect of this phase is its break with the concepts and propositions of Phase I, the primary danger here is the uncritical acceptance or undue romanticization of ways of the past.

It needs to be stressed that there is a danger of falling into the trap of exalting traditional virtues and racialising culture without at the same time condemning faults. To accept totally the values of Third World traditional cultures without simultaneously stamping out the regressive elements can only lead to "a blind alley" as Fanon puts it, and falsification of the true nature of culture as an act or agent of liberation. Therefore, unless this phase, which predominates in Third World film practices today, is seen as a process, a moving towards the next stage, it could develop into opportunistic endeavours and create cultural confusion. This has been brilliantly pointed out by Luis Ospina of Colombia in his self-reflective film *Picking on the People,* in which he criticises the exploitative nature of some Third World film-makers who peddle Third World poverty and misery at festival sites in Europe and North America and do not approach their craft as a tool of social transformation. An excellent case in point is the internationally acclaimed film, *Pixote* by Hector Babenco. According to a *Los Angeles Times* correspondent in Rio de Janeiro, Da Silva, the young boy who played the title role of the film was paid a mere $320. The correspondent writes, "In a real-life drama a juvenile judge in Diadema, a suburb of Sao Paulo, last week released Da Silva, now 16, to the custody of his mother after his arrest on charges of housebreaking and theft." According to Da Silva's mother, who sells lottery tickets for her living, "after a trip to Rio when he got no work, he told me, 'Mother, they have forgotten me, I am finished.' " In the meantime Mr. Babenco, the now famous film director is about to shoot his next feature, *The Kiss of the Spider Woman,* in collaboration with producers in Hollywood.[2]

The style: Some attempts to indigenise film style are manifest. Although the dominant stylistic conventions of the first phase still predominate here there appears to be a growing tendency to create a film style appropriate to the changed thematic concerns. In this respect, the growing insistence on spatial representation rather than temporal manipulation typifies the films in this phase. The sense of a spatial orientation in cinema in the Third World arises out of the experience of an "endless" world of the large Third World mass. This nostalgia for the vastness of nature projects itself into the film form resulting in long takes and long or wide shots. This is often done to constitute part of an overall symbolisation of a Third World thematic orientation, i.e., the landscape depicted ceases to be mere land or soil and acquires a phenomenal quality which integrates humans with the general drama of existence itself.

Phase III—The combative phase

The industry: Film-making as a public service institution. The industry in this phase is not only owned by the nation and/or the government, it is also managed, operated and run for and by the people. It can also be called a cinema of mass participation, one enacted by members of communities speaking indigenous language, one that espouses Julio Garcia Espinosa's polemic of "An Imperfect Cinema,"[3] that in a developing world, technical and artistic perfection in the production of a film cannot be the aims in themselves. Quite a number of social institutions of cinema in the Third World, some underground like Argentina's "Cine Liberacion" and some supported by their governments—for instance, "Chile Films" of Allende's Popular Unity Socialist government—exemplify this phase. Two industrial institutions that also exemplify this level are the Algerian L'Office National pour le Commerce et l'Industrie cinematographique (ONCIC) and Cuba's institute of Film Art and Industry (ICAIC).

The theme: Lives and struggles of Third World peoples. This phase signals the maturity of the film-maker and is distinguishable from either Phase I or Phase II by its insistence on viewing film in its ideological ramifications. A very good example is Miguel Littin's *The Promised Land,* a quasi-historical mythic account of power and rebellion, which can be seen as referring to events in modern-day Chile. Likewise, his latest film *Alsino and the Condor* combines realism and fantasy within the context of war-torn Nicaragua. The imagery in *One Way or Another* by the late Sara Gomez Yara, of an iron ball smashing down the old slums of Havana not only depicts the issue of women/race in present-day Cuba but also symbolises the need for a new awareness to replace the old oppressive spirits of *machismo* which still persist in socialist Cuba. The film *Soleil O,* by the Mauritanian film-maker, Med Hondo, aided by the process of Fanonian thesis, comes to the recognition of forgotten heritage in the display of the amalgam of ideological determinants of European "humanism," racism and colonialism. The failure of colonialism to convert Africans into "white-thinking blacks" depicted in the film reappears in a much wider symbolic form in his latest film, *The West Indies,* where the entire pantheon of domination and liberation unfolds in a ship symbolic of the slave-ship of yesteryear.

The style: Film as an ideological tool. Here, film is equated or recognised as an ideological instrument. This particular phase also constitutes a framework of agreement between the public (or the indigenous institution of cinema) and the film-maker. A Phase III film-maker is one who is perceptive of and knowledgeable about the pulse of the Third World masses. Such a film-maker is truly in search of a Third World cinema—a cinema that has respect for the Third World peoples. One element of the style in this phase is an ideological point-of-view instead of that of a character as in dominant Western conventions. *Di Cavalcanti* by Glauber Rocha for instance, is a take-off from "Quarup," a joyous death ritual celebrated by Amazon tribes.[4] The celebration frees the dead from the hypocritical tragic view modern man has of death. By turning the documentary of the death of the internationally renowned Brazilian painter Di Cavalcanti into a chaotic/celebratory montage of sound and images, Rocha deftly and directly criticised the dominant documentary convention, creating in the process not only an alternative film language but also a challenging discourse on the question of existence itself. Another element of style is the use of flashback—although the reference is to past events, it is not stagnant but dynamic and developmental. In *The Promised Land,* for instance, the flashback device dips into the past to comment on the future, so that within it a flash-forward is inscribed. Similarly, when a flash-forward is used in Sembene's *Ceddo* (1977), it is also to convey a past and future tense simultaneously to comment on two historical periods.

Since the past is necessary for the understanding of the present, and serves as a strategy for the future, this stylistic orientation seems to be ideologically suited to this particular phase.

It should, however, be noted that the three phases discussed above are not organic developments. They are enclosed in a dynamic which is dialectical in nature; for example, some Third World film-makers have taken a contradictory path. *Lucia,* a Cuban film by Humberto Solas, about the relations between the sexes belongs to Phase III, yet Solas' latest film, *Cecilia,* which concerns an ambitious mulatto woman who tries to assimilate into a repressive Spanish aristocracy, is a regression in style (glowing in spectacle) and theme (the tragic mulatto), towards Phase I. Moving in the opposite direction, Glauber Rocha's early Brazilian films like *Deus E O Diabo Na Terra Do Sol* (literally "God and the Devil in the land of the Sun," but advertised in the United States as "Black God, White Devil!") and *Terra em Transe* (The Earth Trembles) reflect a Phase II characteristic, while his last two films, *A Idade da Terra* (The Age of Earth) and *Di Cavalcanti,* both in their formal properties and subject matter manifest a Phase III characteristic in their disavowal of the conventions of dominant cinema. According to Glauber Rocha, *A Idade da Terra* (which develops the theme of *Terra em Transe*) and *Di Cavalcanti* disintegrate traditional "narrative sequences" and rupture not only the fictional and documentary cinema style of his early works, but also "the world cinematic language" under "the dictatorship of Coppola and Godard."[5]

The dynamic enclosure of the three phases posits the existence of grey areas between Phases I and II, and II and III. This area helps to identify a large number of important Third World films. For instance, the Indian film, *Manthan* (The Churning), the Senegalese film *Xala* (Spell of Impotence), the Bolivian film *Chuquiago* (Indian name for La Paz), the Ecuadorean film *My Aunt Nora,* the Brazilian film *They Don't Wear Black Tie* and the Tunisian film *Shadow of the Earth* occupy the grey area between Phase II and III. The importance of the grey areas cannot be over-emphasised, for not only do they concretely demonstrate the *process of becoming* but they also attest to the multifaceted nature of Third World cinema and the need for the development of new critical canons.

Components of Critical Theory

From the above it can be seen that the development of Third World film culture provides a critical theory particular to Third World needs. I would like to propose at this stage an analytic construct consisting of three components that would provide an integrative matrix within which to approach and interpret the Three Phases drawn out from the Third World's cultural history. The components of critical theory can be schematised as follows:

Component 1: *Text*

The intersection of codes and sub-codes; the chief thematic and formal characteristics of existing films and the rules of that filmic grammar.

And, the transformational procedures whereby new "texts" emerge from old.

Component 2: *Reception*

The audience: the active interrogation of images versus the passive consumption of films. The issue of alienated and non-alienated identity and the ideal/inscribed or actual/empirical spectatorship illustrates this component of critical theory.

Component 3: *Production*

The social determination where the wider context of determinants informs social history, market considerations, economy of production, state governance and regulation composes this stage of the critical constructs. Here, the larger historical perspective, the position of the institution of indigenous cinema in progressive social taste, is contexted. The overriding critical issue at this juncture is, for instance, the unavoidable ultimate choice between the classical studio system and the development of a system of production based on the lightweight 16mm or video technology. The pivotal concern and the single most significant question at this stage, therefore, is, "Precisely, what kind of institution is cinema in the Third World?"

Confluence of Phases and Critical Theory

Each Phase of the Third World film culture can be described in terms of all the three components of critical theory, because each Phase is necessarily engaged in all the critical operations. For instance, Phase I is characterised by a type of film that simply mirrors, in its concepts and propositions, the *status quo,* i.e., the text and the rules of the grammar are identical to conventional practices. The consequence of this type of "mimicking" in the area of "reception" is that an alienated identity ensues from it precisely because the spectator cannot or is unable to find or recognise himself/herself in the images. The mechanisms of the systems of "production" also acknowledge the *status quo*—the reliance is on the studio systems of controlled production and experimentation.

If we apply the components of critical theory to Phase II only a slight shift in the Text and the rules of the grammar is noticeable. Although the themes are predominantly indigenised, the film language remains trapped, woven and blotted with classical formal elements and remains stained with conventional film style. In terms of "reception" the viewer, aided by the process of memory and an amalgam of folklore and mythology, is able to locate a somewhat diluted traditional identity. The third level of critical theory also composes and marks the process of indigenisation of the institution of cinema where a position of self-determination is sought.

Finally, the three components of critical theory find their dynamic wholeness in Phase III—the Combative Phase. Here, the Text and subtexts go through a radical shift and transformation—the chief formal and thematic concerns begin to alter the rules of the grammar. Another film language, and a system of new codes begin to manifest themselves. With regards to "reception" we discover that the viewer or subject is no longer alienated because recognition is vested not only in genuine cultural grounds but also in an ideological cognition founded on the acknowledgment of the decolonisation of culture and total liberation.

The intricate relationships of the three *phases* of the evolution of African film culture and the three analytic constructs for filmic institution help to establish the stage for a confluence of a unique aesthetic exchange founded on other than traditional categories of film conventions (see fig. 1).

This new Third World cinematic experience, inchoate as it is, is in the process of creating a concurrent development of a new and throbbing social institution capable of generating a dynamic and far-reaching influence on the future socioeconomic and educational course of the Third World.

I contend that the confluence obtained from the interlocking of the *phases* and the critical *constructs* reveals underlying assumptions concerning perceptual patterns and film viewing situations. For instance, with respect to fiction film showing in Third World theatres, rejection on cultural grounds forces incomplete transmission of meaning. That is, the intended or inscribed meaning of the film is deflected and acquires a unique meaning of its own—the mode of address of the film and the spectator behaviour undergo a radical alteration. Therefore, what has been presented as a "fiction" film is received as if it were a "documentary." The same fiction film screened in its own country of origin, however, claims an ideal spectatorship because it is firmly anchored in its own cultural references, codes and symbols. A classic example of how films from one culture can be easily misunderstood and misinterpreted by a viewer from another culture is Glauber Rocha's *The Lion Has Seven Heads* (Der Leone Have Sept Cabezas). Extensively exhibited in the West, one catalogue compiled in 1974 credited Rocha with bringing:

> the Cinema Novo to Africa for this Third World assault on the various imperialisms represented in its multilingual title. Characters include a black revolutionary, a Portuguese mercenary, an American CIA agent, a French missionary, and a voluptuous nude woman called the Golden Temple of Violence.[6]

Again, a most recent compendium of reviews, *Africa on Film and Videotape, 1960–1981,* dismisses the film completely with a one-liner, "An allegorical farce noting the bond between Africa and Brazil."[7]

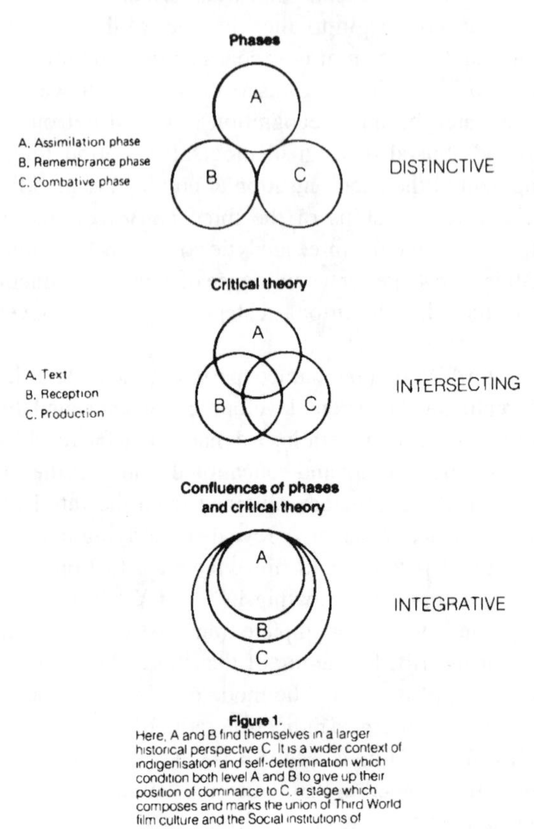

Figure 1.
Here, A and B find themselves in a larger historical perspective C It is a wider context of indigenisation and self-determination which condition both level A and B to give up their position of dominance to C, a stage which composes and marks the union of Third World film culture and the Social institutions of cinema

Yet, Glauber Rocha in an interview given to a prominent film historian, Rachel Gerber (author of *Glauber Rocha, Cinema, Politica e a Estetica do Inconsciente*) in Rome, February 1973, and in a discussion with this author at UCLA in 1976, said that the film is a story of Ché Guevara who is magically resurrected by Blacks through the spirit of Zumbi, the spiritual name of the late Amilcar Cabral. To Rocha, the film is in fact a homage to Amilcar Cabral. Thus, while the West looks at this film as an offering of clichéd images and an object of curiosity, the film-maker is only trying to affirm the continuity of the Third World's anti-imperialist struggle from Ché to Cabral (and beyond), to initiate an awareness of their lives, and the relevance to us today, of what they struggled and died for. To the extent that we recognize a history of unequal exchanges between the South and the North, we must also recognise the unequal "symbolic" exchanges involved. The difficulty of Third World films of radical social comment to Western interpretation is the result, a) of the film's resistance to the dominant conventions of cinema, and b) of the consequence of the Western viewers' loss of being the privileged decoders and ultimate interpreters of meaning.

The Western experience of film viewing—dominance of the big screen and the sitting situation—have naturalised a spectator conditioning so that any communication of a film plays on such values of exhibition and reception. The Third World experience of film viewing and exhibition suggests an altogether different route and different value system. For instance, Americans and Europeans hate seeing a film on African screens, because everybody talks during the showings; similarly, African viewers of film in America complain about the very strict code of silence and the solemn atmosphere of the American movie-theatres.

How the system of perceptual patterns and viewing situation varies with conditions of reception from one culture to another, or how changes in the rules of the grammar affect spectator viewing habits is part of a larger question which solidifies and confirms the issue of cultural relativism and identity.

The confluence of the phases and the constructs also converges on the technologically mediated factors of needed production apparatuses, productive relations and the mechanisms of industrial operations. It needs to be stated outright that "technology" as such does not in itself produce or communicate meaning; but it is equally true to say that "technology" has a dynamic which helps to create ideological carry-overs that impress discourse language, i.e., ideological discourse manifests itself in the mechanisms of film discourse. By way of an example, it is possible that a film-maker might have the idea of "filmic form" before having "a content" to go along with it. Third World films are heterogenous, employing narrative and oral discourse, folk music and songs, extended silences and gaps, moving from fictional representation to reality, to fiction—these constitute the creative part that can challenge the ideological carry-overs that technology imposes.

From the needs of Third World film criticism, contemporary film scholarship is criticised on two major fronts: first, contemporary film theory and criticism is grounded on a conception of the "viewer" (subject or citizen) derived from psychoanalytic theory where the relation between the "viewer" and the "film" is determined by a particular dynamic of "familial" matrix. To the extent that Third World culture and familial relationships are not described through psychoanalytic theory, Third World filmic representation is open for an elaboration of the relation "viewer"/"film" on terms other than those founded on psychoanalysis. The Third World relies more on an appeal to social and political conflicts as the prime rhetorical strategy and less on the paradigm of oedipal conflict and resolution.

Second, on the semiotic front, the Western model of filmic representation is essentially based on a literary or written conception of the scenario which implies a linear, cause/effect, conception of narrative action.[8] However, Third World oral narratives, founded on traditional culture, are held in memory by a set of formal strategies specific to, repeated, oral, face-to-face, tellings.

It is no longer satisfactory to use existing critical criteria, which may be adequate for a film practice (Western in this case) now at a plateau of relevance,[9] to elucidate a new and dynamic film convention whose upward mobility will result in a totally new cinematic language. The Third World experience is thus raising some fundamental concerns about the methods and/or commitment of traditional film scholarship. The Third World filmic practice is, therefore, reorganising and refining the pictorial syntax and the position of the "viewer" (or spectator) with respect to film. The Third World cinematic experience is moved by the requirements of its social action and contexted and marked by the strategy of that action. We need, therefore, to begin attending to a new theoretical and analytic matrix governed by other than existing critical theories that claim specific applications for universal principles.

Cultural contamination is a deeply-rooted human fear: it smells of annihilation. Spiritual and traditional practices have a terrific hold on the Third World rural populace. This reminds us of the maxim which was enunciated by Confucius in the sixth century BC and still prevails: "I'm a transmitter, not an inventor." To the Third World, spirits, magic, masquerades and rituals, however flawed they may be, still constitute knowledge and provide collective security and protection from forces of evil. Unknown forces for the rural community can only be checked or controlled if they can be identified.

One way of readily understanding what Third World culture, therefore, is, is to distinguish it from what it claims not to be.[10] We call at this juncture for a thorough and comparative analysis of "oral" or "folk" art form and "literate" or "print" art form to situate the foregoing discussion on critical theory into focused attention. I propose here to examine the centrifugal as well as the centripetal cultural forces that might determine not only film, but also the media, in the Third World. This dialectical, not differential or oppositional, conception of cultural forms takes into account the dynamics of their exchange.

Several factors ensue from the examination of the two modes of cultural expression. While, for instance, the community issue is at the heart of Third World traditional culture, the issue of the individual is at the base of Western or print culture. With regards to performatory stage presentation, a Western actor interacting with the audience breaks the compact or marginal boundary. Because a special kind of magic enters a playing space, Western stage performance does not allow cross-over. While, therefore, a Western person feels his privacy violated with interactive drama, in the Third World context the understanding between the viewer and the performers is that their positions are interchangeable without notice.

Awe for the old in the Third World culture is very much in evidence. Several films reflect it. The old or the aged as repositories of Third World history is well documented in such films as *Emitai* from Senegal, *They Don't Wear Black Tie,* from Brazil, *Shadow of the Earth,* from Tunisia and the *In-Laws* from the People's Republic of China. The issue of the aged in Third World culture is beautifully illustrated in Safi Faye's film *Fad Jal,* where the opening sequence of the film states: "In Africa, an old man dying, is like a library burning down."

A major area of misunderstanding (if we take into account the "Cognitive Characteristics" of the "Folk/Print Art" dichotomy in Table 1) is the definition and replacement of "man," the individual, within Third World societies. For any meaningful dialogue centering on Third World developmental schemes the issue of "man/woman" in a society must be carefully debated. As Julius K. Nyerere of Tanzania puts it, "The Purpose is Man,"[11] and as the Wolof saying goes, "Man is the medicine of man."

A cultural orientation of "man," the individual, as changeable and capable of effecting change is a condition that reverberates in all advanced societies of the world, be they of capitalist or socialist persuasion. The idea that man, both in the singular and in the plural, has the capability of controlling his/her own destiny and effecting change by his/her own will is a dynamic force which can alter both the thought patterns and work habits of a people. This concept, it must be stated, is not the opposite of the Third World ideal of the primacy of the community over the individual. An excellent example is the film, *Beyond the Plains* (where man is born) by Michael Raeburn, in which a young man from the Masai tribe in Tanzania was able to change his peoples' negative attitude towards education by not only doggedly pursuing it to the university level, but also never losing contact with his people. As he grew up he made sure he performed all the customary rites and fulfilled all the obligations demanded by his people, thus demonstrating that Western and tribal cultural education were not incompatible. From this, it can be seen that the major difference between the Third World and the West with regards to changing the community from a passive to a dynamic entity is one of approach. Whereas the former aims at changing the individual through the community the latter wants the community changed by the individual. Only time will tell which of the two approaches makes for sustained, beneficial social progress.

Table 1: Comparison of Folk and Print Art Forms

Folk (or Oral) Art Form	Print (or Literate) Art Form
Conception of the Value and Evaluation of Art	
Deeper meaning of art held by cultural group or community. Interpretive device: one needs to belong and/or understand cultural or folk nuances.	Deeper meaning of art held as the sole property of the artist. Interpretive device: the artist proclaims 'it is for me to know and for you to find, or art is what you mean it to be.'
Recognises general level of excellence, hence, emphasis on group competence in the aesthetic judgement of art.	Recognises exceptions, hence, emphasis on individual achievement and individual responsibility.
Master artist concept—gifted but normal, and so conforms to the group.	Master artist concept—gifted but eccentric and essentially nonconformist.
Art as occasion for collective engagement.	Art as occasion for 'escape' from normal routine.
Emphasis on contextual relevance.	Emphasis on conceptual interpretation.
Art defined in terms of context.	Art defined in terms of aesthetic.
Performatory Presentation	
Held in fluid boundaries, churchyards, fields, marketplaces—operating in a 360° dimension.	Boxed-in theatres and elevated to a stage—operating in a 180° dimension.
A scene flows into another. Cyclical progression linked thematically.	Each scene must follow another scene in linear progression.
Performatory Effect	
Expects viewer participation, therefore, arouses activity and prepares for and allows participation.	Discourages viewer participation. Puts an end to activity. Inhibits participation.
Multiple episodes that have their own centres.	Singular episode extended through detail.
Cognitive Characteristics	
Man defined as 'unchangeable' alone. Change emanates from the community.	Man defined as 'man', changeable and, by virtue of his person, capable of effecting change and progress.
Individual interlinked with total social fabric. Concept of human rather than concept of 'man' as such.	Individual perceived primarily as separated from general social fabric.
Strong tradition of suggestion in the cultural symbol and in the use of linguistic formulae.	Strong tradition of detail and minute (graphic) description.
Time assumed to be a subjective phenomenon, i.e., it is the outcome of conceptualising and experiencing movement.	Time is assumed to be an 'objective' phenomenon, dominant and ubiquitous.
Wisdom is a state of intellectual maturity gained by experience. Cumulative process of knowledge, derived from the past. Characterised by slowness to judgement.	Wisdom is characterised by high degree of specialisation in a particular field or discipline. Characterised by quickness of judgement based on a vast accumulation of data and information.
Earth is not a hostile world; e.g. the cult of the ancestors is an attempt at unification with the past, present and future.	Earth is a hostile world and has to be subdued. Paradise is in the future or elsewhere.

Manipulation of Space and Time In Cinema

A child born in a Western society is encased, from the initial moments of birth, in purposive, man-made fabricated objects. The visual landscape he experiences is dominated by man-made forms. Even the child's dolls reflect the high technology of the environment. Nowadays, a child who is beginning to learn to spell can have a computer that can talk to him and interact with him in a human way. All of these developments are based on the insistence of a society that puts a high price on individualism, individual responsibility and achievement as most necessary.

A child in a rural Third World setting is born in an unrestricted natural landscape. From the day he/she is born the child is dominated by untampered natural forms. Even the interior of the dwelling where the child is born is made to look like the natural environment: it is not unusual to see fresh grass and flowers lending nature's colour to the child's initial world setting. The child grows in this vast universe where his place within the family and in nature is emphasised. A child born and raised in this situation is taught to submerge his individuality and show responsibility to his extended family and his community. His accomplishments are measured not only by his individual achievements but by the degree to which they accomplish and contribute to the social good.

Culture, the terms on which films are based, also naturally grows from these environmental factors. An examination of oral and literate culture in terms of film brings to light two very crucial elements of cinema, namely, the concepts of "space" and "time." All cinema manipulates "time" and "space." Where Western films manipulate "time" more than "space," Third World films seem to emphasise "space" over "time." Third World films grow from folk tradition where communication is a slow-paced phenomenon and time is not rushed but has its own pace. Western culture, on the other hand, is based on the value of "time"—time is art, time is money, time is most everything else. If time drags in a film, spectators grow bored and impatient, so that a method has to be found to cheat natural time. In film, this is achieved in the editing. It is all based on the idea that the more purely "non-dramatic" elements in film are considered "cinematic excess," i.e., they serve no unifying purpose. What is identified as "excess" in Western cinematic experience is, therefore, precisely where we locate Third World cinema. Let me now identify those essential elements of cinematic practice that are considered cinematic excess in Western cinema but which in the Third World context seem only too natural.

The long take: It is not uncommon in Third World films to see a concentration of long takes and repetition of images and scenes. In the Third World films, the slow, leisurely pacing approximates the viewer's sense of time and rhythm of life. In addition, the preponderance of wide-angle shots of longer duration deal with a viewer's sense of community and how people fit in nature. Whereas, when Michelangelo Antonioni and Jean-Luc Godard use these types of shots it is to convey an existential separation and isolation from nature and self.

Cross-cutting: Cross-cutting between antagonists shows simultaneity rather than the building of suspense. The poser of images lies, not in the expectation we develop about the mere juxtapositions or the collision itself, but rather in conveying the reasons for the imminent collision. Where, therefore, conventional cinema has too often reduced this to the collision of antagonists, on a scale of positive and negative characters, Third World films doing the same thing make it more explicitly an ideological collision.

The close-up shot: A device so much in use in the study of individual psychology in Western film-making practice is less used in Third World films. Third World films serve more of an informational purpose than a study in "psychological realism." The isolation of an individual, in tight close-up shots, seems unnatural to the Third World film-maker because (i) it calls attention to itself; (ii) it eliminates social considerations; and (iii) it diminishes spatial integrity.

The panning shot: Since a pan shot maintains integrity of space and time, the narrative value of such a shot renders "the cut" or editing frequently unnecessary. The emphasis on space also conveys a different concept of "time," a time which is not strictly linear or chronological but co-exists with it. My own observation indicates that while Western films tend to pan right on a left-right axis, Middle Eastern films, for instance, tend to pan generally toward the left, as in *Alyam Alyam* (Morocco) and *Shadow of the Earth* (Tunisia). It is quite possible that the direction of panning toward left or right might be strongly influenced by the direction in which a person writes.

The concept of silence: The rich potential for the creative interpretation of sound as well as the effective use of its absence is enormous in Third World films. For instance, in *Emitai,* there are English sub-titles for drum messages, and a rooster crows as Sembene's camera registers a low-angle shot of a poster of General de Gaulle. A neat visual pun! Silence serves as an important element of the audio track of the same film. It is "a cinema of silence that speaks." Silences have meaning only in context, as in the Ethiopian film, *Gouma,* and the Cuban film, *The Last Supper,* where they contribute to the suspension of judgment which one experiences in watching a long take. Viewers wonder what will happen, accustomed as they are to the incessant sound and overload of music of dominant cinema.

Concept of "hero": Even if a Western viewer cannot help but identify and sympathise with the black labour leader in *They Don't Wear Black Tie,* the lunatic in *Harvest: 3000 Years*, the crazy poet in *The Chronicle of the Years of Ember* and the militant party member in *Sambizanga,* the films nevertheless kill those characters. This is because wish-fulfillment through identification is not the films' primary objective; rather, it is the importance of collective engagement and action that matters. The individual "hero" in the Third World context does not make history, he/she only serves historical necessities.

In summary, Table 2 brings into sharper focus the differences between the film conventions of the Third World and the West and shows the dynamics of their cultural and ideological exchange.

Conclusion

The spatial concentration and minimal use of the conventions of temporal manipulation in Third World film practice suggest that Third World cinema is initiating a coexistence of film art with oral traditions. Non-linearity, repetition of images and graphic representation have very much in common with folk customs. Time duration, though essential, is not the major issue because in the Third World context the need is for films, in context, to touch a sensitive cultural chord in a society. To achieve this, a general overhaul of the parameters of film form is required. Should the reorganisation be successful and radical enough, a rethinking of the critical and theoretical canons of cinema would be called for leading to a reconsideration of the conventions of cinematographic language and technique. The final result would tend towards a statement James Potts made in his article, "Is there any international film language?":

> So, far from there being an international language of cinema, an internationally agreed UN Charter of conventions and grammatical rules, we are liable to be presented, quite suddenly, with a new national school of film-making, which may be almost wholly untouched by European conventions, and will require us to go back to square one in thinking about the principle and language of cinematography.[12]

Film-makers in the Third World are beginning to produce films that try to restructure accepted filmic practices. There is now a distinct possibility of James Potts' perceptive remarks coming true and it is in anticipation of the emergence of the "new national school of film-making . . . untouched by European conventions" that this paper has been written.

Already, certain reactions from film critics may be regarded as a sign of this "emergence." For example, a general criticism levelled at Third

Table 2: A Comparison of Filmic Conventions
(These are tendencies, not absolutes)

	Western Dominant Conventions	Non-Western Use of Conventions
Lighting	High contrast and low key, mostly Rembrandt lighting in drama while comedy uses low contrast and high key lighting.	Lighting as a convention in Third World films is less developed with the exception of Cuban films, whose use of lighting as a language is manifest in *Lucia* and *The Last Supper*.
Camera Angle	Mostly governed by eye-level perspective which approximates to our natural position in the world. Use of angle shots primarily for aesthetic look.	Deliberate choice of low/high-angle shots for purposes of political or social comment. Low/high-angle shots show dominance and power relations between the oppressed and oppressing classes.
Camera Placement	Distance varies according to the emotional content of the scene. Emotion, e.g. anger, is portrayed in close-up.	There is minimal use of the convention of close-up shots. This is perhaps due to lack of emphasis on psychological realism.
Camera Movement	Mostly a fixed perspective (tripod operation), promoting exposition and understanding. Often the camera moves to stay with the individual to study character development and psychological state.	Fixed perspective in African films. A moving perspective (hand-held camera) in Latin American films promotes experiential involvement and dramatic identification. If the camera moves it is to contain a scene or a sequence as a unit and not in response to individual psychology.
Set Design	A studio set. Tightens manipulatory controls, enhances fictional reality.	A location set. Location shooting relaxes manipulatory controls, and enhances documentary reality.
Acting	A Hollywood convention, actor as icon.	Mostly non-actors acting out their real-life roles.
Parallel Montage	Shows the relations of conflicting characters/forces for dramatic and expository narratives purposes, i.e., suspense.	Cross-cutting serves an ideological purpose and denotes ironical contrast and class distinction. Consider the film, *Mexico: The Frozen Revolution*.

Point of View	
Actors avoid looking directly at the camera. Actors are usually positioned or blocked so that their emotional state is easily observed by the camera.	It is not uncommon to see a look directed at the camera, hence a direct address to the audience. A shift to the conventions of oral narrative is evident. Consider the Algerian film, *Omar Gatlato*.

The sum total of what is listed above as technique or elements of the film-making process is what expresses ideology. Films that hide the marks of production are associated with the ideology of presenting 'film as reality', the film that announces its message as an objective reflection of the way things are: whereas, films which do exhibit the marks of production are associated with the ideology of presenting 'film as message'. Predominant aspect or point of view in Third World film is film announcing itself as a polemic comment on the way things are in their 'natural' reflection.

Films, therefore, in their point of view and stylistic choices, are structured to evoke a certain ideology in their production. A consequence of this, quite logically, is their different use of the conventions of time and space in cinema.

World films is that they are too graphic. This spatial factor is part of a general rhythm of pictorial representation in most Third world societies. It is, therefore, precisely because graphic art creates symbols in space that it enables Third World viewers to relate more easily to their films. In the Chinese case, for example:

> The spiritual quality achieved in the supreme Chinese landscape and nature paintings is a feeling of harmony with the universe in which the inner psychic geography of the artist and the outer visual reality transcribed, are fused through brush strokes into a new totality that . . . resonates with the viewer.[13]

Both the Chinese contemporary photographers and cinematographers have attempted to create similar syntax and effects to enhance the people's appreciation of their art.

Again, the most inaccessible Phase III film, the one African film that drops a curtain in front of a western audience, and at the same time a most popular and influential film in Africa is *Emitai* (The Angry God). Shot in social space by the Senegalese film-maker Sembene, the film explores the spiritual and physical tension in a rural community. To begin with the film carries its viewers into the story without any credits, only for the entire credit to be provided some twenty-five minutes later. Spectators have been known to leave the screening room at this point conditioned to read the credits as signalling the end of the film. What Sembene has provided before the credits is essentially the preface of the story like an African folktale. In addition, the ending of the film an hour and half later is anticlimactic and this occurs at the moment the film is truely engaging—the film simply stops—what we hear is the staccato of bullet sounds against a screen gone dark. In this film the film-maker is forcing us to forget our viewing habits and attend to the film in context instead of the experienced, framed as artistic package. A lesson is thus learned; concern should be with the language of the "film text" in its own terms and not with the skeletal structure and chronology of the film.

Cinema, since its creation, has beguiled spectators by its manipulation of time—it expands, contracts, is lost and found, fragmented and reassembled. The resultant multiple time-perspectives have conditioned film appreciation as pure entertainment. There is perhaps some justification for this objective in a society whose stabilizing conditions can afford the use of the film medium solely for entertainment. The Third World, on the other hand, is still engaged in a desperate struggle for sociopolitical and economic independence and development and cannot afford to dissipate its meagre resources and/or laugh at its present political and historical situation.

The Combative Phase, in which the historical determinants of Third World culture occur, provides us with the final horizon of a cinema oriented toward a peaceful coexistence with folk-culture. That oral tradition reasserts itself in a new medium is not only a contribution to Third World societies but to the cinematic world at large.

Film is a new language to the Third World and its grammar is only recently being charted. Its direction, however, seems to be a discursive use of the medium and an appeal for intellectual appreciation. Tomas Gutierrez Alea, perhaps, best exemplifies the new awareness when he says:

> . . . if we want film to serve something higher, if we want it to fulfill its function more perfectly (aesthetic, social, ethical, and revolutionary), we ought to guarantee that it constitutes a *factor in spectators' development.* Film will be more fruitful to the degree that it pushes spectators toward a more profound understanding of reality and, consequently, to the degree that it helps viewers live more actively and incites them to stop being mere spectators in the face of reality. To do this, film ought to appeal not only to emotion and feeling but also to reason and intellect. In this case, both instances ought to exist indissolvably *(sic)* united, in such a way that they come to provoke, as Pascal said, authentic "shudderings and tremblings of the mind."[14]

Notes

1. F. Fanon, *The Wretched of the Earth,* New York: Grove Press, 1963, pp. 207–48. See also, A. Cabral, *Return to the Source*, New York: African Information Service, 1973, pp. 42–69.

2. J. DeOnis, "Pixote role proves all too real," *Los Angeles Times,* Part VI, Tuesday, 5 June, 1984, p. 4.

3. J. Espinosa, "For an imperfect cinema," in M. Chanan (ed), *Twenty Years of the New Latin American Cinema,* London: BFI and Channel 4 Television, 1983, pp. 28–33.

4. R. Gerber, *Glauber Rocha,* Cinema, Politica e a Esthetica do Inconsciente, Brasil: Editora Vozes, 1982, p. 34 and *passim.*

5. G. Rocha, *Revolucao do Cinema Novo*, Rio de Janeiro: Alhambra/Embrafilme, 1981, p. 467.

6. From a film catalogue entitled, *Films about Africa Available in the Midwest,* Madison: African Studies Program, University of Wisconsin, 1974, p. 37.

7. *Africa on Film and Videotape, 1960–81: A Compendium of Reviews,* East Lansing, Michigan: African Studies Center, Michigan State University, 1982, p. 219.

8. It must be freely acknowledged that the future of art criticism and appreciation no doubt lies in the domain of semiotic inquiry. Presently, while its greater virtue lies in the attention it gives to the role of the reader, its greatest weakness is its cultural fixation with Western thought. Third World aesthetics and cultures have been ignored, making it impossible to occupy its premier place in a unified human science. Since the works of Lévi-Strauss and various essays and a book by Roland Barthes nothing of substance regarding semiotic inquiry into cultural studies has been offered. For a general reading on the topic, see Edith Kurzweil, *The Age of Structuralism: Lévi-Strauss to Foucault,* New York: Columbia University Press, 1980 and R. Barthes, *Mythologies,* translated by Annette Lavers, New York: Hill and Wang, 1970. For the various contending factions in the semiotic camp: structuralists, deconstructionists, reader-response critics, theories of intertextuality and narratology, the following books will serve as introductions: R. Scholes, *Semiotics and Interpretation,* New Haven: Yale University Press, 1982 and J. Culler, *The Pursuit of Signs,* Ithaca: Cornell University Press, 1983.

9. Recently Western film-makers, in a bid to revitalise their film world have made "realistic" forays into Third World themes: *Gandhi* on India's struggle for independence, *The year of Living Dangerously* on Sukarno's fall from power, *Under Fire* on the Sandinista revolution in Nicaragua and *Circle of Deceit* on the Lebanese civil war. The statement by one of the characters in *Circle of Deceit:* "We are defending Western civilisation," is an ironic but true epigram for all the films. Far from being radical or new, therefore, these productions give us no more than Hollywood's version of the Third World. For an interesting and illuminating discussion on this recent fascination with "the other," see John Powers, "Saints and Savages," *American Film,* January–February 1984, pp. 38–43.

10. Various sources were consulted, including but not limited to H. Arvon's *Marxist Esthetics,* Ithaca: Cornell University Press, 1973, p. 71 and *passim* and K. Gotrick, *Apidan Theatre and Modern Drama,* Goteborg: Graphic Systems AB, 1984, pp. 140–163. For an elaboration of culture in the context of Third World films, see my book *Third Cinema in the Third World: The Aesthetics of Liberation,* Ann Arbor, Michigan: UMI Research Press, 1982.

11. J. K. Nyerere, *Ujamaa: Essays on Socialism,* London: Oxford University Press, 1968, pp. 91–105.

12. J. Potts, "Is there an international film language," *Sight and Sound* 48 (2) Spring 1979, pp. 74–81.

13. A. Goldsmith, "Picture from China: the style and scope of photography are changing as outside influences mix with traditional values," *Popular Photography,* February 1984, pp. 45–50, 146 & 156.

14. T. G. Alea, *Dialectica del Espectador,* Ciudad de la Habana: Sobre la presente edicion, 1982, p. 21. The first part of the book has been translated by Julia Lesage and appears under the title, "The Viewer's dialectic," in *Jump Cut: A Review of Contemporary Cinema* (29), February 1984, pp. 18–21. This quote is from that publication.

II
The Black Diaspora

Continental Africa

The Artist as the Leader of the Revolution

The History of the Fédération Panafricaine des Cinéastes

Manthia Diawara

The history of the Fédération Panafricaine des Cinéastes (FEPACI) is crucial to an understanding of the development of African film production in general. However, it is in Francophone Africa that the political leverage of filmmakers has met with more success. The efforts of FEPACI in Francophone Africa contributed to the creation of national film centers in the different countries, to the setting up of an inter-African film distribution center (CIDC), to production (CIPROFILM), and to the creation of the Ouagadougou festival as a way of promoting African films. This chapter is a study of the FEPACI with an emphasis on the role of Francophone filmmakers in the development of film in their countries.

African filmmakers are directly affected by the lack of national and international industries that include the structures of production, distribution, and exhibition. In the absence of such money-generating facilities, and because film distribution and exhibition are monopolized by foreign capitalists whose primary concern is to make a profit, Francophone directors are not only forced to depend on the French Ministry of Coopération and similar offices for the production of their films, but they also face problems of programming the films in movie theaters in their own countries and distributing them elsewhere. Such African classics as *La noire de . . . , Mandabi, Le retour de l'aventurier, Concerto pour un exil,* and *Cabascado* were never shown in the movie theaters in the countries of the artists who made them. African directors are also producers of their own films, and, as if that were not enough, they are forced, as in the earlier

Manthia Diawara, *African Cinema: Politics & Culture* (Bloomington: Indiana University Press, 1992), pp. 35–50. Reprinted by permission of the publisher.

days of the invention of motion pictures, to carry their films from place to place for exhibition.

Since the mid-fifties, Francophone filmmakers, organized in national and international associations, have been fighting to change this situation, which Tahar Cheriaa calls "les écrans colonisés," and to set up new plans for the development of film industries in Africa. The strategy consists of denouncing the block-booking system of the foreign capitalists who prevent African films from being seen in Africa. The filmmakers also point out the negative influence on people of American, European, and Indian films that are unchallenged in Africa. Their advice to African governments is to nationalize film distribution, help fund African films by raising taxes on the import of foreign films and ticket sales, and encourage private investment in the production of African films.

Early Efforts to Create an African Cinema

Although the FEPACI is the most significant organization to have influenced governments and international associations in the development of African film, it is important to look first at the efforts of individual filmmakers and/or small organizations of filmmakers who paved the way for the FEPACI. In my chapter on Frances's contribution to the development of Francophone film, I described how Le Groupe Africain du Cinéma attempted to conceive of an African cinema even before African independence. In 1958 Vieyra, the leader of this small organization of filmmakers from Senegal and Dahomey (Benin), put forth the plans for a future African film industry for the Francophone African countries that were dependent on France (only Guinea-Konakry was independent at the time). However, he argued that the countries of the Communauté Francophone should get together to set up an international film center that would have its headquarters in Dakar (Senegal). This center was to serve for the production of educational, instructional, and feature films and was to be funded in the beginning by either the governments of the individual countries or by the central government of the Francophone community, or by both.[1]

After the independence of most African countries in the early sixties, the filmmakers still could not have access to production facilities and to the movie theaters for the projection of their films in their own countries. They consequently increased the political pressure on their governments to intervene and restructure the organization of film activities in a manner that would encourage African productions. Blaise Senghor, ex-member of Le Groupe Africain du Cinéma and director of *Le grand magal à Touba* (1962), criticized the situation by stating that although there were filmmakers in Africa who were authors of some films, there was no such thing as African cinema because the sectors of production, distribution, and exhibition were controlled by foreigners. Senghor believed that African countries, too, must create government-sponsored organizations such as the Office National du Film Canadien, which produces and distributes several short films on culture, education, and research, or the French Centre National de la Cinématographie (CNC), which controls film distribution and exhibition in France and disposes of funds to help produce French films. Similar organizations were necessary for the creation of an authentic African cinema that could be comparable to the New Wave and the French films of *Art et Essai,* all sponsored by the CNC. Senghor writes,

> . . . it is necessary that serious measures be taken to enable the countries to begin a politics of film production, distribution and exhibition that is real and that benefits all. The first measure from which all the rest follows is the creation of an autonomous administrative office, which is like the Centre National de la Cinématographie, and which covers all the issues concerning films.[2]

Timité Bassori, a filmmaker from Ivory Coast, also wrote in the early sixties to criticize the state of film activities in Africa. Bassori argued that African film could not grow when its future depended upon such organisms as the African ministries of information, which produced only newsreels; the Compagnie Africain Cinématographique et Commerciale (COMACICO); and the Société d'Exploitation Cinématographique Africaine (SECMA), which distributed foreign films and saw in the production of African films an interference with their profit making. African directors were forced to work as government employees in the making of newsreels at the ministries of information because the COMACICO and the SECMA were not interested in producing and/or distributing their films. Bassori argued that the imagination and the creativity of filmmakers were stifled at the ministries of information, which were little more than a photo service. Clearly, it was important that governments in Africa change the structure of the film industry. Bassori believed that such a change was all the more crucial because an authentic African cinema could be used to diffuse national culture and enhance national pride and to represent African culture on other continents.[3]

From the mid-sixties on, other filmmakers such as Sembene and Med Hondo became vocal in denouncing the monopolistic practices of the COMACICO and the SECMA and in applying pressure on their governments to restructure film activities. Between 1965 and 1967, national and international associations of filmmakers argued the case of African film at the Colloque de Gène (1965), the Premier Festival Mondial des Arts Nègres de Dakar (1966), and the Table-Ronde de Paris

(1967). The filmmakers at each of these conferences and festivals gave advice to their governments on how to restructure the film market. An elaborate plan for change was submitted, for example, at the Festival Mondial des Arts Nègres in Dakar. The filmmakers recommended the creation of an inter-African film office that would gather and disseminate information about African cinema (film catalogs, statistics on numbers of exhibitions, inventories of the tools of production, and lists of film technicians in Africa).

The new office would propose legislation for the creation of national cinema that would be economically and culturally oriented. It would accommodate different departments on educational films, commercial films, and film of art and essay. There would be an effort, as much as possible, toward a complete transfer to Africa of the facilities of production and postproduction. The film market would be restructured so that a system could be put in place to account for the box-office revenues in each country and on the African level. The office would also reorganize the block-booking system that was used to discriminate against African films. In this vein the current programming system, which allowed theater owners to show two feature films, one after the other, would be changed to one film per sitting. The theater owners would be required to improve the conditions of the theaters, and more African films would be on the agenda.

Other recommendations of the filmmakers at the Festival Mondial des Arts Nègres included the creation of funds to sustain African film production, the building of cinémathèques and archives for the preservation of films, and the organization of festivals to promote African cinema and allow artists to exchange their views. Schools and institutes were also needed in Africa, where filmmakers, technicians, actors, and critics could be trained. The filmmakers believed that the funds for these proposals could come from taxes off the import, production, and exhibition of foreign films. Some additional funds could come from the ticket sales.[4]

The filmmakers' intervention at these conferences and festivals was significant because it underscored the need to control film activities, which in the eyes of the filmmakers constituted an industry for the development of Africa. The filmmakers' plan to restructure and to base film activities in Africa as an economic and cultural factor in the development of the continent gained support from many heads of states and specifically from Niger's president, Hamani Diori, who was then acting-president of the major economic grouping of Francophone Africans, the Organisation Commune Africaine et Mauritienne (OCAM).[5] Furthermore, the filmmakers used the conferences and festivals as platforms for pressuring African governments to stop compromising with the exploitative measures of the COMACICO and the SECMA. They also posed an ethical problem for the French government, which, on the one hand, was producing individual African filmmakers and organizing festivals to promote their films and, on the other hand, was not making an effort to stop the French-owned COMACICO and the SECMA.

The FEPACI and African Film

Clearly, the ground was prepared for an organization such as the FEPACI, which was created in July 1969 in Algiers. The governments were sympathetic to the idea of national cinema, the OCAM had demonstrated its intention of creating an inter-African film industry, and the monopoly of the film market by the COMACICO and the SECMA was weakening as countries such as Upper Volta and Mali made moves to nationalize distribution and exhibition in 1969. Filmmakers needed an organization such as the FEPACI in order to coordinate their efforts and be more effective. Because they were for the most part leftists and idealists who were committed to the notion of Pan-Africanism, the new members of the FEPACI believed their prophetic mission was to unite and to use film as a tool for the liberation of the colonized countries and as a step toward the total unity of Africa. It was in this sense that in its early days the FEPACI sought to be affiliated with its sister association, the Organization of African Unity (OAU).

In 1969, at the Festival Panafricain de la Culture in Algiers, African filmmakers gathered to create an inter-African organization. In 1970, during the third meeting of the festival, Journées Cinématographiques de Carthage (Tunisia), the filmmakers inaugurated the organization and called it La Fédération Panafricaine des Cinéastes. Clearly, the filmmakers in sub-Saharan Africa were benefiting from the experience of their counterparts in Algeria and Tunisia, who had already nationalized their film industries. By 1970, both Algeria and Tunisia had well-defined policies of production, distribution, and exhibition. The Office National pour le Commerce et l'Industrie Cinématographiques (ONCIC) in Algeria, and the Secrétariat aux Affaires Culturelles et à l'Information (SACI) in Tunisia were established for the restructuring of the film industry in favor of national cinema. The national film industry in Algeria had already produced such war epics as *L'aube des damnés* (1964) by Ahmed Rachedi, *Vent des aurès* (1966) by Mohamed Lakdhar-Hamina, and *La voix* (1968) by Slim Riad. In Tunisia, Journées Cinématographiques de Carthage was the first and only film festival devoted to Arab and African cinemas.

The FEPACI was designed to be an association of national filmmaker organizations. Unlike Le Groupe Africain du Cinéma, it was not for individuals who had no base in the national associations of their countries.

Filmmakers were first encouraged to form national organizations and to affiliate them with the FEPACI. In this manner, in order for a filmmaker in Senegal, for example, to become a member of FEPACI, he must first become a member of L'Association Sénégalaise des Cinéastes. This was generally the rule except for cases in which the filmmakers were from an occupied country such as South Africa or from a colonized one such as Mozambique in 1970. Because it had become the association of associations, the FEPACI hoped to be more legitimate in the eyes of governments and other organizations. From its inception, therefore, the FEPACI was recognized by the OCAM, the OAU, and UNESCO, and it increased its membership to thirty-nine countries between 1970 and 1975.[6]

At the inaugural meeting in Carthage, the FEPACI set as its purpose to be committed to the political, cultural, and economic liberation of Africa, to fight the Franco-American monopoly of film distribution and exhibition in Africa, and to encourage the creation of national cinemas.[7] The commitment to the liberation of Africa meant for the filmmakers the creation of aesthetics of disalienation and colonization. Filmmakers were told to use semidocumentary forms to denounce colonialism where it existed and to use didactic fictional forms to denounce the alienation of countries that were politically independent but culturally and economically dependent on the West.

The second purpose of the FEPACI was to fight against the Franco-American monopoly of the film industry in Africa. This was supposed to be achieved by pressuring the governments to nationalize the sectors of distribution and exhibition. At this stage of the development of film in Africa, the filmmakers did not trust private businessmen, whom they suspected of collaborating with the French companies of COMACICO and SECMA and with the M.P.E.A.A. in the United States. The FEPACI could only work with governments to achieve the goal of breaking foreign monopolies and giving African films the chance to be seen in African theaters. However, the filmmakers could not trust the governments enough to ask them to nationalize the sector of production as well. They needed the freedom to express themselves in manners that were not always flattering to the governments. For this reason, they wanted to keep control over the small production houses, such as Film Domirev with Sembene and Soleil O films with Med Hondo. For economic reasons, they wanted the governments to take control of the distribution and exhibition so that they might have the chance to recoup the cost of their films (Boughedir, p. 63).

The last goal of the FEPACI's inaugural meeting, which was to create national cinemas, was to be achieved by raising the level of political consciousness of the filmmakers so that they might increase their lobbying power with their governments and ensure the emergence and maintenance of African cinemas. The filmmakers were also to persuade the governments to create more meeting places like the JJC and the Festival Panafricain de Ouagadougou (FESPACO) in 1969, which could permit them to meet regularly to exchange ideas and promote their films.

Between 1970 and 1975 the FEPACI achieved significant progress in the application of its inaugural resolutions. In Francophone Africa, national film industries were emerging. In 1970, Upper Volta reacted to the unfair monopoly of the COMACICO and SECMA and decided to nationalize its film distribution and exhibition by creating the Société Nationale d'Importation-Distribution (SONAVOCI). The same year, Mali followed suit with the creation of l'Office Cinématographique National du Mali (OCINAM). In 1974, Senegal and Benin got their own distribution houses, the Société Nationale Sénégalaise de Distribution, and the Office Beninois de Cinéma (OBECI). Senegal went as far as to create a Société National du Cinéma (SNC), which is famous for producing four feature films in 1974 alone.[8] Finally, in 1975, Madagascar nationalized the movie theaters and created the Office du Cinéma Malgache.

The French government, too, was forced to readjust its attitude toward the issue of African film development. Afraid that all the countries would nationalize their film import and therefore develop an anti-French attitude (Boughedir, p. 155), the government in France intervened with the COMACICO and SECMA, which forced it to adapt its colonial structures of film distribution to the present realities of Africa. In 1972, the COMACICO and SECMA were bought out by the French Union Général du Cinéma (UGC), and the Société de Participation Cinématographique Africaine (SOPACIA) was created to oversee film distribution in Africa (Boughedir, p. 156.)

In 1975, the FEPACI met in Algiers again to discuss the future of African cinema. This meeting, which is now referred to as the Second FEPACI Congress of Algiers, was concerned with the role of film in the politico-economic and cultural development of Africa. The filmmakers decided that, given the need to raise the consciousness of the African masses and liberate them from the economic and ideological domination of the imperialist countries, they could not indulge themselves in manufacturing films of purely commercial value. They said that African filmmakers should unite with the progressive filmmakers in other countries and join the anti-imperialist fight. It was time to emphasize the instructional value of films. The filmmakers should question the images of Africa and the narrative structures received from the dominant cinema. The question for the FEPACI was how to insert film as an original fact in the process of liberation, how to put it at the service of life, ahead of "art for art's sake"—in other words, how to film African realities in ways that could not be absorbed by the dominant cinema. It was in this light that several filmmakers condemned *Le bracelet de bronze* (1974) by Tidiane Aw and *Pousse Pousse* (1975) by Daniel Kamwa for being overwhelmingly spectacular and less committed to demystifying neocolonialism. On the other hand, the films of Sembene, Med Hondo, and Mahama Traoré were praised for deemphasizing the sensational and commercial aspects and emphasizing the instructional values.

This Second FEPACI Congress also addressed the issue of censorship. The filmmakers argued that the governments had to let them work in freedom in order for them to contribute to the development of the countries in creative and critical manners.

> The State must play a promotional role in the edification of a cinema free of the shackles of censorship and other means of coercion that can affect the creative freedom of the filmmaker and repress the democratic and responsible exercise of this profession. The filmmaker's freedom of expression is, in fact, one of the indispensable conditions of his contribution to the development of the critical senses of the masses.[9]

After the Second FEPACI Congress in 1975, the filmmakers did not meet again until 1982 in Niamey (Niger), where they proposed what is known as "Le Manifeste de Niamey." [See Part III of this book, "Documents," for complete text.] This meeting was necessary for several reasons. Although FEPACI membership was growing and there were more films by Africans, there were no African industries of production, distribution, and exhibition. This was ironical because in 1979 the Francophone countries grouped around the OCAM had created the Consortium Interafricain de Distribution Cinématographique (CIDC) and the Centre Interafricain de Production de Films (CIPROFILM) to replace the Union Africaine de Cinéma (UAC), which was criticized for representing French neocolonial interest (Boughedir, p. 157). A meeting of the FEPACI was therefore necessary to remind the CIDC of its role in the promotion and distribution of African films. Another reason for the meeting was to reassess the role played by governments in film activities. The FEPACI had noticed that most African governments had doubled or tripled their taxes on the revenues of film distribution and exhibition. Some countries even took as much as half the cost of the ticket. Since these tax revenues were spent on activities that were outside the film industry, the FEPACI felt they obstructed the growth of national cinemas. The FEPACI also realized that in those countries where film activities were completely nationalized, the creativity of filmmakers was stifled by governments who only sponsored propaganda films. Finally, in countries such as Senegal and Ivory Coast, the governments, in a reversal of rules, cut the funds set up to encourage the creation of national cinemas.

A third reason that the FEPACI had to convene as soon as possible was the dissension, at the 1981 Ouagadougou Festival, of young filmmakers who called themselves Le Collectif l'Oeil Vert. The young filmmakers charged that the FEPACI did little to help the filmmakers other than to involve them in administrative red tape and visits from the officials of FEPACI in foreign countries.[10] It was generally agreed that the FEPACI had lost its dynamism after 1975. The young filmmakers, by creating Le Collectif l'Oeil Vert, had hoped to bring back this dynamism. Headed by Senegalese director Cheikh N'Gaido Bâ, Le Collectif l'Oeil Vert wanted to take "an immediate and empirical action" toward the solution of African film production. The filmmakers therefore decided to begin by locating one another's equipment and compiling a national inventory. They would also become less dependent on France and other European countries from whom they had been renting equipment. They called their strategy a cooperation "South-South," as opposed to the "North-South" cooperation that existed between the developed and developing countries (Boughedir, p. 65).

Even before the Niamey Congress in 1982, which reunited the FEPACI, the veteran of African cinema, Sembene, had said that the solution to the problems was not in the creation of a new organization, but in convening a meeting and discussing the issues.

Thus, whereas the 1975 Congress of Algiers emphasized the need for the filmmakers to put the instructional values of films ahead of their commercial values, the Niamey Congress was to emphasize the economics and the survival of the FEPACI itself. The Niamey Manifesto introduced new economic clauses in the development of African film that, as Boughedir put it, "contradicted the radical views of the 1975 FEPACI Congress."[11] The expression "Opérateur économique" was used, for example, to designate businessmen and private capitalists. While in the Second FEPACI Congress of Algiers, businessmen were seen as allies of the imperialist countries that were not to be trusted, the Niamey Manifesto described the "Opérateur économique" as necessary to the growth of African film and asked governments to introduce protectionist laws that could guarantee the investment of the "Opérateur économique" and thus encourage him/her to produce more films.

Another important measure in the Manifesto was to link film production with the four other major elements of film activities, namely the distribution, the exhibition, the means of production (equipment, laboratories, and studios), and the training of the technicians. Without the technicians and the means of production, films made by African directors will continue to depend on European technicians and postproduction facilities. Without an

African distribution and exhibition industry, the filmmakers will not only have to depend on European governments and organizations for financial support, but once they have finished films, they will not have a market in Africa to show them.

A third proposal in the Manifesto was to go beyond the idea of national cinema in order to promote inter-African and/or regional cinemas. This idea, too, was economically motivated. The FEPACI knew that it was not possible for a filmmaker to recoup the cost of his/her film in his/her country alone where the total number of theater tickets sold was less than the number of tickets sold in the Latin Quarter alone in Paris for a comparable time. In Africa, the population of most countries is under 15 million, and because of social and economic reasons, many people do not go to the movies. Clearly an inter-African movie industry offered better chances for filmmakers to recoup the cost of their films. In Francophone Africa, the CIDC is one such industry grouping fourteen countries (see note 5). In the perspective of increasing the chances for filmmakers to recoup the cost of their films, the FEPACI also asked African national television services to work with the filmmakers in coproducing and airing films.

The fourth proposal of the Niamey Manifesto concerned the necessity of implementing national control systems that could account for the number of theater tickets sold at the box offices in Africa. In Francophone Africa, for example, if the offices of the CIDC knew exactly the box-office revenues in each country at any given time, it would be easier to divide these revenues to the satisfaction of the filmmakers, the businessmen, and the government for tax purposes. The filmmakers would not trust the theater owners; they would have no other precise way to measure audiences' responses to their films. The implementation of a structure to control the box-office revenues would therefore help the growth of African film.

Finally, the Niamey Manifesto proposed a tax reform plan that would enable the African film industries such as the CIDC to be self-supporting. In this clause, the FEPACI argued that part of the money spent by African moviegoers, both in foreign and domestic films, should be invested in producing new films, refitting old movie theaters and constructing new ones, setting up film archives and cinémathèques, organizing festivals and other promotional activities, training film technicians, and acquiring new equipment. The governments were asked to reduce taxes as an incentive for theater owners to use the additional revenues to improve the conditions of the theaters and to construct new ones. On the other hand, it was proposed that governments begin using the remaining tax revenues in businesses relating to the development of film. It was in this sense that the FEPACI stated in the Niamey Manifesto that "the funds that go into producing African films, whether they are national, regional or inter-African films, should come from the twin sectors of distribution and exhibition, not from the government's budgets. It is only in this manner of film begetting film that a cinema industry is possible" (Boughedir, p. 172).

A Critique of FEPACI

After this panoramic look at the FEPACI, from its creation to the Niamey Manifesto, what can one say of the organization? The FEPACI is unique because it is a movement of the filmmakers of a whole continent, as opposed to such national cinematic movements as the Cinema Novo and the New Wave. The FEPACI is also unique because it is less a cinematic movement aimed at deconstructing traditional film narratives, as is the case with the new Wave, and more a politico-economic movement committed to the total liberation of Africa. As such, the FEPACI has more in common with its sister Pan-African movement, The Organization of African Unity, than with "purely" cinematic movements such as the New Wave and the Italian Neo-realism. The success and the contradictions of the FEPACI should therefore be measured in terms of achievements as a liberationist movement that is committed to the independence and the unity of Africa.

As the above history of the organization shows, the FEPACI has achieved many of its goals. As a liberationist movement, it has led the African governments in breaking the monopoly of such foreign distribution companies as COMACICO and SECMA. It has influenced the governments to create national production centers and to take film seriously as a means of development. The efforts of the filmmakers led to the creation of the festivals of Ouagadougou (1969) and Mogadishu (Somalia, 1980). One can also say that in addition to the economic advantages, it is the spirit of Pan-Africanism that led the FEPACI in the Niamey Manifesto to go beyond national to regional and inter-African issues.

As a liberationist movement, the film directors of the FEPACI have also made several successful films that raise the consciousness of the audiences. Using documentary and didactic fictional forms to denounce neocolonialism and alienation, the filmmakers saw as their prophetic mission to employ film as "a weapon as well as a means of expression for the development of the awareness of class struggle."[12] In retrospect, one sees at least three types of cinema that developed out of this liberationist movement: the semidocumentary, the didactic-fictional, and film research. The semidocumentary depicts and denounces colonialism and/or settler rule and shows the progress of the forces of liberation. Typical examples of the semidocumentary form are the epics of the Algerian war, Sarah Maldoror's *Sambizanga* (1972), Haile Gerima's *Harvest, 3000 Years* (1974), the films by the Mozambican National Institute of Cinema such as *These Are the Weapons* (1979), and the films on apartheid in South Africa.

The second type of film, the didactic-fictional, opposes good and evil in Africa in a Manichaean manner. The films of this genre show the strong, usually from European origins, taking advantage of the weak, symbolized by Africa. Sometimes Islam and/or African governments replace the West as the oppressor in the films. The genre is more prevalent in Senegal, with Ousmane Sembene as its masterful practitioner. Another general trait of the films is a quest (mostly symbolic) to the West, characterizing alienation, and a return to the sources as a way of solving the problem. Typical examples of the form are the films of Sembene and Mahama Traoré, *Touki Bouki* (1973) by Djibril Diop Mambety, *Kodou* (1971) by Ababacar Samb-Makharam, *Soleil O* (1969) by Med Hondo, and *Le bracelet de bronze* (1974) by Tidiane Aw.

The third form, the film of research, is an outgrowth of the didactic-fictional form. Emphasizing less the Manichaeism of their predecessors, the filmmakers look for combinations of solutions to African problems. They depict social changes, breaks, and continuities in history and culture. They pose as challenges the ways in which tradition and modernity, educated and illiterate can be reconciled. Examples of this genre are the films by Souleymane Cissé, *Lettres paysannes* (1975) by Safi Faye, *Jom* (1981) by Ababacar Samb-Makharam, *Wend Kuuni* (1982) by Gaston Kaboré, the powerful documentary *Zo Kwe Zo* (1982) by Joseph Akouissonne, and *Djeli* (1981) by Lancine Kramo Fadiga.

It is important to notice here that all three genres are still being practiced in African cinema. Ideally, the first type coincides with the inaugural manifesto of the FEPACI, which postulates the need to unite and fight against colonialism and settler rule in South Africa. The second type, too, is symptomatic of the anti-neocolonialist and imperialist slogans of the Second FEPACI Congress, and the third type, including such popular films as *Djeli* and *Finye* (1982), seems to represent the Niamey Manifesto, which emphasizes film more as an industry and less as an anticapitalist weapon. The presence of all three genres at the same time indicates the historical situation of Africa as well as the ideological differences of the filmmakers.

As regards the contradictions of the FEPACI, they typically emanate from an insufficient analysis of the blending of political, economic, and artistic realities in Africa. As Cheriaa put it,

> In a place [Africa] where there hadn't been a structure and a tradition of national cinemas before, they [filmmakers] had believed

> that all that was needed, in addition to their own commitment to making films, was the support of the State, which they thought was determined to create national cinemas. Feeling very strongly about what they had to say and imagining themselves free to express it, they believed the states also were in a disposition to work with them. They did not doubt that a limit could be put on their freedom to act.[13]

The FEPACI, as a Pan-Africanist movement, worked to help individual countries gain control over the channels of distribution, exhibition, and sometimes even production. The FEPACI believed that it was making progress toward an autonomous film industry every time that a country nationalized its film activities. However, it soon became clear that the interest of the FEPACI could not always be reconciled with that of the countries it had helped. There were ideological contraditions that were determined by the fact of the African countries' economic and cultural dependence on the West. Thus, although politically it was prestigious for the governments to have national cinemas, economically they could not afford them without the consent of the Western countries. For political reasons, they exported independently made films by Sembene and Cissé (to cite only two), presenting them as national films, even though they had not been distributed nationally. For economic reasons, they collected taxes on film distribution and exhibition, and they used the tax revenues for other problems affecting the countries.

In those countries where the economic issues were temporarily overcome and national film production centers were created, the governments' politics of production were different from those of the FEPACI. Where the FEPACI was committed to making liberationist and Pan-Africanist films, the governments were interested in propaganda films. Where the FEPACI deemphasized the profit-making aspect of the films, the governments emphasized the need for nonpropaganda films to recoup at least their cost of production. The low box-office revenues of national films led to the closing in the mid-seventies of the government-sponsored production units in Ivory Coast and Senegal.

Clearly, these contradictions between the government's national interest and the FEPACI's radical liberationist interest are serious issues that kept setting back the progress of the movement. In Francophone Africa, for example, since the CIDC took over film distribution from the COMACICO and SECMA, many problems had surfaced that have prevented it from functioning full swing in the fourteen member countries. Some governments have failed to pay part of the tax revenues from the film market as membership fees to the CIDC. Some have frustrated the efforts of the CIDC to distribute African films by finding new customers in Switzerland, Société Commerciale de Films (SOCOFILM), which specialized in distributing American films.[14] Finally, some governments defended the right of theater owners who rejected African films on the grounds that the spectators preferred American and European films. The bottom line was, as Cheriaa put it, that there were African filmmakers and films, which he compared to heads, without facilities of production, distribution and exhibition, which he compared to bodies (Cheriaa, p. 8).

Another point of weakness in the FEPACI's endeavor to develop an African film industry is the overreliance of the organization on governments and foreign countries such as France for the production of the films. Thus, it is doubtful whether the filmmakers had thought as seriously about the means of production as they had about the need to nationalize the film market and to raise the consciousness of the audiences with African films. The needs to reflect more on the tools of production led to the creation of Le Collectif l'Oeil Vert in 1981.

The disillusionment of Le Collectif l'Oeil Vert in the superstructural ideology of the FEPACI helps to clarify the extent to which it had failed to integrate the means of production in the total liberation of African cinema. Thus, on the superstructural level, Francophone filmmakers have put a lot of pressure on their governments and on France to improve the conditions of production and distribution. It is also clear that the filmmakers are dynamic, pioneering, and inventive. To be sure of these superstructural qualities, it is enough to look at Sembene's original use of Wolof in *Le Mandat* (1968) and his attack on Islam in *Ceddo* (1976), Djibril Diop Mambety's editing style in *Touki Bouki,* Dikongue Pipa's camera angles and mise-en-scène in *Muna Mota* (1975), and Cissé's synthesis between tradition and modernity in *Baara* (1978) and *Finye*.

But on an infrastructural level, where the forces of production and the means of production are concerned, it is clear that the filmmakers have not done substantial research. The number of camera operators, electricians, sound-engineers, and editors is not increasing proportionally with that of directors. For the most part, after the directorial duties, French manpower is used to finish the films. A look at the tools of production also shows that the filmmakers usually accept equipment without thinking of their specific ends. For example, the 35mm cameras with color and fast film stock are used in spite of the high cost and relative unwieldiness. In this regard, Sembene's candid response to the crucial issues of 35mm versus 16mm is revealing:

> Now, as far as 16mm or 35mm is concerned, it varies according to the filmmaker. It is true that I have always worked in 35mm, but it was just chance that that happened because I have a 35mm camera. On the other hand, we have found a new method. The young people

> whom we have trained like 16mm color because it is much more mobile, the crew is smaller, and it can be blown up to 35mm.[15]

The young filmmakers Sembene is referring to are none other than Cheikh N'Gaido Bâ (Senegal), Sanou Kollo (Upper Volta), Lancine Kramo Fadiga (Ivory Coast), and many others who got together to create Le Collectif l'Oeil Vert. It is hard to understand why an astute man such as Sembene does not get rid of his 35mm for a 16mm that will cut costs by more than half of his long-awaited film, *Samori.*

The FEPACI is to be credited for the disintegration of such foreign monopolies as the COMACICO and SECMA, the creation of national cinemas, and such inter-African organizations as the festivals and the distribution and production units such as the CIDC and the CIPROFILM. Moreover, the FEPACI has contributed to world cinemas by creating a Pan-Africanist and liberationist cinema toward the total independence of Africa. It is therefore to be hoped that the young, such as Le Collectif l'Oeil Vert, will incorporate the tools of production as a realistic economic factor in film production. In colonial times, L. A. Notcutt, with his Bantu Cinema Experiment, invented an economically liberated cinema, although it was racist.[16] Rouch, too, made stylistic and economic breakthroughs with his 16mm camera in Africa. Finally, in Latin America, Solanas and Gettino give an important place to the 16mm or even much less expensive and practical cameras in their definition of Third Cinema, or alternative cinema. If they want to be liberated, African filmmakers must do more research into the tools of production.

Notes

1. Paulin S. Vieyra, "Propos sur le cinéma africain," in *Présence Africaine,* no. 23 (1958): 114–15.

2. Blaise Senghor, "Pour un authentique cinéma africain," *Présence Africaine,* no. 49 (1964): 109.

3. Timité Bassori, "Un cinéma mort-né?" in *Présence Africaine,* no. 49 (1964): 111–14.

4. Jean-René Débrix, "Le cinéma africain," in *Afrique Contemporaine,* no. 40 (Nov.–Dec. 1968): 6.

5. OCAM is an economic and cultural organization grouping the following Francophone African countries: Senegal, Mali, Mauritania, Guinea, Ivory Coast, Upper Volta, Niger, Benin, Toga, Cameroon, Gabon, Chad, and the Central African Republic. For a critique of OCAM, see Brigette Nouaille-Degorge, "OCAM: An Outdated Organization," UFAHAMU 5, no. 2 (1974): 135–47.

6. Férid Boughedir, *Afrique noire: Quel cinéma?,* 63.

7. Pierre Pommier, *Cinéma et développement en Afrique noire francophone* (Paris: Editions A. Pedone, 1974), 153.

8. Férid Boughedir, "La (trop) longue marche des cinéastes africains," *CinémAction* no. 26 (1982): 156–57.

9. Guy Hennebelle, "La charte d'Alger du cinéma africain," *Afrique littéraire et artistique,* no. 49 (1978): 165.

10. Férid Boughedir, *Afrique noire: Quel cinéma?,* 65.

11. Férid Boughedir, "Le nouveau credo des cinéastes africains: Le Manifeste de Niamey," *CinémAction,* no. 26 (1982): 168.

12. Teshome Gabriel, *Third Cinema in the Third World: The Aesthetics of Liberation* (Ann Arbor: UMI Research Press, 1982), 107.

13. Tahar Cheriaa, "Le cinéma africain et les 'réducteurs de têtes,'" *Afrique littéraire et artistique,* no. 49 (1978): 8.

14. Férid Boughedir, "La (trop) longue," 158.

15. Teshome Gabriel, *Third Cinema,* 115.

16. Fernando E. Solanas and Octavio Gettino, *Ciné cultura y decolonizacion* (Buenos Aires: Siglo 21, 1973).

A Cinema Fighting for Its Liberation

Férid Boughedir

The African cinema at the time of the Organization of African Unity (OAU): it was in effect in 1963 that for the first time a film entirely conceived and produced by an African made its appearance on the international scene, was seen by a paying audience, and received a prize. That happened at the International Festival of Tours (France), and the prize-winning film *Borom Sarret* by Ousmane Sembene (Senegal) was established as the first film demonstrating great talent made by an African, opening the way to a cinema of fiction which would develop mainly in the "French speaking" countries of the continent (their "English speaking" or "Portuguese speaking" neighbors favoring on the whole the documentary).

The anecdote of *Borom Sarret* was simple but significant: a poor carter saw the tools of his trade confiscated by the police because he had dared to cross the frontier between the Dakar of the poor and the Dakar of the rich. The emotional nature of this tale, the restrained anger of the author compared with the kind experienced by the more humble, the use of monologue which makes us inwardly experience the plight of the carter, and moreover an open end which appeals to humanity and to the indignation of the onlooker, made this film a premature masterpiece by which, straight away, the African cinema seemed to establish the tone of what was going to become one of its major directions.

The application of picture and sound to the service of works of enlightenment arouses a consciousness of the realities of an Africa hardly

From Renee Tajima, ed., *Journey Across Three Continents* (N.Y.: Third World Newsreel, 1985), pp. 22–25. Reprinted by permission of the publisher.

out of colonialism and living the contradictions of freshly acquired independence. All the subsequent works of Ousmane Sembene, autodidact of the camera—who was a fisherman, a mason, a docker, then a writer before exchanging the pen for the camera "in order to speak even to the illiterate"—confirmed the vocation of the African cinema to inscribe itself as "awakener of the people," in total opposition to the cinema which had preceded it on the continent: that western cinema of "escapism" which the African producers accused of literally drugging the African public with its byproducts and of imparting values of foreign domination.

Domination by Foreign Companies

It is however in terms of the rivalry with this foreign cinematographic distribution that the (short) history of the African cinema will be written from beginning to end. For foreign films (often the worst rubbish of world-wide production) were distributed by the big western companies which controlled the cinematographic market of the African countries, and looked unfavorably upon the appearance of a young African cinema which was going to compete with their films and damage their profitability and their influence.

Several African countries such as Tunisia or the Upper Volta, having tried to create a national monopoly on the importation of films to control their market and reserve a "screen time" for their new-born production, were punished by boycotts as regards their supply of films. The big companies, who regrouped in joint responsibility trusts, obliged the states concerned to back pedal by imposing restrictions on their profit. Deprived of screens and of a paying audience, the African film would never be able to recover the cost of returning and giving birth to a second film—the African cinema was condemned to be still-born.

First Success of the African Cinema

In view of this situation, and above all in view of the inertia of their governments as regards the future of the cinema kept for minor and simple "amusement," the African film producers were quite naturally led to regroup their efforts in order to attempt a unity on the scale of the continent. A first opportunity to meet was offered to them with the creation, in 1966 by the Tunisian Minister of Culture, of the Pan-African festival of the "Cinematographic Days of Carthage" (which acclaimed the first full-length film by Ousmane Sembene *La Noire De*), soon followed in 1969 by the "Panafrican Festival of the Cinema of Ouagadougou" (Burkina Faso). The following year saw the creation of the Federation Panafricaine des Cineastes (FEPACI) which united 33 countries of the continent and whose creed, like that of the two festivals, was to be a voice of incitation directed at the African governments in order that they might take protectionist measures which were necessary for the survival of their new-born cinema—a cinema which confirmed its promise with first attempts that were as good as the works of masters.

The festival of Cannes in 1967 awarded a prize to the very beautiful *Vent des Aures* by the Algerian Lakhdar Hamina, which shows a lyrical and unforgettable sight: the sufferings of a mother during the war. The following years brought two new competent African film producers: Desire Ecare from the Ivory Coast with *Concerto for an Exile,* a bittersweet chronicle about the African immigrants in Europe; and the Nigerian Oumarou Ganda with *Cabascabo,* which shows the difficulties of the reintegration into his society of an African who is an ex-serviceman with the colonial troops.

Finally the Festival of Venice 1968 sanctioned conclusively the African cinema by giving an award to the first full-length color film by Ousmane Sembene, *Le Mandat,* which recounts in the tragi-comic mode the confrontation of the common man with the bureaucracy that arose from independence. It is a film which, for a long time, remained the masterpiece of its author, and which has done the most for raising the knowledge of the African cinema throughout the world. After this film, five great artistic tendencies were henceforth to divide the destiny of the African cinema, whether it be in the north or in the south of the Sahara. The colonial fate of yesterday and the necessities of development today have in effect created among all these works an astonishing continuity of themes, and of preoccupations which go beyond the existing frontiers. One rediscovers in the Maugrabin films, as in those of Black Africa, the struggle against colonialism and neocolonialism, criticism of retrograde traditions (notably maraboutism), rejection of western influence and its servile imitation, the division between town and countryside and its consequences, the rural exodus, demystification of the golden dream of emigration in Europe, and the denunciation of the lot of African women. This cinema seems to have made itself, in many respects, the worthy heir of the old tales of oral tradition which teaches us that the pursuit of money doesn't bring happiness and that fidelity to the given word determines the meaning of life.

The next stage in the history of the African cinema was marked in a significant way, by Algeria, the first country of the continent to prove that it was possible to break the stranglehold of the western distribution companies on the African market. Having made ample provision of films before decreeing a national monopoly on the importation of film in 1971, Algeria resisted for five years the boycott of the all-powerful Motion Pictures Export Association of America (MPEAA), which ended up coming to an agreement; by finally "granting" to this independent country its natural right to choose the films that it imported, and to free screen time for its local productions. It soon became one of the most well-provided of the continent.

Nationalistic Awakenings

The success achieved by the Algerian cinema was to give rise to a series of cinematographic nationalizations in several African countries which took control of their screens. Senegal decreed a national monopoly on the importation of films in January, 1974. It was soon followed by Benin, then in 1975 by Madagascar. The Congo followed suit in 1979.

As Senegalese production expanded, Benin made its first full-length film, and Burkina Faso was already on its second. October, 1974 saw the climax of these initiatives during the fifth session of the "Cinematographic Days of Carthage," the first "conference on the distribution and production of African and Arabian films" uniting the leaders of the new national cinema societies. They concluded that it was necessary to regroup the African cinematographic market in order to create a necessary profitability from local films: profitability that had been impossible in each isolated territory. At the subsequent conferences which took place again at Carthage and at Ouagadougou, at Maputo, Mozambique and at Mogadishu, Somalia, the film producers managed to work out a three-point strategy for the liberation and viability of the national cinemas in Africa—a strategy depending entirely on intervention by the state.

The Success at Cannes

The year 1975, which saw the triumph of African cinema on a world-wide scale: the *palme d'or* of the Cannes Festival was awarded to the Algerian film *Chronique des Années de Braise* by Lakhdar Hamina, thus carrying-off the supreme prize of the greatest festival in the world. The same year, a veritable school of Senegalese cinema was born with several films produced by the Jeune Societe National de Cinema. Most notably were *N'Diangane* and *Garga M'Bosse,* by the prolific Mahama Traore (about lost childhood and the dramas of drought), and *Xala* by Ousmane Sembene, a vitriolic parody on the congenital impotence of the new African bourgeoisie who wanted to imitate the west.

The Mauritanian Med Hondo carried off the main prize of the "Days of Carthage" festival with *Les Bicots Nègres Vos Voisins,* in which he deepened his reflections on immigration that he explored in the preceding film *Soleil-O* (1970). Abhellatif Ben Ammar (Tunisia) presented *Sejnane,* a film on the foundation of the struggle for national liberation. The Cameroonian Jean Pierre Dikongue-Pipa and the Senegalese Safi Faye,

Africa's first producer, created a sensation by the originality of their films *Muna-Moto* and *Lettre Paysanne.* All these films are of an international style, and bear witness to a maturity and an astonishing artistic quality, especially in light of their derisory financial means and their lack of future outlets. The African cinema seemed to have attained its "golden age": it was to be short lived.

The Response of the Foreign Companies

The conference of Tunisia in October, 1974, had been attended by, among others, two guests as "observers": they were, in fact, the representative of the MPEAA who reigned over the Maghrib and English-speaking Africa, and the representative of the Union General Cinematographique, (PDG of L'UGC), the French group whose subsidiary company, Societe de Participation Cinematographique Africaine (SPACIA) dominated the cinematographic market of the whole of French-speaking Western Africa, having repurchased the circuits of the former colonial companies SECMA and COMACICO. From then on, the foreign companies radically changed their approach to the problem—the Algerian example having proved that the time of direct confrontation was gone. Instead there had come the time of "amicable" agreements designed to safeguard national sovereignty, which had become a sensitive issue, whilst at the same time preserving the presence of the same number of Western films on African screens. The result was a boom in foreign films threatened for a moment . . . and a clear regression of the number of African films produced.

Continuing its policy of apparent Africanization of its cinematographic market, SPACIA transformed itself into Union Africaine de Cinema (UAC) and started to sell its cinema houses to African individuals, to try and preserve the most important thing: the importation-distribution of the greatest possible number of foreign films, bought at low cost on the world markets (with, at the top of the list, Indian films and Chinese karate films), and making an excessive profit on African soil.

The Birth of the CIDC

Then a new African partner made its appearance: the Consortium Inter-africain de Distribution Cinematographique (CIDC). One of the regional consolidations that the film producers had formed by their votes, and whose general principles had been voted by several African states of the French-speaking west. It was finally set on its feet in 1979, with a film producer at its head, the Nigerian Inousa Ousseini. A "common market" of cinematographic distribution was born stretching over 14 countries—Senegal, Mali, Mauritania, Guinea, Ivory Coast, Burkina Faso, Niger,

Togo, Benin, Cameroon, Gabon, Congo, Chad, Central Africa. In view of the will of the states, the UAC at the end of 1980 agreed to resell its portfolio of films to the CIDC on such conditions that France would remain one of the privileged suppliers of this market of "French-speaking" countries: nevertheless, for the first time, a common market of film distribution in Africa was really controlled by Africans, who could finally introduce African films there in the normal fashion. This is what the CIDC is committed to. It has distributed since its first year some fifty African films. By trying to introduce this paradoxically new product into the market on its own soil, it regularly comes up against the unwillingness of the owners of cinema houses who prefer to put on what they know: Italian westerns, Italian melodramas, or karate films.

However, the African film has slowly begun to take its place, and the public is helping to conquer the oppositions to the system. It triumphed with the two films that carried off the main prizes of the seventh and eighth festivals of Ouagadougou, *Djeli* by Fadika Kramo Lancine from the Ivory Coast and *Finye (The Wind)* from Soulemane Cisse of Mali. The two films, although considered "cultural" and not commercial enough for the owners of the cinema houses, have nevertheless beaten all box office records in their respective countries.

1982: The Manifesto of Niamey

Assembled in conference at Niamey, Niger in March, 1982, African film producers drew up a manifesto in which they readjusted their earlier stance. The experience of the last 10 years had, in fact, shown the shortcomings of total nationalization. (Too many governments considered the "seventh art" as an instrument of propaganda and have only financed films which glorify the government.) The film producers now called for a balance between State-controlled and private enterprise. The State would have control of the distribution market and voting of protectionist laws in favor of national production—above all the "cultural type." The private producers (that is to say more often than not, the film producers themselves) would have the freedom of choice of subjects, in a system which would guarantee their financial downfall.

Now that the governments have begun to curb the appetites of the suppliers of foreign films, the film producers would like to break away from a state protection that would be too omnipresent. Their new creed is as follows: the cinema must be financed not by state budgets but by money from the cinema. What money from the cinema? That which is spent everyday by millions of cinema-goers to see often very bad foreign films, and of which a part—the taxes—must rightfully return to local film production.

While waiting for this to come about in all the countries concerned, there will certainly follow yet many more conferences, many seminars, many manifestos. The (too long) march of the African film producers towards economic viability of their films continues, even if artistically and culturally, the African films have become an undeniable reality.

Excerpts from the paper read at the first Third Eye Festival, London

Sembene, A Griot of Modern Times

Françoise Pfaff

On many occasions, Sembene has drawn a parallel between the traditional storyteller and the modern African filmmaker in terms such as these: "The artist must in many ways be the mouth and the ears of his people. In the modern sense, this corresponds to the role of the griot in traditional African culture. The artist is like a mirror. His work reflects and synthesizes the problems, the struggles, and the hopes of his people."[1]

To fully grasp the importance of these words, one must refer to the essence of Africa's oral tradition of which the griot is an eminent representative. According to Georges Balandier and Jacques Maquet, the word *griot* comes from the Portuguese word *cridao,* which means servant, or from the French corruption of the Wolof word *gewel,* which refers to the members of a caste formerly attached to a family or a clan.[2] "They are to be found either at the courts of chiefs or established on their own in towns and villages, acting as story-tellers, clowns, heralds, genealogists, musicians, oral reporters, or paid flatterers or insulters."[3]

Elolongue Epanya Yondo of Cameroon writes that the griots are professionals who specialize in storytelling, reciting legends, or recounting the valiant deeds of a family's or country's forebears. Griots may be the chroniclers of an important family or of a group of people—like the Bambara hunters' griot—or itinerant poets and musicians who extol the praises of the person who has hired them for a special festivity.[4] Dorothy S.

Françoise Pfaff, *The Cinema of Ousmane Sembene, A Pioneer of African Film,* (Westport, Conn.: Greenwood Press, 1984), pp. 29–42. An imprint of Greenwood Publishing Group, Inc. Reprinted with permission.

Blair, who translated into English *Birago Diop's Tales of Amadou Koumba* stresses in her foreword that, "in the French-speaking territories of West Africa today, especially in Guinea and Senegal, the griot is an important and respected member of the artistic and cultural community."[5]

Yet the griot has an ambiguous status. While he enjoys freedom of expression and respect for his knowledge from both the elite and the masses, he is also stigmatized because he belongs to an inferior caste within Senegalese society. Although Mbye B. Cham agrees with the general definitions of griots he also makes a very important distinction between two kinds of griot found in Wolof society: the gewel and the *lebkat.* The gewel is someone who possesses "mastery over the word and knowledge" by social tradition and training (craft transmitted within a cast from generation to generation), while the lebkat is a raconteur whose storytelling can be performed at any time by any member of a given community—heads of families or women, especially in such countries as Senegal.[6] Also, it often happens that the professional griot looks down upon the lebkat's storytelling. Yet both the gewel and the lebkat are the repositories of Africa's oral treasures, the collective product of a common experience.

The oral tradition embodied by the griot or the conveying of information and memories orally has existed in all societies at one stage or another of history when spoken words and gestures were more important than graphic signs. In all areas of the world, mankind has first expressed itself through the spoken word before inventing a writing system. One remembers that the epics of ancient Greece, such as the *Iliad* and *Odyssey,* were transmitted orally before being codified in writing. Although the oral tradition is by no means limited to Africa, it is there that it probably remains the most popular and the most vital today.

Until recently, the verbal richness of black Africa was found outside of books. It was kept alive in people's minds and expressed in specific circumstances such as festivals or evening storytelling sessions by professional or extemporaneous performers. In Africa, oral literature is still very much a part of a deep-rooted collective custom, adapted to a communal, largely illiterate, and agriculturally based society. Through its myths, legends, epics, tales, or historical poems, it has celebrated the prowess of kings and warriors, perpetuated its past, and preserved its ancestral wisdom while strengthening the sociocultural and historical identity of the group. The story is probably the most common and the most appreciated literary genre of black Africa's oral tradition.

If "in the beginning there was the word," African storytellers have since made ample use of it, and their influence has gone beyond Africa into Europe and the Americas. The findings of the English linguist Alice Werner prove that their stories were the source of Aesop's fables—

whose name derives from "Ethiop," a general term for "black African" in Greek antiquity—which in turn influenced La Fontaine's fables. The folk literature of the New World is also immensely enriched by its African heritage via the slave trade.

It is true that the love of telling and listening to stories is a universal human characteristic, yet equally undeniable is the fact that it is becoming less and less prevalent in industrialized societies. In such societies, television has become a surrogate storyteller presenting children with cartoons and other stories which are but modernized tales. At one time or another, tales full of magic and spirits have been spread all over the world. Yet it is in countries such as Africa that it is still widely appreciated.

"Africa is the land of story-tellers and tales," asserts Georges Balandier,[7] while Elolongue Epanya Yondo underlines that "the tale is the nurturing breast which feeds most of the Negro African literary genres."[8] The presence in African cinema of tales and oral stories adapted to the screen comes as no surprise in films by Momar Thiam or Mustapha Alassane. One also notes that the atmosphere of film screening parallels the traditional time for much of storytelling at gatherings when the family or community meets. Now it is left to determine how and to what extent African film, as a means of expression and communication, is indebted to the oral tradition. According to Alphonse Raphaël N'Diaye from Senegal, " . . . the oral traditions constitute the global expression of the life of our people."[9] Besides, in a Western frame of reference, Joseph M. Boggs also emphasizes the storytelling aspect of film: "Although film is a unique medium, with properties and characteristics which set it apart from other art forms such as painting, sculpture, fiction and drama, it is also, in its most popular and powerful form, a story-telling medium."[10] Thus, because of structural similarities to the medium of the tale, film is the ideal means of conveying the treasures and techniques of the African oral tradition. If indeed most films tell stories, their content and form are determined by a broader cultural context; that is, the traditions and needs of a given society. This is precisely what the Ethiopian filmmaker Haile Gerima implies in his observation: "The oral tradition is a part of African film aesthetics in terms of space, pace, and rhythms."[11]

Most contemporary African artists draw on their heritage while associating their traditions with Western models. For instance, both African writers and filmmakers illustrate African themes through forms and/or tools which are not indigenous to their culture—be it the novel or the camera. They introduce the same symbiosis in art as that which is unavoidable in their transitional African society based on both African and Western patterns. It is natural, therefore, that Sembene, in addition to defining himself as a griot, should provide works whose style and themes derive from Africa's oral tradition.

The griot is physically present in a number of Sembene's films. He is shown performing varied functions as the actor/narrator of *Niaye* and the cart driver's family griot in *Borom Sarret.* In *Xala,* griots are also part of a celebration following the "Senegalization" of the Chamber of Commerce and of El Hadji's wedding festivities. In *Ceddo,* Fara, a griot, follows the princess and her captor.

Sembene's films, like the griot's renditions, provide the immediacy of visual and auditory action. Thus, when he presents and discusses his works with various groups of viewers in Senegal, Sembene arouses a dynamic interchange with the audience, somewhat re-establishing the verbal directness which has always existed between the griot and the spectators.

Indeed, Sembene's works, like those of other African directors, are accessible to popular audiences because they represent a collective experience based on visual and aural elements with characteristics which can be compared to the griot's delivery. Anyone who has attended a film screening in a working-class district of Senegal is struck by the intensely vocal participation of the viewers who comment on the plot of the film, respond to one another's remark, address the actors, and laugh at their mishaps just as they would during the griot's performance in which dramatic mimics and gestures are used to encourage audience reaction.

"Sembene Ousmane is a born story-teller," observed a film critic in the early stages of the Senegalese director's career.[12] No one could refute such an assertion. As a modern griot, Sembene performs the ritual incantation of images and words which link reality to metaphor in the creative process which constitutes a work of art. In terms of style and content, the griot's handling of social satire in African societies is legendary. The songs and tales of traditional folklore make ample use of comic gestures, words, and situations. The griot is known to have the unique capability of presenting the most serious issues in a humorous and entertaining fashion and so does Sembene. Furthermore, the filmmaker's introduction of musical elements in his works can be compared to the griot's frequent use of songs in his delivery. Sembene favors the use of the Wolof guitar in his soundtrack just as the Senegambian griot likes to accompany himself with that same string instrument.

Structurally, the clear linear progression usually found in Sembene's films can be compared to that of the griot's story. *Mandabi* and *Xala* have the freshness and the atmosphere of tales, while *Emitai* and *Ceddo* reflect the epic tone of some of Africa's oral legends and heroic deeds. Sembene's use of African languages, songs, and proverbs confer on his works the same local flavor which can be found in African storytelling. The circular conversations present in *Emitai* and *Ceddo* are reminiscent of palavers. Then, in traditional tales, " . . . the point of the story is often summed up in a proverb-dicton."[13] This is exactly what happens in *Niaye, Mandabi, Xala,* and *Ceddo.*

In the African oral tradition, the word generally suggests more than it actually says because the griot makes use of a core of visual imagery and metaphors common to both performer and listener. Sembene's treatment of visual signs adheres to such a principle, and all of his stories can be considered at various levels of meaning beyond their realistic components. Also, it is often noted that his characters are types rather than individuals. They have an overt and a covert significance just as in African storytelling and rituals people are brought to focus more on types represented than the psychological intricacies of the characters.

Sembene's films deal with a world which is well known to his audiences. This trait is also observed in traditional African stories. In his study of the Cameroonian oral tradition, Philip A. Noss observes that "the world of the tale is normally the immediate world of the people for whom it is told. Rarely is the tale about distant places and persons."[14] Because of their familiar context, African oral stories are readily understood by listeners. They present social types like the pauper, the imam, the merchant, and the trickster, all of whom are easily identifiable. The stories also include the king or the princess, legendary forebears known to all. These characters appear in Sembene's films as well. The trickster, for instance, usually a dishonest individual who embodies antisocial traits, appears as the thief or the corrupted civil servant or bourgeois in *Borom Sarret, Mandabi,* and *Xala.* The beggars and physically deformed people, who are often a part of African oral stories, are present in Sembene's plots. In *Xala,* the jealous co-wife Oumi and the candid peasant who gets robbed as he comes to town are stock characters of African folklore. The tree which figures in countless African tales and which symbolizes knowledge, life, death, and rebirth or the link between heaven and earth is omnipresent in *Emitai.*

In Sembene's films, people are usually what the viewers see them to be or do on the screen. Except in *Borom Sarret, Emitai,* and *Ceddo,* the past of his protagonists is unknown—because of the difference in format in *Black Girl* and *Xala*, the written stories from which they derive often detail their past. Thus, through the immediacy of his typified characters, Sembene reflects collective ideas and attitudes. Soon those protagonists become the parameters of a sociohistorical period, while remaining oral narrative types responding to typical situations. The principal character of *Borom Sarret* has no name and is remembered through his trade and the problems he is not able to overcome. The heroine of *Black Girl* is the victimized black maid rather than Diouana. *Mandabi's* principal character is the illiterate traditionalist rather than Ibrahima Dieng. *Xala's* El Hadji Abdoukader Beye is perceived as the unscrupulous impotent business man and it is as the princess that *Ceddo's* female protagonist remains in audiences' minds.

Because of religious beliefs widespread in Africa, tales include not only humans and physical objects and landscapes, but also supernatural beings, who, as spirits or ghosts, are made to share the world of the living. In *Emitai,* there is no dividing line between the physical and supernatural worlds as the high priest of the sacred wood communicates directly with the spirits.

Thematic similarities also can be drawn from a comparison between Semebene's films and African tales. Male impotence, which constitutes the basis of *Xala,* is in itself a subject which is often included in the storyteller's repertoire. *Xala's* theme of punishment of greed, selfishness, vanity, and waste is likewise highly popular in African folktales. So are topics of the lowly rebelling against the powerful and the rescue of the princess, which are both illustrated in *Ceddo.*

In the realm of storytelling, good or bad fortune and curses are very important. This is by no means neglected by Sembene in *Borom Sarret,* where the cart driver's faith in his good fortune results in failure. In *Mandabi,* Dieng hopes that alms giving will favor him. In *Xala,* the main protagonist is the victim of the beggar's vengeful curse.

In many African tales, heroes leave their villages to venture into the unknown, which can be a strange and mysterious forest, an alien land, or extraterrestrial areas inhabited by living spirits or ancestors. The odyssey of such heroes is a quest aimed at the recovery of an object or the finding of their identity. Whether this search is successful varies according to the story, but in most cases it represents an initiatory rite resulting in new knowledge. All these are recurrent elements found in the tales of Birago Diop and Bernard Dadie, which have been directly inspired by African legends.

If the quest for truth of African morality tales is found in storytelling worldwide, the obstacles the protagonist has to overcome and the manner in which this is accomplished essentially reflects African mores. In their archaic initatory transition from tradition to modernism, as well as through their search for self, most of Sembene's characters also have to face countless obstacles, violated contracts, deceits, and ogre-like unscrupulous individuals. Diouana's journey to a strange land leads her to the land of the dead. The cart driver and Ibrahima Dieng come back to their compound as anti-heroes who have not been able to triumph over the obstacles which hinder their quest in the threatening downtown of Dakar. In the case of *Tauw,* the passage to adulthood is at the core of Sembene's plot. Thus, the cure for El Hadji's impotence necessitates cleansing and follows the pattern of death and rebirth common to many African oral stories.

As African tales, Sembene's films not only represent some kind of initiation and metaphorical rebirth, but also cause an initiation, new awareness, and basic change in the existential world view of both the protagonist and the viewer. But if Sembene uses the structure of the African tale, he knows how to adapt it to fit his didactic needs. And, in Mbye B. Cham's words: ". . . Whereas linearity functions in the traditional narrative to underscore the need to preserve and maintain social order and harmony, it works in Sembene's narrative to inspire a struggle against and rejection of unjust social and political order."[15] In Sembene's disenchanted fables, the *prise de conscience* or consciousness awakening of his main character derives from an acute *crise de conscience* or crisis of consciousness brought about by the juxtaposition of opposites in the context within which they evolve: the old versus the new, good versus evil, the weak versus the powerful, poverty versus wealth, and so forth. Henri Agel, undoubtedly influenced by Levi-Strauss's thesis of antagonisms, stressed that such binary oppositions were determinant factors in myths, tales, and poetry and that they were often included in the plots of films.[16] Sembene's motion pictures provide a good example of this observation since the dynamic force of his films relies on a narrative movement between two opposite poles. Thus, one might rightly wonder if the conflicting elements of Sembene's cinematic art are not more related to African oral storytelling rather than solely, as many critics have pointed out, to the Marxist tendencies of his ideology.

By presenting a microcosm of the communities within which they have been created, African oral stories have a pedagogical and social value understood and appreciated by adults as well as children. Tales which include moral codes sanctioned by the society are called morality tales. They constitute the most common and most popular literary genre in black Africa, similar to the European fables and moral tales of the Middle Ages, which also helped people gain a clearer understanding of some aspects of life at a time when literacy was limited. At the same time as they contain moral statements about life and human nature, those tales are social comments on the faults and follies of man, while criticizing some of the social institutions he has established. In that sense, Sembene's films are indeed morality tales, an expression of wisdom, popular belief, and social criticism. Yet Sembene's works, unlike most traditional African stories, are often tales of disenchantment which do not always ensure the successful revenge of the small and the humble against the great and powerful. Neither do they, as a rule, immediately restore an ideal order of justice and benevolence. They do, however, suggest or advocate their implementation. Moreover, Sembene's motion pictures derive specifically from African dilemma tales, the outcome of which is debated and in a way decided by the spectators. With the open-endedness of most of his plots, Sembene trusts the viewers' imagination to prolong his films. In *Borom Sarret, Mandabi,* and *Ceddo,* he leaves his spectators with a choice between several alternatives as the film comes to a close.

In addition to being compared to tales, the works of Ousmane Sembene also show similarities with Africa's ancient and modern drama, both of which have developed from the oral tradition. Sembene's films can be compared in some respects with traditional Yoruba theater or the Ikaki masquerade. Like traditional African drama, they include satirical comedies and historical re-enactments. Similarly, they use largely non-professional actors from the community they intend to reach and have the "plot unity" of traditional African drama. The following quote by Bakary Traore on Mandingo theater could be applied almost in its entirety to Sembene's cinema, whose purpose is to generate or maintain ethical values: ". . . the essential function of the Kote Komanyaga of the Mandingo theatre is to make a public spectacle of all efforts to violate the group's moral code: conjugal infidelity, theft, jealousy, etc. . . . and thereby ensure their suppression. . . ."[17]

Sembene's works also offer similarities with modern African drama, inspired by traditional African and European theater. *Xala* calls to mind Nigerian dramatist J. P. Clark's 1966 *Song of a Goat* in which the male protagonist has lost his virility, allegedly because of an ancestral curse. The tone of *Mandabi* is reminiscent of G. Oyono Mbia's social comedies. Sembene's protagonists in *Emitai* and *Ceddo* speak sometimes in proverbs in the same style as Nigerian playwright Wole Soyinka's characters often do. In the same manner, both authors' heroes have very specific social roles. Sembene and Soyinka, as well as numerous other African authors use their art to reflect the world around them and influence it. They focus on the social patterns and the political events of present-day Africa caught between tradition and modernism. While written literature is unknown to the bulk of African people, Sembene's cinema, like African drama, is more accessible to them. It can indeed be compared to African theater since "perhaps the most important thing to stress is that theatre is being used in Africa as a means of education, celebration, protest and discovery."[18]

Although identical structures and intentions exist between Sembene's films and African storytelling as well as drama, there are also limits to such a comparison. If in Africa the collective experience of watching movies reminds one of the audience attending the griot's delivery, films remain the same each time they are presented. They do not have the improvisatory nature, the flexibility, and the versatility of the oral transmitter. The griot can, at any time, respond to the audience's reaction by altering his tone of voice, by introducing new gestures, mimes, and repetitions, and by interpolating songs. Films do not allow such emotional and lyrical interaction of the teller and listener. Furthermore, film screening usually takes place in an enclosed room which does not offer the vivacious communal appeal of the village square or the nocturnal atmosphere of evening gatherings in African compounds, which is so propitious to the magic of storytelling. Films can be seen casually and on an everyday basis, whereas some griots perform only on the occasion of cyclical festivals and special celebrations.

"The traditional story-teller no longer exists today and I think that the filmmaker can replace him," says Sembene.[19] These words—stressing the important role of the African filmmaker—should not be taken literally. The griot is still very much a part of African culture even if he lacks the unchallenged appeal he once had in all of Africa. Admittedly, oral tradition is progressively disappearing in Africa's changing societies because of the expansion of education and modern mass media such as radio, television, and cinema. Nevertheless, it could be argued that African filmmaking and the works of Sembene in particular make use of or expand storytelling rather than supplant it. Not only do Sembene's films evolve from storytelling, but they also preserve it. As writer Bernard Dadié once said: "We have treasures to convey: our message must be transmitted, we will do so in expressing our values under new forms."[20] Sembene's cinema undeniably stands proud among such new forms.

According to many authors, tales have nourished most written literary genres in black African culture. Such contemporary writers as Birago Diop (Senegal) and Bernard Dadié (Ivory Coast) have produced written versions of traditional stories—as did the brothers Grimm in nineteenth-century Europe—or they have used them as a basis for new ones. Likewise, the influence of oral tradition can hardly be missed in the novels of Mongo Beti (Cameroon), Aminata Sow Fall (Senegal), or Ousmane Sembene himself. As a writer, Sembene recognizes having been influenced by the griot.[21] As he transfers his talents from writing to filmmaking, Sembene remains faithful to the same source of inspiration: African oral tradition. He is not remote from his people, creating works limited to the intelligentsia. Instead, he continues the role of the griot as the chronicler of his people's history. According to Mbye B. Cham, Sembene is "a blend of the most relevant, useful and progressive aspects of both the 'gewel' and the 'lebkat.' "[22] He is the kind of griot who has maintained an almost complete freedom of expression and should not be confused with the court griot owing allegiance to a powerful family or group. Sembene's objective is not to provide an escape from reality by embellishing it, for "the griot may only embellish reality during victorious times through what people call court songs or festive songs. In times of crisis, however, a griot does not embellish reality. On the contrary, he finds himself in the brutality of surrounding events. I have never tried to please my audiences through the embellishment of reality. I am a participant and an observer of my society."[23]

The constant social criticism present in Sembene's films attests to the fact that his has remained a free and independent spirit. As a subversive artist he is often censored and kept apart from the mainstream of society like the *griot casté,* the slave without master, who, at the same time as he is considered an important transmitter of culture is paradoxically shunned as a helot. "The African filmmaker is like the griot who is similar to the European minstrel: a man of learning and common sense who is the historian, the raconteur, the living memory and the conscience of his people."[24]

It is as a griot that Sembene uses the magical, emotional, and mystical aspect of film experience to create new depths of understanding. For Sembene, "the African filmmaker is the griot of modern times."[25] It is as such that he defines himself, saying: "I am a storyteller and I tell stories. I hope to tell more of them and better ones."[26]

Notes

1. Ousmane Sembene, "Filmmakers and African Culture," *Africa* n. 71 (1977), p. 80.
2. Georges Balandier and Jacques Maquet, *Dictionary of Black African Civilization* (New York: Leon Amiel, 1974), p. 161.
3. Ibid.
4. Elolongue Epanya Yondo, *La Place de la Littérature Orale en Afrique* (Paris: La Pensee Universelle, 1976), p. 101.
5. Dorothy S. Blair, translator, *Birago Diop's Tales of Amadou Koumba* (London: Oxford University Press, 1966), p. ix.
6. Mbye B. Cham, "Oral Narrative Patterns in the Work of Ousmane Sembene," unpublished paper (1982), p. 4.
7. Elolongue Epanya Yondo, *La Place de la Littérature Orale en Afrique,* p. 23. Author's translation.
8. Ibid.
9. Alphonse Raphaël N'Diaye, "Les Traditions Orales et la Quete de l'Identité Culturelle," *Présence Africaine,* n. 114 (1980), p. 9. Author's translation.
10. Joseph M. Boggs, *The Art of Watching Films* (Menlo Park, Calif.: Benjamin/Cummings Publishing Co., 1978), p. 19.
11. Haile Gerima at the symposium "Literature, Film and Society in Africa: Dialectics of Artistic Creativity and Social Consciousness," organized by the African Studies Program at the University of Illinois, Urbana-Champaign, May 1–3, 1980.
12. Michel Capdenac, "Le Mandat, film sénégalais de Sembene Ousmane," *Les Lettres Françaises,* n. 1259 (1968), p. 22. Author's translation.
13. Joyce A. Hutchinson, ed., *Birago Diop—Contes Choisis* (Cambridge: Cambridge University Press, 1967), p. 15.

14. Philip A. Noss in Daniel J. Crowley, ed., *African Folklore in the New World* (Austin: University of Texas Press, 1977), p. 86.

15. Mbye B. Cham, "Oral Narrative Patterns in the Work of Ousmane Sembene," p. 6.

16. Henri Agel, *Métaphysique du Cinéma* (Paris: Payot, 1976). pp. 197–206.

17. Bakary Traore, *The Black African Theater and Its Social Functions* (Ibadan, Nigeria: Ibadan University Press, 1972), p. 66.

18. Martin Banham with Clive Wake, *African Theater Today* (London: Pitman Publishing, 1976). p. v.

19. Noureddine Ghali, "Ousmane Sembene," *Cinéma 76*, n. 208 (1976), p. 89. Author's translation.

20. Joyce A. Hutchinson, ed., *Birago Diop—Contes Choisis,* p. 8. Author's translation.

21. Ousmane Sembene, *L'Harmattan* (Paris: Présence Africaine, 1963), p. 9.

22. Mbye B. Cham, "Oral Narrative Patterns in the Work of Ousmane Sembene," p. 4.

23. From a series of interviews with Ousmane Sembene conducted by the author in Senegal in 1978.

24. Françoise Pfaff, "Notes on Cinema," *New Directions,* vol. 6, n. 1 (1979), p. 26.

25. From a lecture delivered by Ousmane Sembene at Howard University, Washington, D.C., 19 February 1978.

26. Sembene to author, Senegal, 1978.

[South Africa] Independent Cinema

Keyan Tomaselli

No theory of production is neutral.

Peter Anderson, videographer-lecturer, 1983

South Africa offers a unique opportunity for making films about interracial relationships, social problems, class conflicts and political despotism. Images, themes, plots and stories scream out from the environment for cinematic treatment, but they are generally missed by the South African producers. Such subjects have been treated by British television, the films of Peter Davis and the productions made under the auspices of the banned African National Congress (ANC). Most, consequently, are made by foreigners for foreign audiences. Only a few have been made with South African capital for South African audiences. These films are aimed at middle class non-South African viewers and have little in common with the notion of "Tercine Cinema" or radical film-making.

Although Tercine Cinema or Third Cinema emerged in other neo-colonial countries in North Africa, South America and Asia during the 1960s and 1970s, this revolutionary cinema has yet to be realized in South Africa. While the economic systems of other Third World countries such as Brazil, Argentina, Chile, Senegal and Algeria have all suffered from colonial and neo-colonial domination, their filmmakers are sufficiently

The Cinema of Apartheid: Race and Class in South African Film (N.Y.: Smyrna/Lake View Press, 1988), pp. 195–213. Reprinted by permission of the publisher.

conscious of the structural underpinnings of international monopoly capitalism to articulate a radical commentary in ways which few South African filmmakers would understand.

Tercine Cinema, so named by Argentinian Fernando Solanas, is the cinema of the guerrilla by camera units: a helping hand in the rebuilding of an oppressed nation and a way of anticipating events in order to expedite them. Revolutionary in content, polemical in form, questioning bourgeois esthetic canons, it re-defines cinema's relationship with the audience.[1] It is this outlook which separates these practitioners from many of their South African counterparts who only began to expose and exploit the gaps in the repressive state machinery to any significant degree in the early 1980s. The more critical South African filmmaker tends to be far more conservative, both in terms of political position and in the use of the medium.

The concept of "independent film-making" is a vexed one which has yet to be given a theoretical content. The term is generally used to describe practitioners working outside a system or industry. Historically, this would range from amateurs through to "entrepreneurial film and video producers who are not affiliated with the networks or highly capitalized producing companies."[2]

Moving from the reality that independent film-making is likely to be low in cost and produced outside the conforming influences of the market place, we may define this type of production in terms of six basic criteria:

1. *Intention.* Whatever else, profit or earning back its costs is not a major criterion.
2. *Alternative Exhibition Venues.* Independent productions are screened (or not screened at all) on alternative circuits such as film festivals, university venues, church halls, trade union venues, and homes of specifically interested audiences.
3. Independent films may or may not have secured *censorship* clearance. Where exemption has been granted it is more than likely that such films will have been passed against the expectations of the producer.
4. Independent film/video production is of a *low budget* nature and is financed either by the film/video makers themselves, or by organizations which have made known their opposition to the state or the existing order of things. This is the most difficult category to define, for the contradictions of capitalism allow filmmakers to take dialectic advantage of its weaknesses to criticize the status quo. For example, this is presently the case with Brazilian Cinema Novo, which although funded by the state, is nevertheless critical of it. This category would thus embrace co-option of institutional forms of capital where the

filmmaker produces a text which runs counter to the accepted norms of those whose money is being spent.

5. Set against the traditionally high costs of film production within the "industry," the independent film is one which by virtue of its relative cheapness and financial autonomy is best able to exploit the *relationship between cost and content.* As Ross Devenish puts it, filmmakers in South Africa should "explore the freedom of their poverty," accepting gladly its limitations and turning them to advantage. In other words, the bigger the budget, the less likely is the film to deal with social realism and contentious class conflicts from the point of view of the oppressed. Worldwide, it is the low budget film made outside, or on the periphery of the established industry, which generally fulfills this function.
6. Independent filmmakers try to work within-the-possible *to prepare the way for the not-yet-possible.* Whether operating from an exile base like Nana Mahomo, Chris Austin or the ANC, or from within the country, these filmmakers deliberately exploit hegemonic fissures in the course of their filmmaking practice within the state.

This circuit of production, distribution and exhibition which operates separately from, but in parallel with the industry—though it may use some if its facilities—has been variously labeled as "independent," "oppositional" or "radical" cinema. These terms are not wholly interchangeable. The term, *independent,* as used here, describes the general social practice of film and video making which operates outside the system in terms of methods of production, content and funding.

More specifically, *oppositional* film-making accounts for the "practice involving an opposition to the strait jacket imposed on film-making by the profit motive and the ideologies that justify, legitimate or simply fail to engage with capitalist organization of this cultural sector."[3] Oppositional film-making makes visible or draws attention to the structured absences of commercial cinema brought about by the prevailing productive forces and legitimized by bourgeois critical methods.

Radical cinema is used here to describe films which are not only in opposition to the capitalist mode of production, but are aware of their own technique/style/technology/conventions and the way in which these mold the view of the reality portrayed. In other words, radical cinema consistently reminds the audience of its relationship to the film/video crew. Radical cinema thus refers beyond the text and documents not only to the *what,* but the *how* and the *why.* A radical film, furthermore, is one which devises directions for cultural resistance/action against an oppressive social order in *cooperation with the subject community.*

Funding

> In the case of [South Africa an experimental production fund] encourages the discovery of new talent. In . . . other countries, no expenditure has been more effective.
>
> *John Grierson, 1954*[4]

Although a state subsidy for commercial feature films has been available since 1956, not a single cent was granted between 1964 and 1978 to shorts, documentary or experimental film other than the propaganda films made by the National Film Board. The initial suggestion for an experimental film fund was made by John Grierson of Canada's National Film Board, who was consulted by the South African government on the establishment of a national film board. Grierson's report, submitted in 1954, pointed out that experimentation was germane to national cinematic and democratic progress. The proposed film board was to provide the ideal mechanism through which the exploration of film could be fostered in what he described as a vigorous political climate. Set up ten years later, the structure of the Board (NFB) in South Africa differed in crucial ways from Grierson's original proposals. These were devised to stimulate a vigorous political forum for the democratic discussion and dissemination of information within the body politic through film. As constituted by the South African government, however, the NFB subverted Grierson's democratic assumptions and, until its dissolution in 1979, functioned primarily as a production and distribution facility for National Party propaganda.

Oppositional and independent filmmakers have had, therefore, to search out other sources of finance. Other than personal investment, funding has been made available by the National Union of South African Students (*Wits Protest*—1970–1974), the South African Council of Churches *(This We Can Do For Justice and Peace, If God Be For Us),* the Inter-Church Media Programme (*Alexandra, Part of the Process* and *A Film on the Funeral of Neil Aggett*), NOVIB (Holland) and the International University Exchange Fund *(You Have Struck a Rock),* European and British television stations (*Athol Fugard: A Lesson From Aloes* and the Gordimer series), and private benefactors such as the Maggie Magaba Trust *(Awake From Mourning),* financed by an expatriate South African now living in London. Limited funding has come from the Danish anti-apartheid movement *(The Other South Africa),* while substantial amounts have been awarded to Ross Devenish by the Ford Foundation and the BBC (*The Guest* and *Marigolds in August*). Universities teaching film and television production have been responsible for a noticeable upswing of oppositional material since 1981. The French government established a Centre for Direct Cinema under the auspices of the Federated Union of Black Arts and the University of the Witwatersrand, Johannesburg, in January 1984, and five black South Africans were selected to undergo a training course in France. One of these, James Mthoba, was invited to manage the Centre. This funding was part of the socialist government's foreign policy in Africa, as it had previously set up similar centers in Mozambique and Angola. Paradoxically, much of the finance for film stock has come from the International Communications Agency, an American imperialist apparatus. The students of the Centre come from all walks of life: laborers, clerical workers, bus drivers and so on. At any one time as many as ten projects are operating in Johannesburg and Soweto.

The Community Video Resource Assoc. (CVRA) of the University of Cape Town is involved with investigative, documentary and trade union videos. Where English language universities have spearheaded the oppositional movement, Afrikaans-speaking campuses have been less concerned with film and video, and even then, with a basically conventional application.[5] Television studios at the "tribal" colleges remain beyond the access of students and staff wanting to make critical material.

Academic conferences are a significant source of inspiration and limited funding. The History Workshop of the University of the Witwatersrand in 1981 assisted the production of a film on Alexandra township which revealed the effects of apartheid, enforced uni-sex hostel life and environmental degradation. *Kat River—The End of Hope* was made for the 1984 workshop. The Second Carnegie Inquiry into Poverty in South Africa held in April 1984 made available R50 000 for film and video production: *I am Clifford Abrahams, This is Grahamtown; The Tot System; Mayfair; Reserve 4; Place of Tears; Loaded Dice* and many others.

The sums granted ranged from as low as $100 to $1,000 in the case of oppositional films, while the more commercial fare of Fugard and Devenish, shown in cinemas and on foreign television stations, might solicit as much as $40,000 from a single source. The Gordimer series, costing about $1 million, was almost entirely funded by German and Dutch television stations.

Distribution of independently made films is a problem. No national coordinating agency exists. Distribution is done on an *ad hoc* regional basis which is inefficient and disorganized. Each producing body disseminates its own material to film festivals, academic conferences, universities, churches, trade unions and private homes. No central catalogues exist and the titles available are known mainly to the small group of people connected with the production collective. There are two basic reasons for this state of affairs. The first is that few of these films and videos have obtained censorship clearance. The costs of censorship, which is mandatory, must be borne by the producer. It is doubtful that many of the films made would be granted exemption. The distribution of non-censored material has led to police surveillance of filmmakers, detentions and confiscation of copies. Second, the mainly working class viewers of such productions would not be able to afford the kind of hire charges necessary to bankroll a central distributing agency.

The lack of a distribution organization puts the community media at a crucial disadvantage. The money, energy and time put into the production of motion pictures is rarely amortized in terms of audience size and composition. These films and videos are mainly seen by the already converted and it is rare indeed that they will be shown to hostile or uncommitted audiences, though many copies have been confiscated by the security police.

Production Facilities

> Films from Asia, Africa, and Latin America are films of discomfort. The discomfort begins with the basic material: inferior cameras and laboratories, and therefore crude images in muffled dialogue, unwanted noise on the soundtrack, editing accidents . . .
>
> *Glauber Rocha, Brazilian director, 1967*[6]

Although cinema was used by Afrikaner Nationalist elements to fight cultural and economic domination by English-speaking South Africans as early as the late 1930s, it was only after 1970 that a critical and independent film movement opposing apartheid began to assert itself. This spurt was stimulated by a number of interrelated factors. The first was the improvement in Super-8 technology, with which most oppositional films were made during the 1970s. The lifting of the embargo on video technology to facilitate the setting up of a broadcast television service in 1976 made small format cameras and recorders available for the first time. The third and most significant factor was the introduction of film and television studies at English-language universities toward the end of the 1970s. Within a very short time, working with equipment supplied by these institutions, students and staff started producing material critical of apartheid. The makers of many of these films and videos have resisted the lure of the established industry and have forced a space for a progressive cinema not dependent on the industry and profit.

Currently, most of these film or video makers are white. Despite their counter-ideological tendencies, they show a continued allegiance to a culturally ingrained stance on esthetics and form. They often succumb to the mystique of film-making and take too much for granted. Unable to maintain a critical distance from their own ideologies, many have produced confused statements which, paradoxically, lend themselves to appropriation by the dominant ideology. As a result, recurring bewilderment occurs when films like *This We Can Do For Justice and Peace* (1981) and *Awake From Mourning* (1981) are granted censorship clearance. The reasons are generally attributed to the recently "enlightened" attitude of the censors. A more valid explanation, however, is to be found in the relation of the film to its context. Where the context is displaced or obscured, the film is likely to be passed.

Films made to exercise a cinematic urge through the fun of filmmaking are often a precursor to entering the established industry *per se.* Entertainment is the prime concern of these directors who are hostile to "message movies" or social statements. The classic example is, of course, Jamie Uys. His first film, *Daar Doer in die Bosveld* (1951) was a refreshing departure from previous independent (and commercial) Afrikaans film. The spontaneity with which he captured a uniquely Afrikaner humor catapulted him to fame within a very short period of the film's release. The film cost R6 000 and was shot on reversal stock with a basic Kodak camera. Uys did not remain independent for long, for he was soon to be backed by Schlesinger, and later, big Afrikaner-dominated capital. Other independent filmmakers operating at that time generally failed because they had neither the technical skills nor the social understanding of the Afrikaans community to make their films attractive.

Authentic cinema, although dealing with other than white characters, is designed to appeal to a wider than purely black audience. Such films are not intended for a particular population group. Because of their theme, treatment, characters and style, such films actively solicit cross-cultural and inter-racial audiences. Examples are *Jim Comes to Joburg* (1949) and *Magic Garden* (1961), both made by Donald Swanson. The former deals with the experiences of a black ruralite as he tries to cope with the strange city life of Johannesburg. It is a highly humorous film and has a number of scenes with the famous Jazz Maniacs band which developed an African jazz. Of a more serious nature, certainly more traumatically accurate, is Zoltan Korda's *Cry the Beloved Country* (1951), adapted from Alan Paton's novel. Other examples are Fugard and Devenish's *Boesman and Lena* (1973) and *Marigolds in August* (1980), David Bensusan's *My Country My Hat* (1983) and Ashly Lazerus's *e' Lolipop* (1975). Lindi Wilson's *Last Supper at Hortsley Street* (1982) is a moving reconstruction of the last family to leave the District Six coloured township in Cape Town from which they had been evicted to make way for a white suburb.

Jamie Uys's *Dingaka* (1964), based on his earlier documentary, *The Fox Has Four Eyes,* though shot in a Hollywood panavision style and pandering to an American romanticized conception of Africa in places, would fall into this category. Despite this film's ignorance of the pass laws which controlled black migrants and the serious dislocation caused even the more isolated black rural communities, Mtutuzeli Matchoba described *Dingaka* as an:

> . . . honest attempt to represent the controversial traditional theme. A picture may not be especially intended for a particular population group, but because that group recognizes itself authentically represented within the theme, it will respond positively to it.[7]

The dual theme in *Dingaka* is the opposition between white justice and black justice—that of a tribal law, "he who kills must be killed," and that of the "civilized" state. Through the progression of the plot it becomes clear that despite the formal, inflexible trappings of state law, that it is adaptable and can take account of tribal law. White justice is shown to rule, but black justice is portrayed as more humane.

Most independent films are personal explorations. A couple of these films have been made in 35mm, funded by makers who are wealthy. This wealth, more often than not, has been derived from the extraordinary profits made by working within the system, mostly on advertising films. The desire to make one's "own movie" was an objective common to many of the younger filmmakers who worked during the 1970s. This attitude in itself is something new and is regarded with suspicion by the diehards who see film-making as merely a job like any other job which must fit the demands of the market place. An example of this kind of film is Leslie Dektor's *We Take Our Prisons With Us* (1976), a poetic examination of circus people. These films have a limited distribution since they are not the sort of fare that commercial distributors like, even as supporting programs. Unless they are shown at film festivals, they obtain little exposure.

On 16mm are films like *Angsst* (1979) and *Die Moord* (The Murder, 1980) made by Chris Pretorius. Both were partially financed by Pretorius himself, though he had considerable help from a wealthy benefactor for his later effort. His two films overturn accepted conventions with anarchic determination. *Die Moord,* an experimental counter-narrative, is stylistically more successful, though more frustrating, than *Angsst.* While both work at the level of the sexual, the fear of a young man of women, *Angsst,* apparently more coherent, is laden with potentially symbolic and contextual cues referring to apartheid. It is very much like *Die Moord,* constantly alluding to what is not there, consistently negating itself and thwarting audience expectations conditioned by convention.[8] Indeed, the full meaning of these films only becomes apparent on an *auteur* analysis which relates the text to the personality of the filmmaker.

Fugard's People (1982) and *Athol Fugard: A Lesson From Aloes* (1980) reveal the man through a close examination of his work, his directing methods and origin of the ideas which inform his theater. In the former film, clips from his plays and films are intercut with interviews with Fugard. The latter film, made for the BBC, exposes the fears, traumas and trepidations of the actors as they rehearse the play. The documentary becomes itself an allegory of repression and state invasion of individual privacy.

On video is *The Story of Sol Plaaitjie* (1981), a rather static examination of an early black writer. In total contrast is *I am Clifford Abrahams, This is Grahamstown* (1984), an investigative *cinema verité* documentary which breaks stylistic conventions, not least of which is the payment—on camera—of the narrator/subject, Clifford Abrahams. The usual separation of subject and production crew is abandoned, and the result is a cinema of deconstruction. Abrahams leads the crew and the viewer to a unique "street" view of Grahamstown, spanning black, white and coloured geographical areas and social spaces. Abrahams narrates his life in the places where he begged, sold newspapers, slept out under bushes and in old wrecked cars. Beaten by his father, grandfather, police, and employers since the age of ten, he has maintained his humor as a defense against a harsh environment. The camera reveals the paternalism and brutality of white Grahamstown and the pressures on blacks through his interactions with the population of the town.

The films of Andrew Tracey on Chopi music are the most well-known South African ethnographic cinema. More recent productions include those made on Super-8 and three-quarter inch video.[9] Some of these have attempted to document cultural responses by blacks to apartheid and come closest to an understanding of both ethnography and cinematic techniques and styles of production.

Ethnographic film is not radical film because it seeks to document rather than provide strategies for resistance. The documentation may itself deal with those strategies, such as the video, *Shixini December* (1984), which shows how the traditional beer drink has become a means of maintaining social cohesion in the face of the destabilizing effects of migrant labor among the remnants of a subsistence economy in the Transkei. Such films may be perceived as "oppositional" in character because they question conventional wisdom and accepted myth.[10]

Films falling into the category of counter-culture and metaphysical explorations place themselves in a context which is closer to U.S. West Coast counter-culture than anything local. These films lack indigenous symbolism and draw their images and themes from the *Easy Rider*-type cinematic culture. Examples are Cedric Sundström's *Suffer Little Children* (1976) and John Peacock's *A Certain Delegation* (1976), both made in 16mm. *Suffer* is set in a Crown Mines village in 1934, but is a macabre re-enactment of the Biblical martyr, Stephen. Peacock's film deals with the "plastic society" and the violence individuals impose on themselves through technology. A number of others made on Super-8, for example, *Freedom O* (1972) and *The Surfworshippers* (1973), show alienation, death and rebirth. In the former, the characters wrestle with drugs and the "freedom" of the open road, while the latter deals with the compelling, almost religious cult of surfing.

The transformation of short stories into films which explore the South African milieu are seen in Cedric Sundström's *The Hunter* (1974), based on an Olive Schreiner story, and Lynton Stephenson's *Six Feet of the Country* (1977), based on a Nadine Gordimer story. Although made under all the difficult constraints which face the independent filmmaker, the latter film was successful as a pilot on the basis of which a further five Gordimer stories were filmed in 1981/82. Financed by a German consortium and filmed in South Africa and Kenya, they have been screened in several nations but only shown at film festivals in South Africa. One program, *Chip of Glass Ruby,* remains banned, though initially three were assigned this status. Other than the producer and Peter Chappell, who directed *Oral History,* the crews consisted entirely of South African technicians.

The censorship issues which arose with regard to the Gordimer series have already been discussed. The producer had to contend with censorship and security police surveillance. The German executive producer was denied access to South Africa. The production company was registered in Lesotho to obviate the cultural boycott against South Africa and to take advantage of the financial rands incentive existing at the time. This was the only form of subsidy available to the company as the films were not feature length and therefore unlikely to be screened in cinemas even if accepted by the major distributors.

Though a challenging and exciting series, certain recurring themes and character treatments caused considerable unease among foreign film festival audiences. These concerned the portrayal of the lead women characters by the various male directors. *City Lovers,* for example, was accused of being "racist" and "sexist" by certain members of the audience attending the Amiens Film Festival in France in 1983. While it was clear that these viewers had misinterpreted the subtleties of the relationship between the white German immigrant and the coloured cashier, the accusations of sexism were less easy to counter. An analysis of *Country Lovers* will explain the cause of this problem. The white farmer's son falls in love with a black female farm laborer. They have a child. In order to hide his fall from grace, the farmer's son poisons the baby, and his father sends the woman's family away. The narrative is told from the male point of view, the woman is merely the vehicle for the story. Shot mainly in close-up and mid-close-up, the sons's reactions are dominant. Alternatively, the audience is presented with point-of-view shots which originate from the protagonist. The film is shot in flashback, and the first shots established this male perspective. As one female critic puts it:

> If we are male, we could, presumably, easily identify with the male's dilemma: the threat of emasculation (linked with Nationhood through images of the gun, and references to the vote, his father's Afrikaner history) by "lowering" himself by having an affair with an "unworthy" woman. In South Africa, matters relating to colour take precedence over class, so the woman is "unworthy" because she is black. This adds the further dimension of bringing the *Volk* down—miscegenation would lead to the end of the line which the protagonist's father is keen to preserve.[11]

Essentially, the cinematic emphasis is displaced from the interracial love affair to a plot structure which tries to explain the reasons for the now young man's return to the farm. Because the film appears in flashback it becomes necessary to explain the cause of his reminiscences in the empty farm house rather than to examine the human and social relationships set in motion by their love affair. The farmer's son thus takes on a significance beyond himself, as a representation of Afrikaners, rather than as an individual caught up in "deviant" sexual behavior. The characters become vehicles to describe a political situation.

Because the farmer's son falls in love with the black woman, she becomes the object of his desire and in so far as the audience is positioned to view from his perspective, she becomes the object of its gaze. This is a male gaze, as all the sexual or vaguely erotic scenes are filmed to stimulate male desire. For example, when the son first sees her climbing down the ladder, the camera (he and the audience) focus on her leg: in the dam scene, she tucks up her skirt and there are close-ups of her legs. Male viewers are thus placed in a voyeuristic position vis-à-vis the action on the screen. That this male emphasis occurs in films made by different directors does not necessarily locate the problem with the writer. It would appear to rest more with the conventions of film-making where the director is unaware of the sexist nature of many of the techniques so often taken for granted. Garden's accusation that the "lack of latitude, and Gordimer's slavish adherence to form and content of the original stories, has made the films stilted and unnatural" may have some validity at the level of performance, but this does not account for the sexist nature of much of the camera framing and editing selection.[12] It was the directors who did not follow through the raw material provided them by the scripts. Had the camera made ironic use of the male gaze to make an anti-sexist statement, the series might have more faithfully reflected

Gordimer's original written treatments. This kind of problem has more to do with a lack of knowledge about the *ideological* effects of cinematic techniques on the part of the technicians than with Gordimer's alleged interference.

Social documentaries both challenge the *status quo* and reflect the way in which certain organizations are coping with oppression. Most of these films or videos are made within the conventional documentary frameworks, accepting implicitly the raw material presented to the camera without question. Content ranges from highly structured documentaries such as *This We Can Do For Justice and Peace,* which outlines the efforts of the South African Council of Churches (SACC) to combat structurally induced poverty and the appalling consequences of state enforced resettlement, to Grierson-type documentaries of student demonstrations *(Wits Protest)* and on-camera harangues by resettled homeland dwellers *(Place of Tears).*

Wits Protest, made between 1970 and 1974, was shot on Super-8. Music and commentary were added in the post-production stages. At that time single system direct sound cameras were but a dream. Although *Wits Protest* had not been screened on television, it has been seen at a number of local universities and churches and by overseas groups. The intention of its makers was both historical and functional. During the production period, the film served to communicate what was happening on campus in a manner not available through existing channels. As the events of 1972 unraveled, for example, footage of the police/student confrontations was screened in an unedited state within one or two days of their occurrence. *Wits Protest* started from a straight "objective" style but became more expressionistic as the camera took the point of view of the students. In so doing, *Wits Protest* presents a dramatized documentary. Its participatory style is a conscious and deliberate choice on the part of its makers.

Another film dealing with protest is *You Have Struck a Rock* by Debbie May, which follows the actions of black women during the 1950s in their protests against the mandatory carrying of passes. A rather confusing film to audiences not aware of the momentous history of black repression in South Africa, the film does hint at the double repression suffered by black women through the indifference of black men to their efforts.[13]

The Community Video Resources Association (CVRA) of the University of Cape Town is very active in highlighting particular features of ongoing trade union campaigns. One example is *Passing the Message,* a video made by the South African Food and Canning Workers' Union during the nationwide Fattis and Monis strike. Video was employed to pass messages from the workers in Cape Town to their families in the

Ciskei and vice versa. This occurred at a time when many had lost contact with each other and worked to keep channels of communication between the migrants and their families open.

FOSATU: Building Worker Unity documents a non-racial socialist-oriented trade union federation. Though attempting to contribute to revolutionary action, it fails dangerously, as it is a hodgepodge of incompatible codes and muddled techniques. While context is provided, the narrative relies on verbal codes which are seldom connected to the image. The narration, for example, tells of workers on strike, but presents shots of them working. People within the film are not always identified and many who are working towards a socialist economy are paradoxically portrayed as sinister conspirators hiding behind white skins and dark glasses.[14]

The vast majority of films concentrate on urban communities. This geographical bias is inevitable given the urban location of universities, film and video facilities and technicians. It is, however, paradoxical, in that the cities have become the points of (limited) reform and co-option of the other than white working and middle classes. While there is a struggle in the cities being negotiated through the mechanisms of trade unionism and political groupings like the United Democratic Front and National Forum, it is rural dwellers and marginalized black population who are largely engaged in an unequal struggle for survival itself. Videos like *The Tot System,* which shows how coloured wine farm laborers are held in an alcoholic subservience through wine rations in lieu of some wages, *Kat River—The End of Hope,* which exposes the futile efforts of a coloured peasant farming community to prevent dispossession, and *Shixini December* which documents ways of coping with the stresses caused by migrant labor, are under-represented in their concern for a rural and homeland perspective. The videos commenting on urban issues, though appearing to offer counter-ideologies, often build inconsistencies into the assumptions of their texts. What is omitted from the latter videos is an examination of the *relationship* between the urban events documented and the processes operating within the political economy as a whole. The urban-rural connection is ignored as these videos highlight events at the expense of ongoing social processes. *Future Roots* accomplishes the process most successfully, exposing the structural constraints of bottom-up development strategies. By concentrating on a specific scheme in the Ciskei, the producers argue that they are designed to reduce social and labor costs of mining and industry in the white areas by creating the conditions for a viable subsistence economy in the homeland. This video, as with *Shixini December,* is clear as to the relationship between the micro and macro elements of the political economy.

Class analysis is also often lacking. An example is *Mayfair,* which deals with the responses of Afrikaners, Indians and coloureds who live in this multi-racial suburb, to the government's announcement that it is to be declared an Indian Group area. This video never explains causation or context: how did this suburb become multi-racial in the first place? How did the extreme right-wing racist Afrikaners come to agree to living among people they hate and typify as "foreigners"? Why are significant numbers of this integrated community standing together to resist the government decision? And above all, why are whites going to be moved—the first time this has ever happened to an originally white area?

The producers of *Mayfair* should have examined the context of the shifting class structure which created the conditions for unexpected government action on *Mayfair.* The analysis would have had to take account of a maturing economy which needs more skilled labor and professionals. This has led to the co-option by the state of the Indian and coloured "population groups" which, together with the alienation of right-wing Afrikaners from the National Party, has resulted in a new political alliance. In a similar vein, *Awake From Mourning,* which documents the role of black women in building community cohesion and providing much-needed social services following the 1976 Soweto disturbances, misrepresents the economic determinants of apartheid.

The question of context is crucial, particularly where the state film-makers have tried to mystify historical process through engaging the images of oppositional filmmakers, reinterpreting them, and then representing them to both local and international audiences as "communist propaganda." The images of oppositional filmmakers have proved susceptible to co-option by the repressive agencies of the state. Not only have the images of films like Nana Mahamo's *The Dumping Grounds* and *Last Grave at Dimbaza,* amongst others, been turned against the oppressed in state-sponsored films like *To Act a Lie,* but they have been submitted by the security police as evidence to commissions of inquiry designed to intimidate dissident organizations. This occurred with *This We Can Do For Justice and Peace,* which was passed by the Directorate of Publications on the assumption that its "biased" and "one-sided" message would prove to be counter-productive to the South African Council of Churches (SACC) at a time it was undergoing intensive state scrutiny. Furthermore, the security police edited together clips from a number of "anti-South African" films, including *Justice and Peace* and submitted them on an edited videotape which collapsed the various films into one another. The resulting combination presented the appearance of a continuous program. Some accusatory editorial comment was added and the result was presented as evidence against the SACC.

Radical Cinema

> If video is to become cultural action for freedom, its core problem emerges as one of method.
>
> *Peter Anderson*[15]

If the state has outmaneuvered oppositional filmmakers in terms of the conventions of documentary, it becomes necessary to explore dialectical forms which should be designed to minimize the possibility of co-option. Few of the above-mentioned social films provide historical or geographical contextualization. Even fewer employ reflexive techniques which identify the ideological position of the producers in relation to the working class struggle.

One video which is sensitive to dialectical coding, technology and the trap of conventional approaches is *Kat River—The End of Hope.* This video offers a detailed historical-geographical analysis which clearly locates the interviews within their class contexts, and the relationship of that context to capital, the state and dispossession. Another is *I am Clifford Abrahams, This is Grahamstown,* a participatory *cinema verité* documentary which, at the level of appearance, works as a biographical documentary. The presence of the crew is stamped into the image and the sound, and the central character, Cliffie, a down and out alcoholic coloured raconteur was also consulted during the editing phase. At a deeper level, the result is an exploration of interacting social, political and economic conditions which locate Cliffie within a particular class fraction. The video shows how he makes out through a number of survival networks which span the black, coloured, Indian and white areas of the town. The reflexive techniques used by the crew make visible their own assumptions and methodology presenting not a camera determined "truth," but Cliffie's interpretation of it.[16]

Analysis of the fourteen films and videos presented at the Carnegie conference on poverty in April 1984 suggests that not only do the majority of these practitioners have little theoretical cinematic or video understanding (though they do have an understanding of social process), but they also have yet to master the basic techniques and conventions of structured documentary. The films exhibited a minimal awareness of how the conventions and techniques employed could be co-opted by vested interests or how they might lead to confused interpretations on the part of even sympathetic audiences. Only *District Six* offered contextual information with largely matching visual material. Most of the others simply attached historically sensitive verbal information to irrelevant and distracting visual material. If filmmakers are going to engage the state in terms of the dominant documentary form, they must at least have a knowledge of that form.

Notes

1. H. Salmane, S. Hartog, and D. Wilson (eds.), *Algerian Cinema* (London: British Film Institute, 1976), pp. 1–4. For an outline of the term, Third Cinema, see F. Solanas and O. Getino, "Towards a Third Cinema" in B. Nichols (ed.), *Movies and Methods* (Berkeley: University of California Press, 1976). For a general overview of cinema in social change, see L. M. Henry, "The Role of Film Makers in Revolutionary Social Change," *Praxis,* V. 1, N. 2, 1976, pp. 157–175.

2. P. R. Zimmerman, "Public Television, Independent Documentary Producer and Public Policy," *Journal of the University Film and Video Association,* V. 34, N. 3, 1982, p. 9.

3. P. Willemen, "Presentation," in D. MacPhereson and P. Willemen (eds.), *Traditions of Independence: British Cinema in the Thirties* (London: British Film Institute, 1980), p. 2.

4. J. Grierson, The Grierson Report, Union of South Africa, unpublished.

5. For information on approach and course content see *The SAFTTA Journal,* V. 1, N. 2, 1980 and K. G. Tomaselli, "Media Education and the Crisis of Hegemony in South Africa," *Media Information Australia,* Feb./March, 1985, pp. 9–20.

6. Quoted in R. Johnson and R. Stam (eds.), *Brazilian Cinema* (New Jersey: Associated University Presses, 1984), p. 77.

7. Unpublished essay, 1979.

8. For a more detailed criticism see J. Bruwer, *"Die Moord,"* and J. van Zyl "Angst," *Critical Arts,* V. 1, N. 1, pp. 54–56.

9. See K. G. Tomaselli, A. Williams, L. Steenveld and R. E. Tomaselli, *Myth, Race and Power: South Africans Imaged on Film and TV* (Bellville: Anthropos, 1986).

10. See P. McAllister and G. P. Hayman, "*Shixini December:* Responses to Poverty in the Transketi," Carnegie conference paper, 1985.

11. Unpublished essay by Lynette Steenveld, Rhodes University Department of Journalism and Media Studies, 1984.

12. G. Garden, *Rand Daily Mail,* March 23, 1983.

13. For a review, see J. Margolis, *Cineaste,* V. 12, N. 4, 1983, p. 55.

14. For a detailed analysis of this film, *This We Can Do for Justice and Peace* and *Awake From Mourning* in the context of radical film, see K. G. Tomaselli, "Strategies for an Independent Radical Cinema in South Africa," *Marang,* V. 4, 1983, pp. 51–85; and K. G. Tomaselli, "Oppositional Film Making in South Africa," *FUSE,* V. 6, N. 4, pp. 190–94.

15. P. Anderson, "The Tiakeni Report: The Maker and the Problem of Method in Documentary Video Production," *Critical Arts,* V. 4, N. 1, 1985.

16. L. Steenveld, "Who Sees? A Way of Seeing *I am Clifford Abrahams and This is Grahamstown,*" Department of Journalism and Media Studies, Rhodes University, Grahamstown, 1986, 26 pp. Note the introduction by Hayman.

A Mirage in the Desert?

African Women Directors at FESPACO

Claire Andrade-Watkins

Every other year, thousands take to the air from points in Africa, Europe, and the US to head to the Festival Panafricain du Cinéma du Ouagadougou (FESPACO). Braving searing heat and dust rising from the Sahara Desert, the hordes descend on Ouagadougou, or "Ouaga," as it is affectionately called, the capital city of the tiny West African country Burkina Faso (formerly Upper Volta) for the continent's biggest cultural event.

Held in alternate years from the Carthage Film Festival in Tunisia, the eight-day FESPACO showcases productions from sub-Saharan Africa and the diaspora—particularly the Caribbean, US, and Brazil. Burkina Faso was one of the first West African countries to nationalize film production and distribution and maintains a firm commitment to African cinema. The brief term of charismatic President Thomas Sankara, who came to power in 1983, witnessed an unprecedented increase in popular support for African film within Burkina. Sankara threw open the doors to African Americans and others from the diaspora with a formal invitation to the ninth festival in 1985, followed again by a warm reception in 1987. His successor, Blaise Campore, has continued this commitment, which, along with strong international support, particularly from France, has sustained Burkina as the home of FESPACO.

Now in its twelfth year, FESPACO remains the best place to see new African talent and gauge the level of activity and interests of African filmmakers. It is a festival created by African filmmakers for

Reprinted with permission from the author and *The Independent Film & Video Monthly*, a publication of the Association of Independent Video and Filmmakers in New York City, vol. 14, no.9, 1991, pp. 26–29.

two main purposes. The first and more visible side of FESPACO is the international showcase of African cinema. But FESPACO is also the preeminent meeting place for African filmmakers. Day and night, clusters of directors, producers, critics, journalists, scholars, actors, and distributors from all over the world can be seen milling around the pool of Hôtel Indépendence, the festival's unofficial hub. There and elsewhere they debate issues surrounding African cinema, from its financing to its audience and distribution within the continent and abroad.

Films are screened from morning to night in the city and outlying areas, in theaters ranging from the posh, air-conditioned Burkina to the wonderful open-air Raile and Neerwaya theaters where one sits comfortably on bench chairs under the stars and enjoys the lively calls, laughter, and warnings shouted by the African audience to their favorite characters.

The wise and experienced FESPACO visitor comes armed with festival French. Other essential skills are quickly acquired, such as resisting the lure of the open markets where fabulous wares from all over West Africa are sold and sidestepping myriad street vendors. If your foray into the city is successfully negotiated, with enough bottled water for the day (dehydration is a reality when temperatures hover around 100 degrees Fahrenheit), a veritable film feast of new and classic African films awaits you. This year's festival featured 68 films, with 33 African countries participating and close to half-a-million visitors—a quantum leap from the first festival in 1969, when five countries took part and the audience totaled 20,000.

This year's festival opened with *Karim and Sala,* the fourth feature by internationally acclaimed Burkina Faso filmmaker Idrissa Ouedrago. Ouedrago's previous feature, *Tilai,* won FESPACO's grand prize and was later a hit at Cannes, marking the first time a native son of Burkina earned this prestigious award. Other festival highlights attest to the wide range of new faces and work on the scene, with premieres from Cameroon, Mali, Burkina Faso, and Guinea Conakry.

A festival favorite, *Ta Dona,* by Adama Drabo from Mali, received the Oumarou Ganda award for outstanding first work. In it, Drabo weaves Bambara mysticism and ritual—an increasingly recognizable trait of Malian films—into the quest of a young engineer for the lost seventh Canari, the last link in a mystical chain of knowledge. Drabo is part of an impressive line of filmmakers from Mali, whose ranks include Souleymane Cisse and Cheik Cissoko.

Burkina Faso launched several new features. *Laada* (meaning traditional law), Idrissa Touré's first feature, examines a group of young men as they choose between following the path of traditional life or leaving for the city. Abdoulaye Sow's *Yelbeedo,* a Burkina Faso/Togo coproduction, examines the issue of abandoned children and incest through the eyes of a young couple who take in a baby left on the street. Pierre Yameogo's *Lafi* reveals the dilemma and anxiety of a recent graduate who wishes to attend medical school abroad despite administrators' fears that, like many others, he may not return home with his new skills.

Cameroonian novelist Basek Ba Kobhio garnered special mention in the Critic's Prize his second film and first feature, *Sango Malo*. Based on Kobhio's novel, this tight narrative tells the story of a young, progressive teacher who introduces his students to ideas on politics and sex, much to the chagrin of the headmaster. He eventually collides with the village chief and nobility when he tries to start a peasant cooperative.

Allah Tantou, an outstanding documentary by David Achkar of Paris/ Guinea Conakry, is a brilliantly crafted purview of the heady, early days of Sekou Toure, one of Africa's first nationalist leaders. *Allah Tantou,* which won the Telcipro award, is a beautiful mixture of archival footage, home movies of the director's father, who was an official under Toure, and stunning dramatic segments of the father's imprisonment. The work is an absorbing glimpse of a fascinating moment in history.

Tunisian filmmaker Férid Boughedir, a longtime journalist on African cinema, presented his first feature, *Halfaouine,* a charming and sensual portrait of a young boy crossing the bridge to manhood. Among other award-winning films was Felix de Rooy's (Curaco) *Almacita di Desolato.* This netted a Paul Robeson Award, and a special mention went to British director Auguste Reece's *Twilight City.*

Of the 32 films in competition, none were by sub-Saharan women, and those films in the festival by women from the diaspora—Zeinabu Irene Davis and Carmen Coustaut from the US—were scheduled late in the week and hard to find. The invisibility of African women behind the camera is not unusual. Addressing and correcting this inequity has been a priority of the Fédération Panafricaine des Cinéastes (FEPACI) since its conception in 1969. An organization of African filmmakers from 33 countries, FEPACI has served as a powerful lobbying voice for African cinema within the continent and abroad.

The role of women as central characters in films, however, is an honorable and long-standing tradition in sub-Saharan Africa. Powerful female characters have anchored such classic films as Senegalese director Ousmane Sembene's *Ceddo* and *Emitai,* Ivory Coast director Desire Ecare's *Faces of Women*, and Med Hondo's masterful portrayal of the legendary warrior queen *Sarrounia.* More recently themes concerning African women's struggle against female circumcision have been manfully tackled by Cheik Cissoko in *Finzan.*

In part the scarcity of women directors has to do with the scarcity of resources for film south of the Sahara. Only a very small group of filmmakers, male or female, are actively engaged in African film production. Feature filmmakers number no more than 40 or 50, with 20 or 30 having financing at any given time. Still, scarce resources alone do not explain why, in African film's 30-year history, only two women south of the Sahara have achieved the prominence of their male counterparts. In 1972 Sarah Maldoror (Guadeloupe/Angola) became established as the first African women director with her feature *Sambizanga.* She subsequently went on to make close to 20 films. Maldoror was followed by Senegalese Safi Faye, who has made about 10 films, including two features, *Peasant Letter* (1975) and *Fad Jal* (1979), since 1975, and is currently completing her third. This stands in contrast to countries north of the Sahara, particularly Tunisia, Algeria, and Egypt. They have long-standing film traditions and industries from which a respectable cadre of women directors has emerged.

Tucked in the corner of FESPACO was a workshop on Women, Cinema, TV, and Video in Africa, organized in collaboration with the Montreal organization Vues d'Afrique, which hosts a large annual festival of African film. This workshop was the very first gathering of women from Africa and the diaspora. It brought together close to 100 invited guests from the continent, the US, Europe, and the Caribbean. According to the official communiqué, the workshop was intended to identify the problems facing women in their fields and come up with strategies ensuring their participation in media and its development.

At least eight women filmmakers were in attendance. Two, Flora Shelling M'mbugu from Tanzania and Anne Mungai from Kenya, came seeking financing for feature projects they had in hand. Veterans Sarah Maldoror and Miriama Hima from Niger, plus Attia Kehena (Tunisia), Kadiatou Konate (Mali), Grace Kenyua (Kenya), Sepati Bulane (South Africa), and Lola Fani Kayode (Nigeria) also took part. For most participants, it was the first time they had met women engaged in similar work from other countries.

The energy found at the workshop generated the kind of excitement that marked earlier start-up initiatives at FESPACO such as the International Market for African Film and Television (MICA) in 1983 and the International Partnership Day in 1989. If a typical conference had taken place the following sequence would have unfolded: The group would be well met, engage in fruitful dialogue, have a substantive closing plenary, establish an agenda for the next gathering, and top off the whole event with gracious multilingual social events, finally departing with warm feelings of sisterhood and a common sense of mission and purpose for women in cinema, TV, and video in Africa and the diaspora.

It didn't happen. Instead, the workshop unleashed a riptide of emotion, confusion, and animosity which tore across the festival. It triggered often heated debates on a broad range of areas, including the relationship of the diaspora to Africa, relations between the French- and English-speaking regions of Africa and their positions within African cinema, and the appropriateness of FESPACO as the forum for this workshop.

Following a week of informal and cordial meetings between African women and a handful from the diaspora prior to the festival's official start, the workshop opened with an unexpected request. As invitees gathered around the table, the panel chair asked all non-Africans to leave the meeting. Most of the participants were caught by surprise, particularly given the legendary warmth and camaraderie of the festival. Over the hubbub, the request was repeated. Things quickly deteriorated.

Emotional addresses from women of the diaspora had the poor translators at a loss for words, clearly distressed by the messages they were forced to convey back and forth between workshop spokeswomen and participants. Further confusion ensued as a debate arose on what exactly constitutes an African. Women born in Africa but raised elsewhere angrily defended their right to be there, as did women born abroad of African parents. Before the debate was over, most of the "others" had left.

The emotional momentum caught festival organizers off guard. Some Africans saw the incident as a misunderstanding; others, particularly from the diaspora, saw it as a rejection of non-Africans. A call for action came from many fronts to make amends to the diaspora and put the workshop back on course. A formal letter of protest was sent to festival organizers by several women of the diaspora. Subsequently there were many abashed and embarrassed apologies by all parties.

At a glance, the workshop could be perceived as an unmitigated disaster. But over the long run, it could be a catalyst for understanding and growth. During the pre-workshop discussions and workshop itself, patterns emerged that showed many more African women working in television and audiovisual services than in film. Those in cinema were generally actresses, with no ready access to technical training for production. In the television and audiovisual sectors, participants noted that women were often steered toward distribution and editing. Even those with extensive production backgrounds, such as Deborah Ogazuma, a senior producer at the government-owned Nigerian Television Authority, the country's largest television network, feel restricted. Ogazuma is one of the few women in Nigerian television to direct drama. Her projects have included 41 live weekly episodes based on a literary adaptation of the novel *Magana Jari Ce (Wisdom Is an Asset).* Ogazuma notes, however, that women are steered away from directing drama and toward women's magazine and children's programming.

These bits and pieces of different women's experiences are part of a larger pattern of problems that, prior to the workshop, women struggled with in isolation. These were summarized in the workshop's opening statement by chair Annette M'baye D'erneville, who laid out the program's broad objectives: 1) provide a forum for women to exchange and share their experiences; 2) adopt propositions that will help ensure women their rightful place, particularly in the areas of training and production; 3) devise a follow-up structure for dialogue and common action; 4) identify the frustrations of women professionals and produce images that consciously reflect women's realities, social contexts, cultures, and histories; and 5) disseminate that perspective.

The workshop made clear that while African women filmmakers share many of the same obstacles, there are also vast differences. It seemed that participants were looking at the role of women—in cinema, Africa, and the diaspora—through lines of different focal lengths. Their widely divergent expectations reflected the tremendous range of the participants' complex cultural, historical, political, and societal realities.

Most advancements in African cinema have been the result of arduous and painstaking effort over many years. Viewed from this perspective, the women's workshop was a painful birth for what may become a new network of professional peers. Perhaps holding it under the aegis of FEPACI and FESPACO was inappropriate. Anything that occurs within such a context is subjected to a tremendous amount of international attention. For such a fledgling initiative, some privacy for dialogue, growth, mistakes and the formation of a sense of identity might have been better for meeting the workshop's objections and mission.

The workshop did produce some concrete results. An eight-member panel of women from Burkina Faso, Kenya, Nigeria, Ghana, South Africa, Rwanda, Gabon, and Tunisia was established and workshop participants identified four initial projects for the panel to pursue: 1) develop a repository for film and audiovisual works by African women; 2) establish subregional itinerant training workshops; 3) train staff to conduct these workshops; and 4) support the participation of women at film festivals in Africa, the Caribbean, and elsewhere. In addition to these plausible and pragmatic goals, an appeal was made by the panel in its closing communiqué to FEPACI to support the initiative under its aegis, in addition to generally increasing its activism for women in the film profession.

The pattern of the women's workshop is consistent with the vision and weaknesses associated with the development and dissemination of African cinema as a whole. The dominance of Africa's French-speaking areas in the development of cinema is the direct result of a very aggressive program by France on the eve of its colonies' independence in 1960. France distributed financial and technical assistance and expertise through the Ministry of Cooperation in areas ranging from agronomy to cultural expression. Support to cinema was considered important because it coincided with France's intent to maintain the cultural and linguistic bonds that had characterized the colonial relationship. In contrast, France's colonial counterpart, Britain, had no great interest in supporting cinema by Africans, either before or after independence.

The catch 22, however, was that the financial and technical facilities and personnel provided by France were based in Europe. Distribution and exhibition circuits within Africa were also controlled by non-Africans. So, paradoxically, the most prolific region of African film production teetered on a very precarious base—one without an indigenous infrastructure for production, distribution, and exhibition.

In 1980 France began to shift funding from individual African film projects toward experiments in regional infrastructure. When this happened, the filmmaking community was jolted into seeking alternative means of support. The emphasis on strong ideological, Marxist, postcolonial critiques in film during the subsidized 1960s through late 1970s gave way to a focus on building infrastructures, both through South-to-South cooperation and North-to-South collaborations that could help develop Africa's film resources, technical expertise, and trained personnel.

FESPACO is the forum through which filmmakers address such concerns and chart the future course of African cinema. It is regular practice for FESPACO to focus on specific themes through its workshops, colloquia, and panels. In the past, programs on the oral tradition and narrative film structure and international coproduction have anchored significant activities during the week, often launching major continent-wide initiatives. One such venture was the International Market for African Film and Television, which celebrated its fifth anniversary at FESPACO this year. Set up at the George Méliès Center, a beautifully appointed air-conditioned space, the market served as a convenient meeting place for filmmakers and distributors. Both festival films and other works from Africa, the Caribbean, US, and Brazil were available for buyers, distributors, and exhibitors to screen at their convenience.

FESPACO's major business agenda centered on the Second International Partnership Day, an initiative of FEPACI, which built on the foundation laid during the 1989 festival. The idea at this year's International Partnership Day was to look toward aggressively maximizing north-to-south and south-to-south cooperation and coproduction in order to obtain the resources necessary to finance and disseminate African cinema. Over 100 participants came together, including numerous representatives from Europe and the US—including Britain's BBC and Channel Four, ZDF in Germany, Centro *Orientamo Educativo* in Italy, *Agence de Cooperation, Culturelle, et Technique* in France, and the Rockefeller Foundation in the US.

Although African cinema's basic problem of undercapitalization has not been resolved, both dialogue and action on this matter are now moving beyond the rudiments of production to other, more internationally attuned concerns, such as competitiveness in the foreign market and the accessibility of African films to non-African audiences.

Even with the current interest in African cinema at international festivals and the promising prospects of partnership, there still are no guarantees for the individual filmmakers. They must still log thousands of miles on the gypsy trail, hopping between three continents to pursue partnership and coproduction opportunities. Despite daunting obstacles, African filmmakers persevere. It is this vitality and courage that makes FESPACO more than just a film festival. FESPACO showcases a cinema on the move and a movement with potentially unlimited range and impact within the continent and throughout the global community.

Black African Cinema in the Eighties

Roy Armes

Jeune de vingt ans, le cinéma d'Afrique noire est encore à l'aube de son devenir.

Férid Boughedir

This article stems from several years research into African cinema which has taken me to the three major African film festivals (Carthage, Ouagadougou, Mogadishu) as well as various gatherings in Europe.[1] The research owes much to the work of a small number of energetic and committed francophone critics (Guy Hennebelle, Victor Bachy, Paulin Soumanou Vieyra, Férid Boughedir, Charles Lemaire), whose efforts in a variety of books, articles and conference papers have done much to elucidate the issues and to establish the factual basis on which future research can be built. This present article, which forms part of a forthcoming study of Arab and African cinema, is based on a viewing of 36 of the 67 or so feature films which would seem to have been produced in black Africa during the five years 1980–1984. Not all these films have as yet been seen even inside their countries of origin. For example, Wole Soyinka's study of the Shegari era in Nigeria, *Blues for a Prodigal,* was seized by the authorities when due to open at the National Theatre in Lagos, and many other black African features have found no commercial distributors. Since

Screen, 1985, vol. 26, nos. 3–4, pp. 60–73. Reprinted by permission of the publisher and author.

the films themselves are virtually unknown in the United Kingdom, I have tried to set them in the context of the development of Third World cinema and to draw out points of wider interest. In particular the difficulties encountered by African film-makers emphasise the need for studies of the economic organisation and control of western cinemas. Too often such studies as currently exist—and here I include my own histories of British and French cinema—deal with a "national" cinema whose autonomy and unity is simply assumed. Study of cinema in Africa shows this to be a myth. Being a purely commercial operation, distribution has no need to be linked geographically to the studios where films are made or to the film theatres where they are shown. But even when operating from afar, it remains the key sector of the industry.

I

The failure to build a black African film industry over the past decade—despite the valiant efforts of individual film-makers and a number of governments—is hardly surprising in view of the fact that modern industry of any kind in Africa dates only from World War II when, as John Iliffe points out in a recent study of *The Emergence of African Capitalism,* "several circumstances came together to change the old pattern of exported raw materials and imported manufactures": colonial governments seeking to diversify their economies, local European settlers aiming for greater autonomy and foreign firms seeking commercial advantages.[2] As late as 1950 in Nigeria, which was later to become something of an economic giant in black African terms, the manufacturing sector still provided "only 0.45 percent of GNP (the smallest proportion of any country producing statistics)."[3] To the problem of late capitalism was added that of foreign control: since the initial industrialisation occurred under colonial rule, early industrial enterprises tended to be owned by foreign capital. All Third World film industries have been created by indigenous capital attracted by the profits to be derived from catering to the entertainment needs of the new audience composed of those drawn into the cash economy by urban industrialisation and the rural exodus. But black Africa, coming late to industrialisation, missed out on the boom which, in India and Egypt for example, was occasioned by World War II speculation. In these instances cinema came to be seen as an excellent investment for undeclared profits from the illegitimate economy. In black Africa, however, though subsequent programmes aimed at giving indigenous control of foreign firms have certainly enriched local elites in countries like Nigeria and Zaire, there have been far more lucrative and less speculative outlets for reinvestment than a nascent film industry. The industrial infrastructure for cinema—studios, sound and editing facilities, laboratories—is therefore almost completely lacking in black Africa, with the privately owned Cinafric studios in Ouagadougou (capital of Burkina Faso, formerly Upper Volta) standing as virtually the sole realisation of the commonly held capitalist dream of establishing local production facilities as a prelude to venturing forth onto the international film scene. But Ouagadougou remains a paradoxical location for Cinafric (as for most of francophone black Africa's cinema organisations) since Burkina Faso is one of the world's poorest countries and possesses barely a dozen cinemas for its five million inhabitants.

The factors which go to shape cultural production in black Africa constitute only partial grids, each of which implies a different set of divisions. Though traditional beliefs persist, the varying penetration of Islam and Christianity tends to differentiate the North from the South—Muslim Niger, for example from Congo or Zaire. Traditions of capitalist development, in contrast, make a division between East and West. In West Africa entrepreneurs emerged largely from artisanship and trade, whereas in East Africa they have tended to come "through the straddling process of western education and modern-sector employment."[4] This finds its reflection in cinema to the extent that there is no equivalent in East Africa to the individual initiatives, backed by local capital, which have led to the production of fictional feature films in Nigeria and Ghana. There is an extensive and well-organised production and distribution of films in East Africa, through the Kenyan Institute of Mass Communication, for example. But all this local production is of documentaries serving government educational and agricultural programmes and even in the 1980s remains an expression of official views and ambitions: there have been no initiatives for privately funded features. Of course, the most important set of divisions in black Africa is that deriving from colonialism, which even after 25 years of formal independence continues to tie states—and in particular their Western-educated elites—to the former colonial capitals of Europe—London, Paris, Brussels.

European influences on the development of cinema—from the efforts of Belgian missionaries in Zaire to the traditions of "neutral" informational documentary in anglophone Africa—continue into the 1980s. It is the Catholic organisation OCIC (Organisation Catholique Internationale du Cinéma et de l'Audiovisuel) which in the 1980s has done most to bring African film-making to European attention,[5] and undoubtedly given many young African film-makers an orientation towards Europe. The lack of post-production facilities in black Africa means that most films are finished in Europe, which not only increases enormously the cost of African production but also cannot fail to influence the attitudes of film-makers. But December 1980 did see an attempt by certain African states to break free from the most highly developed of these foreign shaping influences, with the ending at their insistence of the system of aid to francophone African film-makers through the French Ministère de la Coopération. This remarkable seventeen year example of "enlightened neo-colonialism" was the main force behind the development of film-making in the fourteen states of what was previously French West and Equatorial Africa. (For example, 125 of the 185 films of all kinds made between 1963 and 1975 received Coopération technical and/or financial assistance.[6]) The method of finance, through the purchase, at a larger than normal fee, of the non-commercial distribution rights in the film by the French ministry, was quite separate from the commercial film distribution system in francophone black Africa, itself controlled until the 1970s by a French commercial duopoly. The result of the Coopération scheme was the production of a great number of films which would not otherwise have been made, but these were ghettoised in Africa (finding showings only in French cultural centres), and more accessible in Paris (through the ministry's archive) than in Africa. The scheme did little to bring African film-making to African popular audiences, but it has given francophone Africa a tradition of personal "art" cinema, often only tenuously rooted in the specificities of the maker's national culture and the demands of a mass audience, which persists into the '80s, long after the cessation of the system.

The cultural importance attached to film by the French Ministère de la Coopération also served to bring cinema to the attention of francophone African governments, so that initiatives to support film-making are stronger in, say, Senegal or Burkina Faso than in Nigeria. But the problems of film-makers struggling to create commercially viable film production remain the same throughout black Africa and point to the need for government assistance in such matters as the regulation of the import and distribution of foreign films, the reduction of taxes on cinema admission (set at extremely high levels by most colonial administrations), the establishment of a national ticketing system (which alone would ensure producers a fair return on local distribution), production assistance at home and promotion abroad. But film fits awkwardly into the state's institutional priorities it cannot be seen as a governmental achievement in modernisation (like the building of a dam or industrial complex), nor is it an aspect of traditional culture to be promoted internationally along with, say, local carpet making or rural handicrafts. For African rulers, usually Western-educated and always sensitive to the image of their country abroad, a speculatively financed local film production designed for the mass audience would be the last type of product to be advertised abroad, while a film which looked critically at local society would be simply intolerable to them. On the rare occasions when Third World governments have become culturally involved with film—as in India with the National Film Development Corporation—the result has usually been rather like the efforts of the French Ministère de la Coopération: the creation of a hybrid product—part local, part Westernised—which no longer corresponds to local audience taste.

One area in which international organisations such as UNESCO and the Paris-based Agence de Coopération Culturelle et Technique can make an unambiguously positive contribution is through support for events and organisations allowing African film-makers to meet to exchange views, see each other's films and debate issues. Africa has two long-established biennial film festivals which have continued to prosper in the 1980s, the JCC (Journées Cinématographiques de Carthage), established in Tunis in 1966 and holding its tenth festival in 1984, and FESPACO (Festival Panafricaine de Ouagadougou) established in 1969 and celebrating its ninth gathering in 1985. These two have now been joined by a third festival, this time in anglophone East Africa, MOGPAFIS (Mogadishu Pan-African Film Symposium) which held meetings in the capital of Somalia in 1981 and 1983. These festivals have provided a crucial context for members of the professional organisation of African film-makers, FEPACI (Fédération Pan-africaine des Cinéastes). Established in 1970 and granted observer status at the Organisation of African Unity, FEPACI has continued to be an active force through the efforts of its individual members, though the congress held in Ouagadougou in February 1985 was the first for a decade. Meanwhile the institutional inertia of FEPACI prompted two complementary initiatives in 1981: L'Oeil Vert, a group of black African film-makers interested by the possibility of collective film-making, and CAC (Comité Africain de Cinéastes) a Paris-based distribution organisation headed by the exiled Mauritanian film-maker, Med Hondo.

The 1980s have been a time of enforced reflection for African film-makers, faced with the fact that fifteen years of struggle have not led to the establishment of a film industry anywhere in black Africa. A major statement of their position is to be found in the manifesto issued after a meeting held in Niamey in March 1982. Of particular interest are the five general principles which underlie the various propositions put forward:

— developments in production must be linked to those in the sectors of exhibition, import and distribution of films, technical infrastructure and professional training;
— the intervention of the state is needed to promote and protect private and public investment;
— measures to promote cinema are not viable on a purely national level but need to have a regional and interafrican dimension;

— future developments in African cinema will need to be made in collaboration with television institutions;
— finance for developments in cinema can be found within cinema itself, in the receipts from the showing of foreign films.[7]

The discussions held at Carthage in 1984 continued in a similar vein, drawing up a balance sheet of developments since the festival was established in 1969 and setting out to consider in detail the impact (or lack of impact) of the various resolutions passed at earlier conferences. In a lucid and informative paper presented at Carthage, Férid Boughedir traced "The Evolution of Strategies for the Viability of National Cinemas in Africa from 1967 to 1984."[8] Taking as his starting point the definition of distribution as the key sector of the film industry made by Tahar Cheriaa, the founder of the Carthage festival, in 1967, Boughedir looks back at the African film-makers' initial cry for total nationalisation. Seeing the crucial problems as foreign control of distribution, national markets too small to support a national cinema and the failure to return money taken from cinema in taxation to foster production, film-makers of the late '60s and early '70s placed their reliance on state control. But the ensuing years have shown the limitations of this approach. National film corporations remain vital to control import of foreign films and to regulate the domestic market, but have generally had little positive impact as production organisations: new in the '80s is the respect offered by African film-makers to private producers.

The ideal of film common markets linking regional groups of countries has also proved difficult to realise. In 1981 CIDC (Consortium Interafricain de Distribution Cinématographique) came into operation after years of patient effort, taking into African hands for the first time film import and distribution in the fourteen states of francophone black Africa, where it had previously been controlled by a succession of French-owned companies. The aim was to create a system which would both allow the commercial showing of African films to the African mass audience and, with the profits from the screenings of foreign films, support a parallel production organisation, CIPRO Films (Centre Interafricain de Production de Films). As Boughedir points out, things have not gone smoothly:

— most states did not reform their tax structures to allow the common market to come properly into operation;
— most did not pay their contributions to CIDC;
— many national cinema organisations saw distribution purely as a commercial operation and were uninterested in a cultural role (such as that of finding an audience for African films);

— many exhibitors refused to show African films (though some of those which were properly distributed achieved remarkable commercial success);
— the lack of national ticketing systems prevented the monitoring and control of the market.

By 1984 the CIDC, which in any case had only 50 African films among the 1,200 it distributed (that is, barely four per cent) had lost the confidence of African film-makers and its operation had come virtually to a halt, leaving the market vulnerable again to US films distributed through a Swiss-registered company, SOCOPRINT, which has shown no interest in handling films made by Africans.

II

Turning to production, the figures set out in the following table eloquently demonstrate the huge difficulties facing the black African film-maker. The 68 feature films produced over the 5 year period 1980–1984 have come from 13 countries in all, with half (33 films) from just two countries (Nigeria and Cameroon) and a quarter (17 films) from 3 others (Senegal, Ivory Coast, Niger). No other black African country has produced even one film a year during the period.

Nowhere is there the possibility of a satisfying career for a film-maker. Some of the major figures in African cinema, such as Ousmane Sembene and Med Hondo, have been unable to make even a single film in the five year period, though both have projects on which they have been working for years. Of the fourteen film-makers who had made at least one feature before 1980 (and only one active director, Ola Balogun, had made more than three before 1980), only five (three from Nigeria and two from Cameroon) succeeded in making more than one film over five years. No less than 39 of the 67 films are first features by new directors, but only four of these managed to complete a second film. Though it is often convenient to consider film-makers in national groupings (since these will often have specific problems and opportunities in common), there is no way in which we can talk meaningfully of a "national cinema" when output is so low.

These African films are personal creations in a way that films can never be once film-making is fully industrialised. The African film-maker (almost invariably a man—no feature films have been made in black Africa by a woman in the 1980s) has to concern himself personally with raising the money for his film from state or private sources. Usually it is his own company that produces or co-produces the film or, if the film is produced by a national film corporation, it is likely that he will be the official in charge of production. He will normally have written the script as

Country	Films	Film-makers			
		Total	Estimated before 1980	Debuts	Two or more films
NIGERIA	22	12	3	9	5
CAMEROON	11	7	3	4	3
SENEGAL	6	6	3	3	—
IVORY COAST	6	6	—	6	—
NIGER	5	5	3	2	—
BURKINA FASO	3	3	—	3	—
MALI	3	3	1	2	—
GHANA	3	3	1	2	—
CONGO	2	1	—	1	1
ANGOLA	2	2	—	2	—
GUINEA	2	2	—	2	—
ZAIRE	2	2	—	2	—
MADAGASCAR	1	1	—	1	—
TOTALS	68	53	14	39	9

well as directed it—sometimes appearing too as actor—and often he will have to take a very direct part in arranging local screenings. He will be his film's sole publicist accompanying his film to foreign festivals. It may be argued that this is the fate of any independent film-maker, but where the African film-maker is unique is that he working in a context where there is no real tradition of film-making, no standard procedure for organising production or conventional source of film finance, no pool of experienced technical or acting talent and virtually no appropriate models for a film's dramaturgy or visual style. Every African film therefore represents an awesome personal effort on the part of its maker, whose rewards will be strictly limited. He and his backers are likely to take years to recover their investment (if at all), the local market is most probably so unregulated that he will never receive the return appropriate to the audience his film has attracted, foreign commercial screenings—in Africa or elsewhere—are rare, and the chances of making another film within four or five years are remote. Nonetheless films do continue to be made in black Africa.

African cinema has grown up largely separate from the African literature and drama in European languages which has been a key element of black African cultural production since the '50s. Apart from Wole Soyinka's recent venture into the cinema, Ousmane Sembene is the only major novelist to turn to film-making and he has spent the first half of the 1980s seeking backing for his major study, in two feature-length parts, of the nineteenth century West African leader, the Almany Samori Turay, who put up a stubborn resistance to French imperialist ambitions. Thanks to sub-titles, African films in local languages are not necessarily impenetrable to outsiders—they can be enjoyed by (literate) audiences throughout the world. The major African films of the '80s—like most of those of the '70s—have been made in indigenous languages, but it is notable that the two countries which produce the most films and possess the film-makers most concerned with purely commercial success—Nigeria and Cameroon—both have a tradition of film-making in the language of the former coloniser. In Nigeria, both Ola Balogun with *Cry Freedom* and Eddie Ugbomah with *Death of a Black President* have worked in English, and the latter film, a ponderous if well-intentioned drama-documentary on the assassination of President Murtala Muhamed, shows the particular problems of this kind of approach: unconvincing characterisation and leaden dialogue sequences. The writing and performing in English of scenes which would inevitably be acted out in indigenous languages gives film-making of this kind an air of contrivance akin to the sense of a "literature in translation" that characterises many of the African novels written in English. In Ghana too, films like *Love Brewed in the African Pot,* directed by Kwaw Paintsil Ansah and *Kukurantumi,* directed by King Ampaw, play out their conventional themes of the clash of generations or the contrast of rural goodness and urban corruption in ways that give fascinating glimpses of Ghana's evolving society, while retaining a "stagey" tone in their key dramatic scenes.

In Cameroon, a country with apparently 427 local languages,[9] virtually all the film-making so far has been in French and much of it has had no more than purely commercial ambitions. Installed in Paris since 1975, Alphonse Béni has directed a series of what Boughedir terms "erotico-disco" thrillers, and his example has been followed by a 1980s newcomer, Jules Takam. The latter's *L'Appât du gain* is essentially a low-budget French production with a black cast. Ostensibly a political thriller about Western exploitation and African corruption, the film's confused plot is little more than a pretext for the series of a dozen or so murders committed by its macho-style vigilante hero. More authentically African in setting are director Daniel Kamwa's commercially highly successful comedies such as *Notre fille* which deals with the contrasting life-styles of a Paris-educated elder daughter and her village chief father, with his eight wives and 30 children. The film's theatrical style and use of the very real problems of African cultural identity for purely farcical ends have provoked the hostility of African critics, but many of the incidental (exaggerated?) details it offers of the life of the Frenchified elite are fascinating (the need to wear a tie to be respected and to enjoy camembert to be considered sophisticated; the preference for local fruit that has been exported, canned and reimported for sale in the local supermarket, and so on). French is also the chosen language for the first two films produced by Burkina Faso's privately owned Cinafric studios, the French director Christian Richard's *Le Courage des autres* and Sanou Kollo's *Pawéogo/L'Emigrant.*

Set against this excessive Western influence in the use of local dramatic forms in certain films from Guinea and Nigeria. The Guinean film *Naïtou l'orpheline,* directed by Moussa Kemoko Diakité, features the national ballet troupe dancing out, in natural settings, a two hour drama of rival wives and an ill-treated daughter. *Naïtou* combines initiation rites and nude bathing scenes, ghosts and scenes of madness. Similarly, in Nigeria, the '80s have seen the production of a number of adaptations of Yoruba folk dramas, written for theatrical performance and set in an idyllic rural world untouched by colonialism or modernity. The series began in 1980 with *Aiye,* directed by Ola Balogun, and since then three of the leading Yoruba dramatists and performers, Chief Hubert Ogunde, Ade Love and Baba Sala, have directed versions of some of their own dramas. Balogun's former assistant Bankole Bello, followed in the same vein with his first film, *Efuntesan,* a period tale of a rich slave-owning woman whose cruelties were stopped by communal action. Perhaps the most elaborate of the series is Balogun's *Orun Mooru,* featuring the Yoruba comic Baba Sala in a story mixing folklore and rural comedy, traditional songs and dances, satire on the modern rich and elaborate trick effects (for a ghostly dream sequence). This kind of interplay between cinema and local dance and drama promises to be highly fruitful, since it offers a genuinely popular cinema which is not derivative of Western film models.

Though to date the resources of film have been used to do little more than record performances already elaborated in stage terms, Balogun's work shows how these elements can be blended with influences drawn from Indian film melodrama to create a cinema of more than regional appeal. At present, however, these productions remain at a largely artisanal level, with Ogunde, for instance, acting as his own distributor and exhibitor and taking the sole print of his films on tour on his theatrical circuit for showing in any available public space.

Since the ending in 1980 of the production system devised by the French Ministère de la Coopération, African cinema has struggled to find alternative local sources of production finance, as yet with only limited success. In the larger countries such as Nigeria and Zaire hundreds of cinemas exist, but many of these are extremely primitive and they are not organised in a way that will support local film production. The very lack of precise statistics indicates the extent to which the market is uncontrolled with dozens of "exhibition outlets" operating illegally to avoid the payment of taxes. Indigenisation measures have done little to improve the situation and in Zaire, for example, only two features have been produced in the'80s, one with equipment from educational television and the other with foreign funding from religious sources in Rome. Nigeria has a well-developed and richly funded television system which could in theory help support film-making, but to date no production finance has been forthcoming and Nigerian producers have been offered only derisory sums for screenings of such films as do exist. In francophone Niger—a country which has produced a number of striking films despite the fact that it has less than a dozen cinemas—the newly established television service (ORTN) has proved more enterprising and the three Niger films shown at the Ouagadougou festival in 1983 were all ORTN productions: Mustapha Alassane's *Kankamba,* Bakabé Mahamane's *Si les cavaliers . . .* and Djingarey Maiga's *Aube noire.*

In many ways African films are defined more by the source of their production finance than in terms of a national identity. Some privately funded films are still highly personal cries of anguish at the effects of cultural disorientation of a kind common ten or fifteen years ago. *L'Homme d'ailleurs,* for example, directed by Mori Traoré from the Ivory Coast, is the story of the disintegration of a black student in Japan. Offering no view of Japanese society or culture, the film concentrates on the slow progress of its hero (played by the director who also wrote the script) towards a predictable suicide. *Suicides,* another self-financed film, directed by Jean-Claude Tchuilen from Cameroon, traces the decline of an African student who witnesses the suicide of a rich French woman.

By contrast, films produced by national film organisations tend to take an optimistic view of human possibilities, laying stress on co-operation.

The archetype of this kind of cinema is *Les Coopérants,* directed by Arthur Sibita and produced in Cameroon by the Fonds de Développement de l'Industrie Cinématographique (FODIC) and telling of a group of well-off students who volunteer to assist in a programme of rural development. Similarly, in *Jours de Tourmente,* directed by Paul Zoumbara and produced by the Centre National du Cinéma in Burkina Faso, the young people of a village get together to confront the lethargy of their elders and dig a much-needed well. *La Chapelle,* directed by Michel Tchissoukou for the Office National du Cinéma (ONACI) in the Congo, is a tale of popular resistance to the intrusion of Christianity in the 1930s. Tchissoukou was one of the few newcomers of the 1980s to complete a second film, a study of traditional wrestling, *Les Lutteurs. Wend Kuuni,* directed by Gaston Kaboré and produced by the CNC in Burkina Faso, combines a simple tale of a young boy found in the bush who is brought up by a friendly village family with a look at pre-colonial Mossi culture. More than simply a semi-documentary account of African rural life, *Wend Kuuni* is marked by a subtle interplay of image and music.

The films produced by a mixture of personal, private company and official organisation funding divide into studies of tradition and looks at contemporary society. The former category contains a number of interesting explorations in the '80s. *L'Exilé,* director Oumarou Ganda's last film before his death at the age of 46 and the first to have adequate financial backing, tells a legendary tale with clear relevance to the present. *L'Exilé* captures the full flavor of the fable: the arbitrary happenings and flat characterisation, the passages of suspense and ritual repetitions. The story itself—told to guests by an ambassador in exile in Europe (Ganda himself)—turns on the importance of the word and verbal agreement in traditional African society, but in the telling a rich ambiguity emerges. Mustapha Diop's *Le Médecin de Gafire,* co-produced by the director's own company, Niger television and the national corporations of Mali (CNPC) and Burkina Faso (CNC), deals with a fascinating aspect of the clash of modernity and tradition by contrasting attitudes to medicine. *Jom,* the Senegalese director Ababacar Samb Makharam's second feature, co-produced with German television and made ten years after the director's first feature, focuses on a key West African figure, the *griot* or popular story-teller, celebrant of social values and preserver of tradition. The director uses this timeless figure to link a variety of stories, all dealing with human dignity (*jom*) ranging from a contemporary strike to early instances of resistance to the coloniser. It marks the return of one of black Africa's earliest film-makers, who made his first short in 1964.

Recent film representing contemporary African society include *En résidence surveillée,* directed by veteran Senegalese film-maker and historian of African cinema, Paulin Soumanou Viey. Here the tactical manoeuvring among members of the ruling elite is set against a coup which reinstates the ousted president in a new guise. *Pétanqui ou le Droit à la vie,* directed by Yéo Kozoloa from the Ivory Coast, contrasts its view of the ruling elite with television images of drought and ecological disaster and combines idealism and corruption, modernity and witchcraft. *Seye Seyeti,* directed by Ben Diogaye Beye, examines polygamy and the attitudes of present-day youth in Senegal. *Comédie exotique*—a film with partial French financing and a French cameraman—directed by the young writer Kitia Touré from the Ivory Coast—is a relatively expensive feature which returns to this contrast of traditional and modern in a story dealing with the making of a Western TV documentary about Senoufo masks.

Djeli, an Ouagadougou prize-winner made as a first feature by the Paris-educated Lancine Kramo Fadika, examines the continuing problem of caste in the modern Ivory Coast. Here again the claims of tradition and modernity are counterposed and the film ends with a freeze frame at a moment of possible reconciliation. Malian director Souleymane Cissé's third film—and sole production so far in the 1980s—*Finyé,* is the first to look seriously at the workings of power in an African society under military rule. Its young student heroes live in a recognisably modern world, marked by the clash of generations, a frustrated anger when confronted by evidence of corruption and a mix of betrayal and commitment when their solidarity is tested. Though the bulk of the film is shot with a close and precisely focused realism, it's protagonists are placed in a context of traditional African beliefs and values in a style which shifts effortlessly into moments of literal unreality.

In the 1980s the established traditions and styles of West African film-making have received a challenge from the cinemas of Portuguese-speaking Africa. Two features from Angola underline the potential of a style of film-making rooted in almost ten years of documentary practice and in a very conscious concern to redefine the social function of cinema in an emerging African society. Both films base their structures on the patterns and rhythms of oral story-telling techniques, so as to reach a popular audience for whom a cinema of social and political awareness is inevitably quite novel after the long years of colonial rule. Both were shot in black and white, *Memoria de um dia* opening with a selection of photographic images of colonial rule, set against a quietly lyrical, meditative text. It contains depictions of the customary range of aggressions by the coloniser—beatings, forced labour, torture—but these are presented through simple re-enactments performed by ordinary peasants from the region concerned. These scenes are illuminated by the direct testimony of witnesses who relate in personal, localised terms the history of the recent past (naming the colonial aggressors and their victims). To bind the film together Orlando Fortunato uses the figure of an old peasant who moves ceaselessly through the landscape of oppression, eventually becoming the mute witness of the tragic aftermath of a 1960 massacre of Agostinho Neto's supporters.

The concern with a people's history and with using cinema to demonstrate truths about the past to a knowledgeable but unsophisticated audience is shown too in Rui Duarte's *Nelisita.* This takes the form of a fable dealing with the last two families left alive in a time of drought. One man discovers a fantastic store of food in a warehouse guarded by mysterious "ghosts" (easily recognisable by their dark glasses). Eventually he and his companion are captured and their families turned into "ghosts" as well, though the oppressors' greed causes them to leave alive one of the women who is pregnant (so they can capture her son too). But the boy, Nelisita, grows instantly to manhood, survives all the tests and trials to which he is submitted, and turns the tables on the families' tormentors. This simple fable—again presented without the need for elaborate props or sophisticated acting—is commented on at intervals by the story-tellers, depicted as sitting among the players and representatives of the audience. Though in itself quite simple, this narrative device points to the subtle rethinking of the relation of film to audience which underlies current Angolan film practice. While one path for African cinema in the 1980s—exemplified by the Moscow-trained Cissé—leads to a greater fluency and mastery of cinema in a largely traditional form, the Angolan work points to a quite different potential, that deriving from a rethinking so as to give African cinema a new social role and relation to its audience.

Notes

1. This research has been made possible by grants from the British Academy and the Leverhulme Trust, to whom I extend my thanks.

2. John Iliffe, *The Emergence of African Capitalism,* London, Macmillan, 1983, pp. 64–65.

3. Ibid, p. 65.

4. Ibid, p. 67.

5. See the series, *Cinémas d' Afrique noire,* edited by Victor Bachy and published by OCIC in collaboration with L'Harmatt in Paris (five volumes to the end of 1984).

6. Jean-René Debrix, interview, in Guy Hennebelle and Catherine Ruelle (eds), *Cinéastes d'Afrique noire,* Paris, *CinémAction,* no 111 and *L'Afrique Littéraire et Aristique,* no. 49, 1978, p. 153.

7. The text of the Niamey manifesto is reprinted in Guy Hennebelle (ed), *Cinémas noirs d'Afrique,* Paris, *CinémAction,* no. 26, 1983, pp. 168–172.

8. Férid Boughedir, "De l'Idéal à la Pratique: L'Evolution des Stratégies pour la Viabilité des Cinémas Nationaux en Afrique de 1967 à 1984," conference paper presented at the Journées Cinématographiques de Carthage, 1984.

9. Victor Bachy, "Panoramique sur les Cinémas Sud-Sahariens," in Guy Hennebelle (1983), op. cit., p. 31.

Toward an African Cinema

N. Frank Ukadike

Chief Eddie Ugbomah has emerged as the most prominent independent filmmaker in Nigeria. His rise is not merely due to the fact that he has made more films than his counterparts; Ugbomah has the proven ability to survive the test of Nigeria's austere economic times. His career spans three problematic decades of Nigeria's history: the buoyant "boom-years" of the 1970s; the economic stagnation of the 1980s; and the entrenchment of poverty in the 1990s.

Where many of his contemporaries have failed, Ugbomah and his production company, Edifosa Films, have produced more than thirteen films without turning to outside funding agencies. He envisions a thriving African film industry that requires no assistance from external sources. This insistence, as well as his demotic flair and socially committed art, makes him a uniquely compelling antiestablishment figure in African film. It has also put him on a collision course with Nigeria's political hierarchy, his fellow filmmakers, and African filmmakers in general. Ugbomah is Nigeria's most outspoken and most controversial filmmaker.

N. Frank Ukadike: You have made more feature films than any other black African filmmaker. How do you overcome the financial problems that plague African filmmaking?

Chief Eddie Ugbomah: The most important thing is that I have relied on my personal finances, money I made while in England and when I

N. Frank Ukadike, "Toward an African Cinema," *Transition,* no. 63, 1994, pp. 150–63. Reprinted by permission of the publisher.

came back home. I invest all the earnings from my films in new projects. In this way I have been able to overcome the major crises and financial headaches of making films in Nigeria.

NFU: How is your situation different from less prolific African filmmakers, who complain that their films do not yield enough earnings to invest in new projects?

EU: I am able to minimize my costs because I've been trained in most aspects of film production. I am one of those people who participates in the production, acting, directing, and writing of his or her own films. This strategy allows me to cut four people from my budget. Secondly, unlike most Nigerian filmmakers today, I do not have a huge family of sixteen wives and a hundred children, so that is great savings. Additionally, my non-ostentatious lifestyle allows me to reinvest my profits into my next project.

Getting funding from entrepreneurs or the government is quite difficult. I would rather reinvest my money than waste my time lobbying, begging for funding to make another film. We do not have entrepreneurs in Nigeria willing to invest in the film industry. There are no foundations or banks that fund film production.

NFU: What are you trying to accomplish as a filmmaker?

EU: I am a committed filmmaker. I know the power of communication, the harm foreign films have done to our society. I see how the various African cultures are gradually disappearing before the onslaught of foreign influences. What I am interested in is film and its impact on society, in the context of enormous, visible changes taking place in our society. I produce films for television and the cinema; sometimes I support stage productions. Most of my films deal with current events and are statement films about societal issues.

I also believe that there is no reason that the Nigerian film industry should not be lucrative. There is a large audience demand for entertainment. We do, however, have one big disadvantage. Foreign films have ruined the audience's appetite for African films. Africans have become addicted to violence, to what they have come to expect from Hollywood and other escapist, alien films. Furthermore, most academicians who call themselves filmmakers wait until they can get foundation funding to make art documentaries, not commercial features. I think it is important in Nigeria today that we use film to entertain people while informing them. For the Nigerian film industry to survive, our films must make money.

NFU: How do you manage the responsibilities of your many roles—producer, director, editor, exhibitor . . .

EU: I do what is necessary to survive. I have to live with smaller budgets, so I assume those duties for myself. It is not an enviable position—it's too stressful and it affects the end product. But if you are committed to a profession, you just have to do what is necessary to survive. Since there is a shortage of skilled technicians, you find you have to get involved in all aspects of production in order to ensure that things get done the way you want them to.

NFU: Many francophone filmmakers receive technical aid from the French Ministry of Cooperation. Can you comment on this, as well as on other financial aid that is provided to African filmmakers?

EU: Francophone filmmaking is not a commercial enterprise. The filmmakers are subsidized either by their own governments or by France; many of the experimental or art films they produce are not released. The difference between anglophone and francophone filmmakers is that the latter do not spend their own money to make films. Even before a francophone film is completed it has already made money from the foundation that is funding it. For me, I invest my hard-earned money. I deny myself a Rolls Royce and Mercedes Benz in order to make films.

In Nigeria and Ghana we make commercial films. Nigeria has a massive market—it is a country of about 100 million people. In order to make a profit, a Nigerian filmmaker has to capture only a fraction of that population, say one million people paying ten naira each to see a film. We do not endorse extravagant filmmaking, we do not have to spend $20 million to make a film in Nigeria. We can afford to be independent.

At the same time we do not want to compete with Hollywood filmmakers who can spend $15 million on a film and are covered by insurance, so that if it flops the filmmaker can borrow money again to make yet another film. We have no such privilege. Nigerian filmmakers cannot afford to produce movies like *Terminator, Basic Instinct, Deep Cover,* or *Lethal Weapon.* These films are very bad for Third World societies who want to preserve their own cultures.

In October of 1991 we proposed a law to the Ministry of Justice which would have required theaters to show African films for at least four days a week, and imported films only three days. It shouldn't be impossible to run African films four days of the week. But sadly, ever since independence in 1960, governmental policy has hindered the development of a viable national film industry. For example, the government uses the Federal Film Unit exclusively to produce self-aggrandizing documentary films that nobody sees.

Nigerian filmmakers are free agents. We do not have to clear our scripts with foreign sponsors or anybody before we shoot a film. Although, since we invest our own money in our films, we do consider the criteria of the Nigerian censorship board.

This brings us to another problem. Where francophone filmmakers are subsidized by foundations and governments, anglophone filmmakers are subject to heavy taxation. This is killing us. The Nigerian government is not investing this money into independent film production. Furthermore, every state has the power to determine its own taxes on admission fees, what they call "entertainment taxes." We have been fighting this outrageous system of taxation to no avail. So filmmaking in the francophone world does not involve the same personal risk that it does in the anglophone, where a filmmaker like Kwaw Ansah had to borrow 22 million cedis from about four syndicated banks to make a film.

Only after intensive lobbying did the Nigerian government agree to build a film industry, the Nigerian Film Corporation (NFC). They have now built a color laboratory, but who is using this 88-million naira lab now? Nobody—because the government has not been able to convince investors that film is profitable. The NFC ought to be run as an economically viable enterprise, which means appointing a reliable management staff, among other things. The government should also establish a commercial bank that would lend money to filmmakers.

NFU: African filmmakers have been fighting for a long time to get African governments to subsidize or sponsor film production. But how can that goal be achieved when African governments have mortgaged African economies to the International Monetary Fund and the World Bank; when many African countries are struggling to feed their people? When governments are concerned with feeding their populations, do you think African governments are going to commit their resources to film production?

EU: This is where people make a mistake. When there is frustration, depression, that is when entertainment pays the most. This is when you have to use entertainment to cool down people's tempers. Now is the most promising time to invest in the film industry. In Nigeria we used to charge a three-naira or five-naira cover fee, but because of the depression, we now charge fifteen naira. And yet the theaters are overcrowded. In a country of nearly 100 million, there are not even two million video machines. The masses still want to see movies, they want to go out. The government has to invest in communication and entertainment now, so it can communicate and address the people's concerns. The media has enormous educational and propaganda potential. The government can use the media to tell the people to buy Nigerian, to be proud of Nigeria. It can communicate that it is cheaper to grow their own rice rather than importing it. Through movies, we can start eliminating those canker worms, tribalism and religion, as I did in my film, *Apalara.* Those are the kinds of things you can achieve through the power of the image. Economic depression should not be an obstacle to the development of the film industry.

NFU: Despite the fact that you have made so many films, you are not well known outside of West Africa. Why is that?

EU: People don't realize that the film industry is a powerful medium protected by governments in Europe and America. It is big money. The industry will do everything it can to frustrate a Third World filmmaker. They condemn films that do not conform to Hollywood formulas. So why should I waste my time? There is an enormous African market to explore. Why should I pay 35,000 naira to fly to America to campaign for the distribution of my films only to discover it is a futile endeavor? I have made thirteen movies and I am happy about it, and I am not worried about whether they are distributed outside Africa. In America, even if I do find a theater to screen my film, the owner expects to keep most of the proceeds from the screening. They show my films and refuse to charge admission. At Howard University I had to pay to show my film.

NFU: Have you considered showing your films at Pan-African Film and Television Festival of Ouagadougou (FESPACO)? African films are discovered by distributors from all over the world who go there to view African films. When I was in FESPACO in 1991, only one filmmaker from the anglophone countries, Kwaw Ansah, came to show his film *Heritage . . . Africa,* which won an award. Wouldn't FESPACO be a good place to show your films?

EU: FESPACO is a francophone business and for the past six years only francophone films have won the three major awards. The year Kwaw Ansah won, the anglophone filmmakers had met before the award and agreed that if anglophone films were not duly recognized we would not return to FESPACO. Yes, *Heritage . . . Africa* was awarded the grand prize, which it deserved. But anglophone filmmakers go to FESPACO merely to promote Pan-Africanism, not to advertise or sell their films. Some anglophones do go—Ade Love, Babasala, and Ogunde have all shown their films at FESPACO. But for me to go and sit in FESPACO, where only French is spoken twenty-four hours a day—it's too frustrating. I do not know of one film that managed to be distributed internationally after having won the coveted FESPACO grand prize.

NFU: There are African films, such as *Yaaba*, that have received much international attention.

EU: They may be screened internationally, at the New York Film Festival, the San Francisco Film Festival, or in Montreal where audiences applaud them, but after the festivals, do they get international distribution? What is crucial at this point in the development of African cinema is for Africa to create a united front to confront the problem of distribution. Individual effort simply is not enough. We must seek a united effort to promote African films outside of Africa. The Nigerian government should engage in publicity or propaganda or whatever. Judging from the few places that I have shown my films, there is a demand, people are inquisitive, they want to see Africa from a new perspective. The days of Tarzan are gone.

NFU: You have emphasized film and entertainment. How do you integrate film and politics?

EU: This is a very delicate issue. Filmmakers have to be very subtle about bringing their political perspective into their work. It is imperative that they do not push it down people's throats. Look at the popular Nigerian television programs like "The Village Headmaster" or "The Masquerade" which use comedy and satire to make political statements. These programs are entertaining, but while people look at them and laugh, they also understand the political message. My film *Oil Doom* (1988) applied the same strategy to warn Nigeria that the oil boom would one day end in doom. Film is a powerful political tool. It is very powerful because you really do not have to receive any special education in order to understand the image and its political implications. There are a lot of filmmakers now in Nigeria who use film as a medium for political criticism. So also are playwrights who have become filmmakers: Baba Ogunde, our famous playwright, dramatist, and filmmaker, uses his plays, some of them adapted to film, as political tools. You really cannot separate film from politics. During the Nigerian-Biafra war, Ojukwu effectively employed the media as a weapon of propaganda, constantly drawing the world's attention to the genocide against his people. So powerful was his manipulation of the media that images of rotting bodies were broadcast all over the world. Filmmakers could be considered harbingers of change. Sometimes they can make and unmake with their stories.

NFU: Could you comment on your first feature, *The Rise and Fall of Dr. Oyenusi,* and its depiction of contemporary urban problems?

EU: *The Rise and Fall of Dr. Oyenusi* was made when armed robbery was just beginning to flourish in Lagos. It was a film I made to show that crime does not pay and to show that most of these criminals have the support of "godfathers," influential individuals with links to the armed forces. I exposed a lot of these political intricacies, all the tricks and maneuvers, and at the same time I entertained people with the film. People have asked why I did not make a film about Zik, Awolowo, or Balewa. I did not make a film about them because these individuals are what I call "Biro-armed robbers." The fact that these figures do not carry guns and shoot people does not mean that their crimes are any less destructive than those of the armed robbers in my film. This film is a reflection of national consciousness, the dichotomy between the oppressed and the oppressor. That is why I made a film about Oyenusi's life and how he was living in glamour at the expense of his victims. But what was his end? Was it worth it? That was the question I wanted to ask, and the answer is that crime does not pay.

NFU: You exhibited courage and audacity in the production of *The Rise and Fall of Dr. Oyenusi.* What are the trials, dangers, and obstacles involved in making such a gangster movie in Nigeria?

EU: The problems I had were not only due to the fact that it is a gangster movie. I have gone through hell making all of my films. Making *The Rise and Fall of Dr. Oyenusi* was no harder than making *The Death of a Black President,* or *The Mask,* or even *The Boy is Good.* I cannot tell how many times my life has been threatened for exposing societal ills. There is even a script, *The Jennifer Connection,* for which I nearly got killed. We were going to shoot this film about a young girl and her connection with drug barons. As soon as it was announced that I was looking for five million naira to make this film, I was targeted by the drug lords. But this time I was not willing to gamble with my life because I had had enough of people's threats. I had to jettison the whole idea. This predicament has destroyed all to many African film pioneers. A lot of trained Nigerian academics and technicians who would have made important contributions to Nigerian cinema have been forced to abandon filmmaking. Also, corruption has taken its toll. Francis Oladele, who made *Things Fall Apart,* was loaned four million naira to make a film called *The Eye of the Man.* Instead of making the film, Oladele sank the money into a farm. This misuse of funds has set a bad precedent, causing the government to abandon about twelve such loans that were earmarked for filmmaking.

I myself have since changed course and adopted the Yoruba film tradition. Since this tradition is oriented towards cultural plays and dramas and remains largely apolitical, a filmmaker is spared the danger of making enemies. The Yoruba theater/film genre does not satisfy my creative impulse, but I have been making money from it. My last four Yoruba language films have been big box office hits. It is very frustrating that the quest for survival has forced me to relinquish my original style of filmmaking.

NFU: Nigerian cinema seems to be a derivative of the Yoruba traveling theater, what might be called "theater on the screen" . . .

EU: What the Yorubas have been able to do through their mentor, Ogunde, was to just lift the stage onto celluloid. Because this tradition originated in the popular Yoruba theater there is a ready-made audience. There are ready-made stars and ready-made stories. Since cinematic aesthetics are secondary to the popularity of the actors, the filmmaker does not need to advertise much. The only people with cinema, stage, and art culture are the Yorubas. Unfortunately, mediocre dramatists will now stage anything, film it, and take it to the hinterlands to show an unquestioning Yoruba audience. The industry is still alive not because of the production quality, but because of the huge, enthusiastic audience.

Still, this process should not simply be condemned. Producers are bound to make mistakes. But some of them will make innovations that will map new territories for African cinema. We cannot wait for Oliver Stone to come and shoot an epic in Lagos. We have to continue producing to sustain the audience's interest and to keep the film industry growing, however gradually.

NFU: Do you see the Yoruba film/theater genre reaching international film markets and functioning as a vehicle for the exportation of Nigeria's cultural heritage? Do you see Yoruba films competing with other well-made films?

EU: Absolutely. The world today is waiting for African films. I would argue that audiences throughout the world are tired of the *Terminator, Lethal Weapon* genre. If we had the means to launch extravagant productions, Africa would take Hollywood by surprise. That means taking Hollywood technology to Africa, shooting films relevant to Africa. Not jokes like *Allan Quartermain, King Solomon's Mines,* or *Out of Africa,* to name a few. We don't need more jungle-melodrama, where zebras and tigers run up and down on screen. The kinds of films that would have a meaningful impact are what we are interested in—epics depicting, for example, the Benin Massacre, the origin of the Oduduwa Dynasty, the true story of the Yoruba people, and many other African stories of epic proportion that remain untold.

NFU: But what about aesthetics? While the Yoruba theater is known for its excellence all over the world, some critics argue that most of these Yoruba films look no different from Hollywood-style jungle melodrama. In the words of one critic, they are not cinematic enough. They are not the sort of artworks which make people proud to say, this is an African aesthetic.

EU: I would argue the opposite. I am proud and happy that Africans are finally appreciating African films. Most Africans do not appreciate the African aesthetic. You never see an African dip his hands in his pocket and buy a painting for $2,000. The filmmakers are lucky that the ordinary citizen appreciates films and comes to watch them. There are many trained filmmakers who can make films better than Hollywood directors, but they do not have Hollywood money. I keep saying it: money and technology. If I had Hollywood money I'd shoot a film that would go to Hollywood or to the Cannes Film Festival and win. Hollywood films have neither substance nor story line, they have only flying helicopters and missiles exploding. I can fill this vacuum with African culture and captivate audiences.

NFU: *The Mask* is an important film. What statement were you trying to make and to whom is the statement directed?

EU: *The Mask* was made primarily as a way of getting even with Britain. In 1976–77, Nigeria hosted the World Black Festival of Arts and Culture (FESTAC). We wanted to use the famous Bini mask as the emblem of FESTAC. But the mask was in the possession of the British. They claimed it was their own, forgetting that the mask had been stolen from Benin City by the British. The most outrageous statement made by the British authorities was that the mask would not be safe in Nigeria, and that it was also too fragile to be returned to Nigeria. Even when the Nigerian government offered Britain 10 million pounds to buy back the mask, they still refused to let us have it. So in the face of this British intransigence, I made a film which suggested that if the Nigerian government cannot legally retrieve the mask, they should take it back the way it was taken by the British. The implication was that 300 billion pounds worth of the British Museum would be destroyed in the process; they did not take kindly to that. The British summoned the Nigerian ambassador to discuss the issue. I am very proud to this day that I made that film because it did send shock waves through the British government. The British media was unsympathetic, condemning *The Mask* as a rough film. They never thought it was possible for a Nigerian to be bold enough to shoot a film inside the British Museum, let alone indicate how the mask could be retrieved. Philosophically, *The Mask* devastated the mind of the British critic. Any time Africans fight for their rights the West automatically claims they are communists, Marxists, revolutionaries, etc. But direct and sometimes forceful confrontation is the only language the British, still in their colonial mentality, can understand.

NFU: When *The Mask* was released, some critics thought that it was rough. But if you look at the history of Third World cinema, Latin American cinema, it is not necessarily the film's "aesthetic" quality that matters, but rather the message and its historical and cultural specificity.

EU: That is exactly my policy, I create my films in the service of my people. I want to see the result of this "rough" film. I want to know if people have listened to the message, if things have changed or if I have educated my people. But I have to tread a fine line because films which are too provocative risk being banned.

NFU: Perhaps the most controversial of your films is still *The Death of a Black President.* The title alone is enough to galvanize or to provoke even the most quiescent observer. Can you talk about the inspiration, what you hoped to accomplish, the reflection on society?

EU: *The Death of a Black President,* like most of my films, was inspired by an actual event. This is a film that focuses on our former President, Murtala Muhammed who, unfortunately, did not live long enough for Nigerians to know who he really was. During the two hundred days he lived as president, Nigeria changed. People began to feel they belonged to the society, irrespective of hierarchical divisions. This sense of belonging connoted a new awareness of their rights as human beings. But this euphoria dissipated abruptly with Murtala Muhammed's assassination. He was killed because he was of that rare breed, a dynamic Third

World leader who Western powers feared would free his country from dependency. His policies were considered anti-British, anti-Western. A stooge, Colonel Dimka, his own friend killed him. This has become a pattern marking the downfall of African leaders. This was how Patrice Lumumba, Kwame Nkrumah, Thomas Sankara, and many more were killed. These were African leaders who took a stand to defend African interests.

Murtala Muhammed openly supported liberation movements in Angola, Mozambique, and Zimbabwe. We all remember, for example, when Britain was not fully supportive of Zimbabwean independence, how Murtala nationalized two British companies in Nigeria, British Petroleum and ESSO, as a signal that favoritism toward the white minority population in Zimbabwe would not be tolerated. This bold move determined the outcome of the Lancaster Talks in London which led to the independence of Zimbabwe. His dynamism was unprecedented and this is exactly what Western imperial governments felt compelled to extinguish. The assassination of African leaders who have refused to be puppets of the West has been orchestrated through blackmail, bribery, and collaboration with the leaders' close, and sometimes most trusted, associates. The film immortalizes this great leader, Murtala Muhammed, for his works, his failures, and his achievements—his place in history. He was almost like a dictator, yes, a dictator, and today Africa needs a dictator, not necessarily Western-style democracy. We need a leader who prioritizes the interests of the people as opposed to exploiting the people under the guise of democracy. Africa needs leaders, not rulers. To this I will add that General Idiagbon, Muhammed's Chief of Staff, his right-hand man who embraced his agenda will remain as one of Nigeria's best leaders, but we did not give him a chance to lead.

NFU: You played a leading role in *The Death of a Black President,* as in many of your films. What motivated you to play General Odongo, which is really a fictional character of General Gowon?

EU: Well, first off, we look alike. I am trained as an actor. And again, most Nigerians did not want to partake in the film. I had to bring in an all-foreign crew and about six Hollywood actors. It was only at the last minute that Nigerians started coming out of their shells to act. And as you can see, the characterization was very good. I did my best in that movie. And I thank God that it was a big box office success in Nigeria.

NFU: In fact, this film was well acted. Rudy Walker and many others gave great performances. Could you comment on the problem of finding professional actors and how the use of nonprofessional actors affects the development of African cinema?

EU: The biggest headache about professional acting is that neither the entertainment nor the art industry in Africa is lucrative. Although almost twenty-six universities in Nigeria have departments of theater arts, few students go into acting because there is no future for them. The film industry is not organized to ensure regular and continuous employment and there are no good television stations.

The people in theater groups have monopolized acting. They have been in the business for a long time, so moving from the stage to film was an easy step for them. They discovered that their presentations could reach a much wider audience more efficiently on film than on stage. Furthermore, it is easier for them to carry the stage play via celluloid to the hinterlands than for the people in the villages to go to the cities to see the stage performances, which they cannot afford anyway. Stage performances used to be very popular in the Yoruba land, but film has killed the stage in Nigeria. Even the Yoruba popular theater, which was once excellent, has lowered its standards because of the impact of film.

NFU: Until recently, you were the chairman of the Nigerian Film Corporation. Before your tenure, that corporation had squandered millions of naira without producing a significant film. What did you accomplish there?

EU: As I have said before, our people do not appreciate what is available to them. Filmmakers often go abroad seeking post-production services when they could easily complete their films in Jos. During my tenure at the NFC, the film laboratory was established, and since then four documentaries have been produced here, including *Ruwanbagaja* by Alhaji Ramalan Nuhu, a former Kano state television producer. The last one was shot to promote the First Lady's (Miriam Babangida) program, *Better Life for Rural Women.* This film went on to win an award in China for idea and content. Also during this period, the corporation was able to loan over three and a half million naira to filmmakers. Furthermore, my film *Toriade* was processed and edited there, the sound track was laid there, and it was only sent to America for the release print to be made. Now the NFC is also sponsoring a 35mm film being made at Ahmadou Bello University called *Kulba na Barna.* The NFC has also contributed to the making of about ten documentaries for state governments, the most well known of which is probably *Women in Development.*

NFU: Let me refer back to 1983 when you said, "As far as I am concerned, the Nigerian Film Corporation looks like another of those government parastatals established to fool the nation." Do you still subscribe to that view as ex-chairman of that corporation?

EU: Yes, I still do. The corporation functioned as a viable entity only when I ran it. Being professionals, we were not happy with how the corporation was managed so we decided to make changes. But some people were adamantly opposed to those changes. As soon as the Minister of Information and Culture, Toni Momoh, who supported the changes I was making, was replaced by Alex Akinyele, the board was quickly dissolved. There was no justification for doing this except for the fact that I did not give them the opportunity to embezzle the corporation's funds as they had expected me to do. And since I left, the corporation's activities have come to a standstill because there is no one there to provide effective leadership to the organization. Some of the NFC staff have put in twenty-eight years of service and they are waiting for their pensions; productivity is not a matter of concern for them so long as their salaries are guaranteed. There was space, there was money, and as chair of the organization, I enjoyed the collaboration of other professionals (there were about three filmmakers on the board). Furthermore, the president gave me a free hand as a professional to do what I thought was best for the corporation. He said to me, "Go and do this thing. You talk too much. Let's see you do it." So I delivered. But unfortunately, you cannot just walk up to the president every day and complain. The civil servants are afraid to talk, afraid to act, and to give constructive criticism. Every month their salary is guaranteed, so why should they make noise? The place has come to a total standstill. I think most industries should be privatized, including the NFC. The laboratory has been stagnant since I left. The lab is supposed to function, it is there to be used.

NFU: You are always at odds with film critics. Some of them have compared your films with the James Bond 007 series pointing out that your films replicate the Hollywood formula of sex and violence. They think you have not been able to generate what might be called an African aesthetics.

EU: When you are going to war, you don't use a sling shot, you use guns. When the British stole the mask from Africa, they shot our people. We are trying to recreate historical events and we are living in a very violent society. Furthermore, the Nigerian film goers are used to seeing Hollywood films, Kung Fu films, Italian spaghetti Westerns, and Indian melodramatic films, which are inundated with sex and violence. When it comes to the question of making an independent film, one has to treat it as a commercial venture because it is usually funded with borrowed money that has to be paid back whether you like it or not. This is a fact any filmmaker has to constantly contend with. And this is why, while you are being patriotic and nationalistic, you cannot avoid being commercially minded.

I wonder if any critic could tell me what this thing that we call African cinema really is anyway. We certainly do not have one yet. There is not one unified film style that distinguishes African cinema. We have to create an African film style before we can talk about African cinema. At this stage, we have a cinema of diverse states.

NFU: That is exactly where African critics fit in. They are trying to look at African films and then talk with the filmmakers to jointly formulate a discourse on African aesthetics.
EU: I would like to emphasize one last thing, that it is an achievement and a record for one African man to have single-handedly made thirteen feature films with hard earned money. It is unfair for critics to sit in their ivory towers and run filmmakers down without considering the situation of the filmmaker. Until the critic comes down to earth from that ivory tower of his and talks to the man in the field, the man on the battlefront, he will never be able to digest African cinema. Sometimes I wonder if the critics are not envious of the filmmaker. I think the role of the critic in the development of the cinema should be to give constructive, contributive advice, instead of picking up his Biro and writing trash. That trash does not move me. I can look into that critic's face anytime and say, go to hell with your writing, I am making more films. If I die tomorrow, people will remember Chief Ugbomah as a filmmaker who made such-and-such films, as the former chairman of the NFC and secretary general of the Nigerian Film Society. Those are good achievements and I am very grateful to God. Critics should take the trouble, as you have done, to develop a dialogue with filmmakers and collaborate in the spirit of Pan-Africanism, if only for the sake of encouraging a critical thought system that would interrogate the diversity of African cinema.
NFU: With the release of your fourth film, *Toriade,* you moved away from your own tradition of filmmaking.
EU: No! No! No! My first films were in English; only later did I start making films in Yoruba. I tried to strike a balance such that my films would appeal to both literate and illiterate audiences. Right now, I'm working on two movies: *Akoba,* a Yoruba film, and *America or Die,* an English film. I serve two audiences. I do it to survive. Luckily for me, the two audiences I am serving are paying. The English films I make sell to my diverse ethnic audience and the Yoruba films I make sell to the Yoruba audience.
NFU: How does this affect your film style?
EU: It kills it. It's pure murder. I'm not free to have an identity. But most of my films, like *The Rise and Fall of Dr. Oyenusi, The Death of a Black President, The Mask, Oil Doom,* and *Bulos 80,* are mainly whodunit films. They are about the problems of the society.

My film *Omiran* was one of the biggest box office successes in Nigeria. I compare myself to Agatha Christie or Alfred Hitchcock. They are the people I try to emulate. When I shot *The Death of a Black President,* many people tried to discourage me from shooting a film about Murtala Muhammed. But I shot the film. I went through a lot of horror, but I shot

the film and today it is one of the biggest successes in Africa. Now ABC TV in America wants to buy the film. To me, this is quite an achievement.
NFU: Do you see oral tradition playing any role in the Africanization of the film language?
EU: American movies have their own identity. British and European countries have their own identities. I think oral tradition is the only true African identity for African movies because the oral tradition is a critical element in our pattern of communication. It seems that the only identity we have now, to which we should hold tight, is the interpretation of our folklore. African film must have its own identity, and its identity is located in the oral narrative technique just as America's lies in the violent, *Star Wars, Terminator* type.
NFU: In other words, is there a possibility for an African film language?
EU: Yes, a very, very big one. It might sound stupid, but it is really true, the world is waiting for African cinema. Someday Africans must deliver. We must redefine an African image that is not Tarzanistic. We have not delivered. African cinema owes the world an answer.

Portuguese African Cinema

Historical and Contemporary Perspectives, 1969 to 1993

Claire Andrade-Watkins

The thirty year history of film production in sub-Saharan Africa, and of films by African filmmakers in general, must be considered in the context of an acute shortage of technical and financial resources, as well as a lack of viable circuits of distribution and exhibition. These difficulties, in turn, have been compounded by colonial and post-colonial traditions and policies regarding cinema: first, cinema targeted for Africans during the colonial period, where it existed, was integrally linked to administrative, military, religious or educational objectives; second post-colonial—either European or African—film policies, and filmmakers' initiatives aimed at ameliorating colonial conditions have been unable to generate an economically viable and stable film industry in the region.

A major stumbling block is the bottleneck created by European and American conglomerates who own and operate the lucrative distribution mechanisms for cinema throughout Africa. For these companies, the continent is merely a commercial market, a dumping ground for foreign films of dubious merit. The continued lack of control by Africans and their governments over the distribution process means that revenues are being drained from the continent, rather than redirected to building and supporting cinema production and its related industries.[1]

Given this scenario, the context of film production and distribution, specifically, the manner in which financial and technical structures of

Claire Andrade-Watkins, in *Research in African Literatures,* vol. 26, no. 3, 1995. Reprinted by permission of Indiana University Press.The essay in this book is a slightly revised version of the original journal article.

film production have a major impact on the ideological perspective, form, content and purpose of cinema in post-colonial Africa, assumes added significance in the history of cinema in Africa. Historically, the dominance of Francophone Africa, in film production was due in large measure to France's two-pronged post-colonial film policy. Financial and technical assistance was provided to African governments for production of newsreels and documentaries, and to aspiring African filmmakers to explore and expand their "cultural expression in film."[2] Technical facilities, however, were not created within Africa, perpetrating the need for African governments and filmmakers to go to Paris to complete the film production process.

Mozambique, on the other hand, insisted on merging ideology with form, content and context, pioneering a successful model of "guerrilla" cinema that embraced a Marxist conception of the engagement between film and society. More importantly, the film industry was nationalized, so that infrastructures of production, distribution and exhibition were created and supported by a government that viewed cinema as a vital force in post-colonial development and education.

This scenario in Mozambique in particular and lusophone Africa in general was fueled by a united, revolutionary desire for independence from 500 years of harsh Portuguese colonial rule. This created an unprecedented hegemony among the farflung members of the Portuguese African empire—Angola, Cape Verde, Guinea Bissau, São Tome and Principe. This unity was reinforced by the leadership of Agostinho Neto of Angola, Eduardo Mondlane followed by Samora Machel in Mozambique and Amilcar Cabral of Cape Verde and Guinea Bissau. The intellectual, political and ideological leadership that challenged, fought and overturned colonial rule, was provided by this vanguard of senior African statesmen and revolutionaries spearheading independence movements throughout Africa.

Conceived ideologically and thematically in the spirit of the liberation struggle against the Portuguese during the 1960s and the 1970s, films from the region—particularly Mozambique and Angola—comprise an important chapter in the history of African cinema in general and the genre of "guerrilla," or liberation cinema in particular.[3]

For the purposes of this study, the period of liberation cinema begins in 1969 with "Towards a Third Cinema," a pivotal manifesto written by Argentinean filmmakers Solanas and Gettino. The subsequent development of "guerrilla" or liberation cinema in Mozambique during the 1960s and 1970s, is examined in the context of film production in sub-Saharan Africa, with particular reference to francophone Africa. The study concludes with the transition in Mozambique in the late 1980s from state to private sector production, international co-productions and financing. The discussion of change in Mozambique and Portuguese-speaking Africa is placed in the wider context of film production in sub-Saharan Africa. The historic and contemporary participation of Cape Verde, Angola and Guinea Bissau is also highlighted.

Portuguese African Cinema: Part I The Origins and Development of Liberation Cinema, 1969–1978

The decade of the 1960s witnessed an explosion in cinema that cut a swath through the "third world." Fueled by Argentinean filmmakers Octavio Gettino and Fernando Solanas's pivotal 1969 manifesto, "Towards a Third Cinema," and leaders of African independence movements, waves of revolutionary ideology swept across Latin America and Africa, leaving in wake a cinema that confronted dominant historical, colonial, cultural and ideological norms in society and cinema. Latin America, Cuba, North Africa, French-speaking, and Portuguese-speaking Africa—especially Mozambique—became major centers for the theoretical and practical development of cinema.

Only Ghana in English-speaking Africa and Guinea Conakry in French-speaking Africa came close to the production potential of post-independent Mozambique. The government of Guinea Conakry shared Mozambique's commitment to a functional and educational "third cinema," going as far as nationalizing part of the distribution and exhibition film sectors in the country.[4] However, the British had no interest in the post-colonial development of cinema, although a colonial legacy of documentary traditions in English-speaking Africa is visible in the strong television and government networks in Ghana and Nigeria.[5] While Ghana inherited full 16mm and 35mm capabilities from John Grierson's Colonial Film Unit, the government unfortunately neither espoused the ideological significance of cinema nor entertained the vision of Ghana becoming a regional center for production.[6]

In contrast, vestiges of colonial cinema were extremely faint in Portuguese-speaking Africa.[7] In the decade preceding independence from the Portuguese—1974 for Guinea Bissau and 1975 for the other colonies—film production was galvanized by two revolutionary forces. The first was an internal, newly awakened sense of unity, purpose and collaboration among the colonies, and the second was external support from the international community for the revolutionary war efforts and governments—in Mozambique, FRELIMO (Frente de Libertação de Mocambique, 1962); Angola, MPLA (Movimento Popular de Libertação de Angola, 1965); and in Cape Verde and Guinea Bissau, the PAIGC (Partido Africano Pela Independencia de Guinè e Cabo Verde, 1956).[8]

Both forces—the African liberation movements and their foreign supporters—viewed cinema as a powerful force in the liberation struggle and a vital component in the documentation, education and dissemination of information about the war. Consequently, films produced both informed the international community of the armed struggle against the Portuguese and contributed internal information, educational and cultural programming for the African populations.

The revolutionary governments of Mozambique and Angola, the most active centers for film production, supported landmark films made by pioneering filmmakers in the nascent African cinema. Sarah Maldoror was a major contributor to both the cinema emerging from within the lusophone region, and the revolutionary cinema of the era. A Guadeloupean by birth who was trained in Moscow, Maldoror was a long time supporter of the independence struggle. *Sambizanga,* her first feature length film made in 1972, was also the first and only fiction film devoted to the liberation struggle in Angola.[9]

The story was based on Angolan novelist Luandino Viera's, "The Real Life of Domingos Xavier," and adapted to screen by Mario de Andrade. Set in the 1960s during the war for independence from the Portuguese, the story follows Maria in her search through the prisons of the capital for her husband, an organizer for the MPLA independence movement.[10] The film celebrates the comrade and his wife's sacrifice and loss while exhorting supporters to continue the struggle. The last line of dialogue is a call to arms for February 4, 1961, the day hundreds of Africans attacked the police and military in Luanda and launched the armed struggle in Angola. This ending clearly reflects the political tone, theme and focus of films made during this period.

Prior to filming *Sambizanga* Maldoror directed a short film of 18 minutes, entitled *Monangambee.* Filmed in 1970 and financed by the Comité de Coordination des Organisations nationalistes des Colonies Portugaises, *Monangambee* illustrates the total lack of understanding between Portuguese and Africans through a dramatic confrontation between an African prisoner, whose comment about a national dish made during a visit by his wife, is totally misconstrued by the Portuguese officer, who orders the prisoner beaten. While not renowned for its cinematic quality, this production reaffirms the themes of revolutionary struggle.

Films made by a distinguished group of international filmmakers and activists from countries as diverse as France, Sweden, Yugoslavia and Cuba comprise the second, congruent movement of revolutionary cinema. Efforts from the United States were spearheaded by Afro-American Robert Van Lierop, a lawyer turned filmmaker. These films chronicled the struggle against Portuguese domination, and were extensively shown abroad, resulting in a ground swell of international support for the liberation struggle.[11]

Neither Cape Verde nor Guinea Bissau were engaged in production. Rather, growing political and revolutionary stirrings against Portuguese colonialism hovered over the Cape Verde islands, a tiny archipelago of drought-stricken islands lying 200 nautical miles off the coast of Senegal. A modest yet significant intellectual movement stimulated by the cinematic and revolutionary ideals of the period, began as early as the late 1950s.

Cape Verdean intellectuals studying at the lycée on the island of São Vicente were profoundly impacted by the fervor radiating from the intellectuals and young revolutionary leaders studying in Lisbon, the Negritude movement, the writings of its leading proponents, Léopold Senghor of Senegal, and Aimé Cesaire of Martinique, W. E. B. DuBois, PanAfricanism and the writings of Afro-American literary giants such as Richard Wright, author of *Native Son.* Imbued with the spirit of African nationalism through books clandestinely brought into Cape Verde from Senegal, the young intellectuals of the 1950s were closely attuned to the activities on mainland Africa and throughout the diaspora. Encouraged by the emergence of leftist protest in Lisbon against the fascist government of Antonio Oliveira Salazar, a small group of Cape Verdean intellectuals, following the example of the ciné-clubs in Lisbon as forums for intellectual dialogue, debate, and artistic exhibitions, formed a ciné-club in the capital city of Praia, on the island of São Tiago.[12]

The first meeting of the ciné-club was held in 1960 at the ciné-teatro Municipal da Praia. Open to the public, this cultural program featured Cape Verdean poetry. Plans were made to develop and continue a range of cultural programming, cinema, music and poetry. A list of possible films was proposed for future exhibitions although the only concrete activity in cinema was a regular radio commentary presented by the president of the ciné-club on social and cultural issues raised in commercial films being shown at the theatres.[13] However, by April, 1961 the PIDE (Polìcia Internacional de Defesa do Estado), the Portuguese political police, fearing collusion or support for anti-colonial revolutionary movements, brought the activities of the fledgling ciné-club to a halt with the arrest and imprisonment of two leaders of the association, A. F. Correia e Silva and Alcides Barros, and the deportation of others from São Tiago.[14] No further efforts were made to revitalize the ciné-club until independence.

Immediately following the coup in Lisbon, Portugal, which overthrew the fascist government of Antonio de Oliveria Salazar, colonial rule ended for Guinea Bissau with a defiant, unilateral declaration of independence in 1974. Angola, Mozambique, Cape Verde, São Tome and Principe followed suit in 1975. The pinnacle of cinematic endeavor in lusophone Africa began at independence with the creation of the Institute of Cinema in Mozambique. The architect of that reality was a legend in the history of world cinema, Ruy Guerra. A Mozambican by birth, Guerra was the leading figure in Brazil's Cinema Novo movement. His return to Mozambique after independence to head the Institute of Cinema was a major factor in the cultural ascendance of Mozambique in southern Africa, sub-Saharan Africa and the lusophone diaspora.

This early national period of cinematic activity was a time of experimentation in the form, direction and format for cinema and television in Africa: a harbinger of subsequent developments in lusophone Africa. In a rare convergence of ego and talent, progenitors of three major movements—Ruy Guerra, Cinema Novo, Jean Rouch, cinema verite and Jean-Luc Godard of the new wave—converged on Mozambique in 1978. The institute had invited Rouch to explore the possibilities of super 8mm film, and Godard had a contract with the government to do a study about the possibilities for television and video in Mozambique.[15] The critical and even acrimonious interactions between these cinematic giants and the forces that shaped the formal, social, and technical development of Mozambique's cinema illuminate a dilemma intrinsic to African cinema in general, and merits a digression in the narrative of the region.

The experimental phase of Mozambique cinema was preceded by the post-colonial initiative in cinema launched by France. While lacking the revolutionary fervor of Portuguese speaking Africa, the French shared a spirit of adventure and experimentation and were optimistic about the prospects for cinema within Africa. In 1961 France created the Ministry of Cooperation with the express purpose of providing financial and technical assistance to her former colonies. The Bureau of Cinema, created within the Ministry in 1963 under the direction of Jean-René Debrix, provided technical and financial assistance to Africans to foster cultural expression through cinema, allowing Africans from francophone Africa to launch the embryonic movement of sub-Saharan African cinema.

In retrospect, African cinema as envisioned by Debrix through the Bureau of Cinema was flawed, if not doomed, from the outset. First, the administrative and operational procedures imposed by the Ministry on the Bureau were heavy and incompatible for production. Government bureaucracy hampered rapid deployment of resources or technical support. Instead of a lab order taking one call, in a non-governmental production center, technicians at the Bureau had to wait days for the processing of a request for the same service through the Ministry.[16]

The French film professionals looked askance at the Bureau, viewing the whole operation as unprofessional and financially and technically inadequate. The inability of the Ministry to provide effective and appropriate administrative mechanisms for production exacerbated increasing tensions between the filmmakers and the Bureau on one hand and the upper echelons of the Ministry and African governments on the other hand about the films, their content, form and distribution.[17]

Jean Rouch, a pioneering ethnographer and filmmaker and controversial figure in African cinema, clashed both with African filmmakers and Debrix on technique and themes. Rouch and the African fillmmakers had distinctly different philosophies; he dreamed small, i.e. 8mm or 16mm, and the Africans dreamed large, 35mm format, the standard for professional, commercial cinema.[18] Rouch was reproached by the filmmakers for trying to institutionalize a level of technical underdevelopment by advocating the use of the smaller formats.

In Rouch's view however, the 35mm format was neither pragmatic nor cost-efficient and the Africans' emphasis on it amounted to a mystification of technology, where "The tripod was the beginning of a temple, an altar."[19] In Rouch's capacity as the director of research at the Centre National de Recherches Scientifiques in Paris, he was involved in innovative experiments with the super 8mm format, which he applied to the 1978 experimental pilot project in Mozambique, supported in part by France's Ministry of Cooperation.

Rouch believed that super 8mm was an expedient, cost efficient format that would help developing countries catch up to the more technologically advanced countries. Further, Rouch felt that the super 8mm format demystified the process of filmmaking, and was a format that was accessible for use by more people since the cameras and editing equipment were cheaper, lighter and smaller than 16mm or 35mm.[20]

While Rouch argued for super 8mm, Godard, on the other hand, was fascinated with the possibilities of video and television, and the creation of the images for that medium. For both men, Mozambique was in some ways a laboratory, an opportunity to identify or select the tools of production to build and shape a national cinema and television.

Neither Rouch nor Godard's experiments came to fruition. Both were perceived as too costly and were canceled.[21] Although unproductive, nevertheless, these efforts in Mozambique exposed the ideological and theoretical implications of production methods and technological choices. These differing perceptions among and between the Europeans, on the one hand, and the Africans on the other, also reflected a wider struggle in post-colonial Africa between neo-colonial and African initiatives. In this case, the arena was the context of production and the concomitant impact on culture, history and identity in southern Africa in general and Mozambique in particular. While Godard and Rouch were arguing the merits of their vision for Mozambique, Guerra was actually realizing his vision in the documentary, educational and feature film projects launched under his direction at the Institute of Cinema. A viable, prolific, engaged cinema—integrally connected to the issues and realities of achieving military, psychological, educational and cultural independence—established Mozambique within Africa as a national cinema of power and force that effectively and triumphantly created a functional "guerrilla" or liberation cinema.

Portuguese African Cinema: Part II Transition from Revolutionary to Free Market Film Production, 1978–1991

Cinema production in Mozambique after 1976 was ascending, due in large measure to the activity of the Institute of Cinema. Launched in 1975 and officially established in 1976 by the revolutionary government of Samora Machel, the Institute of Cinema was the first cultural institution to be set up after independence. By 1978 an ambitious and sustained vision was created in Mozambique that addressed the cultural, educational and informational needs of a people engaged in armed struggle and a socialist reconstruction of society and government.

Despite the internal battles with the South African-backed opposition forces of RENAMO, the Mozambique government continued to support the Institute and their efforts to articulate, document, educate and disseminate cogent and germane films about the crisis in the region and the ongoing destabilisation efforts by South Africa against the Marxist governments in Mozambique and Angola. In Angola, on the other hand, film production after independence dropped off markedly—the result of nonexistent production infrastructures, and ongoing internecine warfare between the Marxist MPLA and the opposing FNLA and UNITA factions.

Angolan filmmaker Sarah Maldoror, a pioneering filmmaker of the previous decade, continued to produce within the intra-regional lusophone community, making two short films in 1979 for the government of Cape Verde, *Fogo, L'île de Feu,* a profile of the environment and culture of the island of Fogo, and *Un carnaval dans le Sahel,* which includes feast day celebrations and a PAIGC rally.[22]

Unlike Mozambique, Angola, which lies parallel to Mozambique on Africa's west coast, never developed a national center or infrastructure for cinema. However, television was established in 1975 and after independence senior Angolan filmmakers including Ruy Duarte de Carvalho and Antonio Ole produced many documentaries for that medium. Portuguese by birth, and Angolan by declaration, Duarte produced five sections of *Sou Angolano trabalho com força,* a major eleven-part 1975 documentary series on the workforce. He collaborated on this series with Ole, who continued on to make *Apprendre pour mieux servir,* and *Le rythme du N'Gola Ritmos* (1977), and *Pathway to the Stars* (1980). Although Duarte produced mostly for television, he also produced the feature film, *Nelista,* in 1982, an elegantly crafted film based on two tales from southeastern Angola. *Nelista* is the story of two families escaping from a great famine, and their efforts to overcome their situation. Nelista, the hero of the film fights against evil spirits, and with the help of animals and his friends, delivers his people.[23]

Television (TNCV), came to Cape Verde after independence in 1974, including productions by local filmmakers on the stories and folklore of their islands. Independence also brought a revival of the ciné-club movement and renewed participation in the dialogue of the nascent African cinema. On May 7, 1975, the Cineclub Popular da Praia was established, including many of the members of the earlier thwarted cine movement of the late 1950s and early 1960s. The objectives of the ciné-club, as delineated in their formal charter, stressed the support of the ciné-club for cinema and TV as vehicles for informing the population about current and foreign events through documentaries, encouraging active participation in national history, contributing to national arts, popular culture, the education of the population—politically and socially—and for creating a national cinema.[24]

Supported and encouraged by the PAIGC, the club resumed its exhibition of films which were rented through the ciné-clubs in Lisbon, and shown on a weekly basis. Fortunately, the artistic films were much cheaper to rent than the commercial ones, giving the members of the ciné-club exposure to many different genres ranging from Italian neorealism, American classics, Cuban, Brazilian and Japanese films. Picking up the slack from the new revolutionary government, which was busy organizing the first elections and facilitating other critical transitions of independence, the ciné-club organized the filming of the first elections and other independence activities.[25] This highly visible, informed and engaged community of intellectuals was committed to the development of cinema and television in Cape Verde and helped disseminate a wider vision of cinematic development throughout mainland Africa.

A delegation from the ciné-club in Cape Verde joined representatives from nine other revolutionary African countries for a historic meeting in Maputo, Mozambique, February 21–24, 1977, when the Conferência Africana de Cooperação Cinematogràfica, or Association Africaine de Coopération Cinématographique (AACC) was formed. The primary objective was to displace the foreign distribution monopoly and create regional, intra-African circuits of distribution for cinema. This initiative failed, due in large measure to a lack of political commitment by the majority of the participating countries.[26] During these halcyon days, the Institute in Mozambique was already recognized as the center for cinema, as evidenced by their convening and hosting this ambitious, but flawed attempt to address the distribution problems within the continent. Fortunately, the internal successes of the Institute were more tangible and long-lived.

The Institute was empowered by its mission and mandate to restructure all sectors of cinema: distribution, exhibition and production. The Institute grew to include a lab, cinémathèque and a training program. Additionally, the significance accorded to cinema in the revolutionary process is apparent in the Institute's allocation of resources and manpower. During the peak years of the Institute, 1976 to 1986, three shifts worked twenty-four hours a day to produce, process and edit the newsreels, documentaries, and later, date, dramatic productions distributed within Mozambique and abroad.[27] Films such as African-American filmmaker Robert Van Lierop's *O Povo Organizado* (1976), a documentary on the challenges facing the reconstruction and development of the newly independent country, helped foster continued support for Mozambique and bolstered the high international visibility of the fledgling Institute.[28]

The Jewel in the crown of the Institute was the Kuxa Kanema. Conceived in 1981, this project was created to answer specific needs of the population for information about the country, and as the first step in the technical training of the staff of the Institute. As the major center for documentary production, Kuxa Kanema produced 395 weekly editions, 119 short documentaries and 13 long documentary and/or dramas before the decline and ultimate collapse of the Institute in 1990.[29]

The military engagement with South Africa, the corrosive influence of the West, internal criticism of Mozambican political structure, the battles against illiteracy, disease and poverty, historical and cultural self-determination—these were the themes that dominated the productions of the Institute.[30] As a result, Mozambique's ability to quickly respond, reflect, document, produce and disseminate documentaries and programs on current events established the country as the ombudsman of the region. Vertically integrated infrastructures of production, a cadre of trained personnel and the Institute's innovative horizontal systems of distribution and exhibition (which included mobile cinema units reaching out to rural areas and urban audiences with little or no previous exposure to cinema) were the marks of a self-sustaining, healthy national cinema; one capable of recouping its production costs through distribution and exhibition. In short, the vision of an ideologically engaged, alternative cinema, appropriate vehicles of distribution and exhibition and the development of trained African technicians—the ultimate objectives of the Institute were achieved.

Mozambique, already assured a leadership role in the cinematic development of the region, was poised to become the model for the future of African cinema.[31]

However, that vision was not to be realized. The revolutionary transformation of the 1960s and 1970s reversed direction in the 1980s: No longer were the ideological demands of liberation struggles the determining force in the form, content and purpose of cinema. A series of external and internal crises accelerated the decline of the Institute and the future prospects of cinema from Portuguese-speaking Africa. First, the assassination of Amilcar Cabral in 1973 and the death of Samora Machel in 1986 weakened the ideological, intellectual and political leadership of lusophone Africa. Second, the hegemony of the lusophone community began to fray after the split in 1980 of the PAIGC, (the political party under which Guinea Bissau and Cape Verde fought for independence) into two separate parties: the PAICV for Cape Verde, and the PAIGC remaining in Guinea Bissau. Third, independence for Angola and Mozambique from the Portuguese was a phyrric victory; a lull before a plunge into protracted, internal guerrilla warfare and the destabilizing maneuvers of then Rhodesia and South Africa.

Activity and production at the Institute began to taper off, especially after the death of Machel in 1986. However, before its demise, four major large scale productions were realized between 1986 and 1991. Zdravko Velimrovic's *Time of the Leopards* (1987), a Yugoslavia/Mozambique/Zimbabwe 90-minute feature co-production, recounts a fictional episode in the armed struggle for the liberation of Mozambique. The primary action of the film takes place during the turbulent, early 1970s when the war weariness of the Portuguese was apparent, and victory was imminent. The story unfolds in the northern plateau's rich and protective cover for guerrilla fighters. A hunt is organized for Pedro, the commander for a FRELIMO detachment whose courageous actions begin to worry the Portuguese military in the area. Pedro becomes the object of a manhunt, is captured and killed. His memory inspires the new generation, who continue the struggle and attack the barracks where Pedro had been imprisoned.

Jose Cardoso's *O vento sobra do norte* (The wind blows from the North (1987), a 16mm 90-minute feature, opens in the north of Mozambique in1968 where the liberation war has been going on for four years. Colonial settlers, unable to comprehend the reality of the slaves' revolt, exhibit an arrogant boldness along with a sense of uncertainty. Rumors of the changes sweeping through the rest of the country create widespread terror and guilt among the colonialists, who fear the vengeance of the blacks, "mainatos," coming to reclaim the land taken from them five centuries earlier.

The third film, *Borders of Blood,* by Mario Borgneth, a 16mm color 90-minute documentary, was shot in 1985 and completed in 1986. This feature film examines South Africa's destabilisation tactics and the effect on Mozambique's reconstruction. Finally, *Devil's Harvest,* a 1988 Institute of Cinema co-production with France, Belgium, Channel 4 in England, and Denmark, directed by Brazilian Licinio Azevedo and Brigitte Bagnol from France, weaves fiction and fact to tell the story of a drought stricken Mozambican village, which is defended by five veterans of the war for independence against the daily menace of harassment by bandits hidden in the surrounding forest. These productions illustrate the capacity of the Institute to produce feature length fiction and documentary films while incorporating the themes of armed struggle, regional destablization, internal cultural and historical change and post-colonial turmoil. Ironically, the Institute halted at the peak of its financial and technical capability to produce, distribute and exhibit politically and ideologically engaged cinema. Unfortunately, instead of being harbingers of a powerful voice within the region and African cinema, these films were public symbols of the end of an era.

The death knell for the Institute was an electrical fire on February 12, 1991. The Institute and its technical facilities were badly damaged: the film equipment depot, sound studio, editing rooms and processing labs were destroyed. As a result, all documentary production halted, training of personnel ceased, and distribution ground to a halt, since all the prints were destroyed in the fire.[32] This devastating loss, compounded by the death of president Samora Machel, changes of leadership within the Institute, the economic toll of protracted internal guerrilla warfare and the declining support and influence of Marxist regimes for Mozambique, effectively brought a close to the fifteen-year history of the Film Institute.

On the other side of the continent, the watershed for francophone African cinema precedes the demise of the Institute of Cinema by a decade, beginning in 1981 with the socialist policies of the newly-elected Mitterrand government. The French Ministry of Cooperation had begun to withdraw and modify its policies toward African cinema following the closure of the Bureau of Cinema in 1979. The early idealism and commitment of France's Ministry of Cooperation to maintaining cultural ties to their former colonies through the Bureau of Cinema soured with the rising cacophony of African filmmakers' demands for more resources to upgrade their work to commercial, 35mm standards, African governments angered by the content of some of the films, long simmering, and increasingly vocal internal opposition to the Cooperation, and french filmmakers protesting the preferential treatment for the African filmmaker.

Finally, the incoming socialist government of Francois Mitterrand in 1981 shifted the emphasis in francophone Africa from financial and technical assistance to individuals to the creation in sub-Saharan Africa of regional centers for distribution (CIDC, Consortium Interafricain de Distribution Cinematographique, 1979), production (CIPROFILM, Consortium Interafricain de Production de Film, 1979) and training (INAFEC, Institut Africain d'Education Cinematographique, 1976).[33]

The void created by the closure of the Bureau was filled by ATRIA (Association Technique de Recherches et d'informations audio-visuelles), a Paris-based, non-governmental private initiative spearheaded by technicians who had worked with African filmmakers at the Bureau of Cinema. This association served both as a production center and liaison for African filmmakers lacking direct connections with the French film industry. For individual filmmakers the situation in acquiring financial and technical resources, always an uncertain process, reached critical proportions.

In French-speaking Africa, the tension was apparent in Africa's premier organization of African filmmakers, FEPACI (Federation Panafricaine des Cineastes), the continent-wide body of African filmmakers created in 1969 to establish policy and direction for African cinema. FEPACI had originally embraced the tenets and philosophy of the ideologically engaged form and content of third cinema. When the socialist government of Mitterrand came to power in 1981, the dramatically restructured assistance provided to African cinema created turmoil in the ranks of African filmmakers.

This upheaval resulted in the Manifesto of Niamey in 1982, an action launched at the 1981 FESPACO film festival in Ouagadougou by a group of young filmmakers, *L'Oeil Vert*. They demanded, among other things, that FEPACI turn their attention toward concrete objectives of production by beginning with an inventory of available resources throughout the continent. The urgency for the inquiry was reinforced by failure of the regional programs—CIDC, CIPROFILM and INAFEC—to develop self-sufficiency and autonomy for sub-Saharan African film production.

For a brief moment, the leadership and hopes of African cinema revolved around the driving force and charismatic leadership of the president of Burkina Faso, Thomas Sankara, and his vision of African cinema. Sankara viewed cinema as an important vehicle for African educational, cultural, economic, historical and political affirmation. The FESPACO film festival, held in Ouagadougou, the capital of Burkina Faso, took on renewed energy and direction under Sankara, who envisioned Burkina Faso as the spiritual center of African cinema. More Importantly, Sankara's government was politically committed to supporting national cinema.

The 1982 production of *Wend Kuuni* by Burkinabe filmmaker Gaston Kaboré, showcased the talent of emerging Burkinabe filmmakers, and was the first feature film to be totally financed by the government of Burkina Faso.[34] Sankara's assassination in 1987 left a huge gap in the momentum of regional initiatives for African cinema. This blow, coupled with the continuing crisis of African cinema and the disappointment in the performance of the regional centers—CIDC, CIPROFILM, INAFEC—made African filmmakers and FEPACI renew the efforts for international support and collaboration.

In the gray period following 1981, hope continued in the African film sector for South to South (African to African) production possibilities. Historic epics launched by leading African filmmakers during this period include Ousmane Sembene of Senegal's *Camp de Thiaroye* (1989), a co-production between Senegal, Algeria and Tunisia; Mauritanian filmmaker Med Hondo's *Sarrounia,* (1986) a collaboration of several west African nations; and Ethiopian-born Haile Gerima's *Sankofa,* (1992) produced with the support of Burkina Faso, Ghana as well as Channel 4 in England and NDR/WDR, German television. These projects, ambitious and principled in theory, were often frustrated by unwieldly administrative processes, horrific field productions and capricious changes in political support by the African partners. North to South collaboration and co-production, while not a panacea, provided an additional partner or an alternative to the oftentimes costly, arduous and precarious nature of South to South productions.

Subsequently, the first International Partnership forum was held at the 1989 FESPACO film festival. The hope was to establish, "a new pragmatic cooperation with practical results."[35] The second major forum on "Partenariat et Cinema Africain," was held during the 1991 FESPACO festival. Major North to South collaborations between European producers and African filmmakers were launched. European co-production partners included German television, NDR/WDR or ZDE, Belgium, Italy, England's BBC and Channel 4, which launched the "South," series, of commissioned films by African filmmakers. France continued to provide support to African cinema through a more selective process administered through a modified Ministry of Cooperation which had been absorbed by the Ministry of Foreign Affairs.

Those funds partially or completely supported veteran filmmakers as well as a new generation of young filmmakers from French-speaking Africa to produce feature films. These films include *Yeleen* (1987), a powerful and beautiful adaptation of a Bambara oral tradition set in thirteenth-century Mali by established Malien director Souleymane Cisse and two films, *Yaaba* (1989) and *Tilai* (1991) by Idrissa Ouedraogo, a former student of Gaston Kaboré and one of a talented cadre of young filmmakers from Burkina Faso. Set in pre-colonial Burkina Faso, *Yaaba* and *Tilai* touch on themes of alienation, friendship, fratricide, love, community conflict and reactions to drunkenness and adultery.

The international acclaim and subsequent interest in these films was not entirely selfless. The international popularity of these films indicated a market potential for African films with crossover marketing appeal. Films with beautiful images, legends, universal human stories or themes of love, friendship, betrayal had a greater chance of being pursued by foreign partners and investors. In short, the audience and the dollars were abroad; the African gaze shifted from within African to without.

This scenario at the end of the 1980s reflects the impact of Mitterrand's socialist policies on francophone African cinema after 1981, and the transition after 1986 in lusophone Africa from state controlled, ideologically and politically motivated cinema to privatized, free-market film production. Film production in francophone and lusophone Africa entered a new period, one shaped by the climate of democratic change in sub-Saharan Africa.

Portuguese African Cinema: Part III
Private Sector Production, 1991–1993

The democratic change sweeping through Africa in the late 1980s was accelerated by the crumbling Soviet and eastern bloc ideological and financial support to Marxist governments. The ascension of a conservative, western-leaning government in 1990 in Cape Verde and movements toward negotiated peace settlements within Angola and Mozambique in 1991 and 1992 presaged subsequent changes in cinema. In Mozambique the bureaucratic and administrative transition from state-controlled production to a free market was already underway, as evident in numerous seminars held on "the democratization of television" for film and television producers.[36] The changes became more concrete after the fire in 1991.

The subsequent shift in Mozambique was both geo-political and economic. Mozambique's geographical, political and economic relationship within the southern African region superseded to a large degree the earlier cultural and political links with the wider lusophone community. The regional realignment of Mozambique with Angola, Botswana, Lesotho, Namibia, South Africa, Swaziland, Zambia and Zimbabwe, however, creates new challenges, not the least of which are diverse historical, political, cultural experiences and differing expectations and traditions for cinema and television.[37]

Hypothetically, the financial and technical potential for regional film production backed by the resources of a stable South Africa are enormous. That stability is moot, given the continuing volatile nature of the political situation in South Africa. Nevertheless, producers from the region are collaborating on a range of ventures, including the production in 1992 of the *Southern Africa Film Television and Video Yearbook and Catalogue* which lists the regional production companies and films available for distribution.

Economically, Mozambique is increasingly linked to the international, competitive, commercial film marketplace. Close on the heels of the "Partenariat" in 1989 and 1991, four privately owned production companies emerged in Mozambique. A leading force is Ebano Multimedia, Lda., an independent production and distribution company established in 1991 by experienced film professionals, including many senior producers and administrators from the former Institute of Cinema. Ebano is the first of the private companies to venture into feature film production with *The Child From the South,* a 1991 co-production with Channel 4 in England. Set in war-torn Maputo, Nadia, a South African woman journalist meets a committed, but weary Mozambican doctor. This elegant, contemporary love story addresses Nadia's feelings of loss and alienation created by her forced exile as a child from South Africa. *Marracuene,* a 43-minute, 1991 Ebano co-production with German television (ZDF) Channel 4 in England, is a dramatic documentary about a village situated in a heavy war zone. Once a bustling stop on the railway line, the village has become a veritable shadow. Every night the remaining villagers flee to the other side of the river to avoid the terror of nightly raids, returning the next morning to the sight of devastated homes and businesses.[38]

Both films—*The Child From the South* and *Marracuene*—modify the treatment of the prevalent war theme of the earlier didactic, revolutionary films of the region, to stories that appeal to the television audience of the international co-producers. In *The Child From the South,* especially, war becomes a backdrop for an intense personal drama. *Marracuene,* while actually set in the village, includes stylized visual cut-aways and dramatized personal accounts of the nightly sieges. Arguably, in both instances, forces of international financing and marketing have resulted in shifts in content and form. A similar trend, to lighter, or stylized touches, is also apparent in contemporary Angolan productions.

A North to South co-production between Belgium and Angola yielded *Mopiopio,* a 52-minute documentary on music and everyday life in Angola made in 1990 by Angolan-born Zeze Gamboa, a veteran of Angolan television. Another recent Angolan co-production between Italian and Portuguese television is *Moia-O Recado das Ilhas,* a 1989, 35mm feature film by veteran Angola producer, Ruy Duarte de Carvalho. A poetic drama, flashing back and forth between the present and an eighteenth-century adaptation of Shakespeare's *The Tempest, Moia* is the story of an Angolan woman of Cape Verdean descent whose return to Cape Verde forces her to confront and question her existence and identity as someone who is neither totally European nor African. Always strong in television, Angola remains relatively quiescent in film production. Unlike Mozambique, Angola did not develop production facilities. Further, the uneasy peace in the country inhibits further television production or wider participation in regional activities. However, interest and hope remains strong among Angolan filmmakers for their future participation.

In west Africa, Guinea Bissau, the tiny mainland neighbor of the Cape Verde islands, has emerged as a major presence in African cinema. The 1989 film, *Mortu Nega,* catapulted native born director Flora Gomes and Guinea Bissau to international acclaim. Told through the eyes of Dominga, the wife of a guerrilla fighter, the viewer witnesses the commitment, tenacity and will for independence that sustains the morale of the soldiers. She follows her husband through the bush as he and his unit engage the Portuguese in unequally matched warfare, providing encouragement, love and unswerving support to her husband and friends. *Mortu Nega* is an unprecedented and unparalleled dramatic and highly realistic portrayal of the high human cost of war against the Portuguese.

Produced solely by the government of Guinea Bissau, *Mortu Nega* affirms the priority of cinema in the country's development plans. Although there are six filmmakers in the country at the moment, the National Center of Cinema is collarborating with the Ministry of Education and the government to: 1) improve commercial importation and exhibition in the country; 2) produce and co-produce films by Guinea Bissau filmmakers; and 3) train personnel for all levels of film production. To reinforce these goals, filmmaker trainees are attached to all productions occurring in Guinea Bissau.

Flora Gomes second feature, *The Blue Eyes of Yonta,* produced in 1991 again brought critical acclaim to the director and his country. The government of Guinea Bissau participated in the production, along with the Institute of Cinema in Portugal, Vermedia Productions, and Portuguese television. The film, set in the capital city of Bissau after the war, is a story about a beautiful girl, Yonta, who falls in love with a war hero. He never learns of her infatuation, nor in turn does Yonta recognize the passion that a young man from the waterfront, Ze, harbors for her. More importantly, the film shows a post-war reality for Guinea Bissau, its people, and their sense of loss, psychological displacement, love, conviction and hope for the future.

In many ways the contemporary movement in Guinea Bissau is similar to the former Institute of Cinema in Mozambique. Unlike the Institute of Cinema, Guinea Bissau and its filmmakers must secure international financing and/or collaborations to survive.

International television is an increasingly important production partner in African cinema. The credits for *Blue Eyes of Yonta* include Portuguese and English Channel 4 television as well as the Institute of Cinema in

Portugal. Experienced Lisbon based producer Paulo de Sousa and his company, Vermedia, have been instrumental in securing international financing for filmmakers from the nascent lusophone sector. Vermedia produced *Yonta* and served in the same capacity for the 1993 production of *Ilheu de Contenda,* Cape Verdean director Leao Lopes feature debut. Based on a novel by noted Cape Verdean author, Teixeira de Sousa, the story takes place in the 1960s on the island of Fogo. Two brothers, united to settle a family estate, struggle with conflicting values and perspectives on emigration and Cape Verdean identity, dominant themes in Cape Verdean literature, history, and culture. Financing for the production was raised from advance television sales and the Institute of Cinema in Portugal. The experience gained in location shooting of *Yonta* in Guinea Bissau and now Cape Verde establishes Vermedia as a leading production partner in Portuguese speaking Africa.

Cape Verde is an increasingly popular location for feature film productions, a development encouraged by the revitalized Cape Verdean institute of Cinema. Founded in 1977, the primary function of the Institute was the distribution and exhibition of foreign films. Since 1988, under the direction of Daniel Spencer Brito, the Institute is successfully broadening its scope to attract foreign productions and train local personnel. Prior to 1988 most productions in Cape Verde were documentaries produced or co-produced for Cape Verdean television. Brito is cautiously optimistic about the future of cinema by and for Cape Verde. He hopes to bring more African films to the screens in Cape Verdean. Language, however, is a major issue. Although Portuguese is the administrative language, Cape Verdean is the language of the people, the music and literature. Further, it varies from island to island, making it difficult for the local distribution of indigenously produced films as well as imported films.[39]

It is clear that Portuguese speaking Africa—Mozambique, Angola, Cape Verde and Guinea Bissau—is in step with changing trends and influences in the production and distribution of cinema in sub-Saharan Africa. How their participation affects, modifies, changes or encourages the development of African cinema remains to be seen.

Conclusion

Mozambique in particular and lusophone Africa in general, constitute a small but vital chapter of the extant body of sub-Saharan and world cinema; an uninterrupted chronicle of the major movements, instrumental if not pivotal, in shaping the first three decades of African cinema. The period between 1969 and 1975 for cinema in Portuguese-speaking Africa was influenced most significantly by: 1) the internal and external movements and productions in support of liberation struggles, 2) the launching of the Institute of Cinema in Mozambique, and 3) broad issues and debates within African cinema—as evinced by the experimental period with Rouch and Godard or the Bureau of Cinema.

Without question, the pinnacle of cinematic achievement for lusophone Africa was the Institute of Cinema. Prior to its demise and destruction in 1991, the Institute had evolved into a mature, successful production center combining theory, practice and implementation: a monument, testament and finally, solitary beacon of sub-Saharan Africa's revolutionary cinema.

Historically, the Institute symbolized the optimism, euphoria and expectations for cinema throughout the lusophone diaspora in the years immediately following independence where, on the one hand, early initiatives such as internal and foreign lusophone collaborations or the cine movement in Cape Verde continued, or in the case of the Institute of Cinema, flourished. Those dreams died, due in large measure to the constant instability of the film production and distribution in sub-Saharan Africa.

Always susceptible to shifting external and internal political and economic trends, sub-Saharan African cinema, is a microcosm, or barometer, of the shifting priorities in socio-political, historical, ideological and economic trends occurring within a broader, continent-wide context. As indicated in the three areas of this study—liberation cinema, the transition from state to free-market production and private sector productions—each shift brought new directions, trends, themes and participants in African cinema.

Sub-Saharan francophone and lusophone filmmakers are today joined in a common, competitive pursuit of the global market. As southern Africa begins to pull together under the aegis of democratization, the senior producers in the region, Mozambique and Angola, prepare to re-enter the global film market on a new footing. The nascent commercial sector is expanding with private production companies developing projects and exploring co-production and collaborative ventures with other African countries and Europe. Cape Verde and Guinea Bissau are also part of this scenario. Francophone Africa also continues to encourage its established and younger generation of filmmakers in the pursuit of new themes and markets.

Although the 1990s have brought increasing visibility and acclaim to African cinema, the reality of an African cinema in Africa remains precarious until the fundamental and perennial problems in distribution, exhibition, and financing are resolved.

Notes

1. Some African countries have successfully nationalized their film sector completely or in part. However, the lack of inter- and intra-regional co-ordination impedes the possibility of creating comprehensive, alternative, continent-wide distribution circuits, creating a void that exceeds the capabilities of individual African states in general or filmmakers in particular.

2. Claire Andrade-Watkins, "France's Bureau of Cinema: Financial and Technical Assistance between 1961 & 1977—Operations and Implications for African Cinema," *Society for Visual Anthropology.* 6.2 (1990). The argument advanced here is that the cinematic development supported by the French in the newly independent French-speaking West African countries was undercapitalized, creating a neo-colonial economic and technical dependence on France that reinforced colonial policies of assimilation through French language, culture and finance.

3. Region in this context refers to historical and cultural ties, not geographic proximity.

4. While capable of 16mm production, the 35mm facilities were never completed, and the lack of laboratory processing facilities made them dependent on technical supports in Europe. Furthermore, the unmitigated anger unleashed by de Gaulle and France at Guinea Conakry's dramatic declaration of independence in 1958 resulted in a brutal economic and political backlash that hobbled Guinea Conakry's potential in all sectors.

5. Ghana has the most sophisticated 16mm and 35mm facilities and laboratories in West Africa, a legacy of William Grierson's Colonial Film Unit. Unfortunately it has never been fully and effectively used within Ghana or for the rest of sub-Saharan Africa. The Bantu Educational Kinema experiment (1937–1939) and the British Film Unit (1939–1945) were short-lived programs aimed, in the first instance at rural education of the Africans, and in the second, propaganda films to mobilize Africans to fight during WWII. Other noted ventures in colonial cinema include films made by Catholic missionaries in the Belgian Congo (CCAC) Congolese Center for Catholic Action Cinema. Finally, in French Guinea Conakry, Sily Cinema was created in 1958, made documentaries, educational films and newsreels. Established basic 16mm production, but 35mm facilities incomplete.

6. Anglophone Africa has major filmmakers who have been pioneers in African cinema, and include Nigerian Ola Balogun, Ghanaians Kwaw Ansah, King Ampew, Kwate Ni Owee and others. However, although they might receive some government assistance, most of the work is produced in the private sector. As noted earlier, dramatic and documentary television programming is really the main activity in Anglophone Africa.

7. Clyde Taylor, "Film Reborn in Mozambique," *Jumpcut.* 28 (1983): 29–31.

8. Lars Rudebeck, *Guinea Bissau: A Study of Political Mobilization.* (Uppsala: Scandinavian Institute of African Studies, 1974.) 7, 12, 21, 71; Hilary Anderson, *Mozambique: A War Against the People.* (New York: St. Martin's Press, 1992); Daniel Kempton, *Soviet Strategy Toward Southern Africa: The National Liberation Movement Connection,* (New York: Praeger, 1989); Barry Munslow, *Mozambique: The Revolution and its Origins,* (New York: Longman), 1983.

9. Guy Hennebelle, *Guide des films anti-impérialistes,* (Centre d'Information sur les luttes Anti-Impérialistes [CILA], 1975) 110.

10. Maldoror also made two short films after independence for the government of the Republic of Cape Verde in 1979, *Fogo, L'île de Feu* and *Un Carnaval dans le Sahel.*

11. Hennebelle, *Guide des films anti-impérialistes,* 111. For Mozambique, films included *Viva Frelimo,* 1969, Dutch, a report on Frelimo and an interview with Samora Machel; *A Luta Continua,* 1971, Robert Van Lierop, of Frelimo and historical analysis of the country; *Dans notre pays les balles comencent a fleurir,* Sweden; *Etudier Produire, Combattre,* a film on a Frelimo school in Tanzania, Guinea Bissau; *No Pincha,* 1971, 70 minute documentary on PAIGC, *Madina Boe;* 1968, Cuba; *Nossa Terra,* 1966, Labanta Negro, 1966, Italy, *Le Cancer de la Trahison, Une Nation est Née,* Sweden 1974, *Free people in Portuguese Guinea,* 1970, Sweden.

12. A. F. da Silva, President of the Ciné-Club, Praia, São Tiago, personal interview, August 22, 1993.

13. Ibid.

14. Ibid.

15. Manthia Diawara, *African Cinema* (Bloomington: Indiana University Press, 1992) 97.

16. Andrée Daventure, personal interview, June 1987.

17. For a fuller discussion of this issue see Claire Andrade-Watkins, unpublished dissertation, *Francophone African Cinema: French Financial and Technical Assistance 1961 to 1977,* (Boston University, 1989), 215–235.

18. This is despite the fact that the productions through the Bureau of Cinema, were 16mm; an issue which became a source of contention between the filmmakers and the Bureau.

19. Jean Rouch, Unpublished Interview, Paris, August 25, 1987. For a fuller discussion of Rouch, see, "Jean Rouch, Un Griot Gaulois," *CinemAction* 17 (1981).

20. *CinemAction,* 20–36.

21. Television came to Mozambique in 1979 without any particular benefit of Godard's participation.

22. Francoise Pfaff, *Twenty-five Black African filmmakers* (Connecticut: Greenwood Press, 1988) 212.

23. Duarte's films, chronologically include; *Sou Angolana, Trablaho com força* (1975), five-part TV documentary; 1976 *Uma festa para viver,* TV; 1977 *Angola 76 e a vez da voz do povo*—three documentaries for TV; 1977 feature film, *Faz a coragem, camarada;* 1979 *Presente angolano, tempo mumuila* (10-part documentary series, TV); 1982 *O balanco do tempo na cena de Angola* (documentary); 1982 *Nelista,* feature; 1989 feature, *Moia, O recado das ilhas.*

24. Unpublished document, *Estatuto,* Cineclube Popular da Praia, Outorgado em 7 de Maio de 1975, Publicado in "B.O." n. 19, de 10 de Maio de 1975.

25. *Estatuto.* In late 1976/1977 the cine-club of Praia suspended its activities. Lack of financial support, internal struggles over direction of the ciné-club contributed to its demise.

26. Pedro Pimenta, letter to Claire Andrade-Watkins, January 27, 1993.

27. Pedro Pimenta, Associate Director of the Institute of Cinema, personal interview, November 1993, New York.

28. For example, the first major benefit in the US for the Institute was a historic national tour in 1981, organized by Positive Productions, Inc. in Washington D.C.,

and spearheaded by Ethiopian filmmaker Haile Gerima. The success of that tour resulted in the purchase of an optical printer and other materials for the Institute of Cinema. Gerima, an early supporter of the Institute, contributed his films to the library and archives of the Institute and encouraged other filmmakers to do likewise. Haile Gerima, personal interview, January 21, 1993.

29. Pedro Pimenta, letter to author, January 27, 1993.

30. Early landmark films of the Institute include; *They Dare Cross Our Borders* (1981 BW 25 min., 16mm or 35mm), South Africa's attacks on Mozambique and the reaction of the government and people; *The Offensive* (1980 BW, 30 min., 16mm), an internal offensive against efficiency and incompetence; and *Unity in Feast* (1980, color, 10 min., 16mm) a film on Mozambique culture, with a particular objective of valorizing and preserving the rich traditions scorned by colonialism. A prime example of regional collaboration, is the documentary *Let's Fight for Zimbabwe* (1981, 60 min.) documentary, co-produced by Mozambique and Angola, documents Zimbabwe's independence and raises questions about the political stability and future of the region.

The first feature length documentary *These are the Weapons* (1979 BW, 50 min., 16mm), a chronicle of the fight for independence, the internal struggles facing the people of Mozambique, and South Africa's strategies of disruption. Bringing the touch of Cinema Novo to Mozambique, Ruy Guerra's *MUEDA: Memorial and Massacre* (1979 BW, 35mm), the first feature by a Mozambican, is a blend of theater and reality. A small village in northern Mozambique, Mueda was decimated by a massacre by the Portuguese in 1960. The theater play, created in 1968, is an annual reenactment by the survivors to commemorate the massacre.

31. Although Angola had a national Film Institute and shared the commitment of Mozambique, their lack of infrastructures limited their production capabilities to collaborations or co-productions. A growing cadre of trained African technicians and administrators found their way to the Institute, such as Pedro Pimenta, who joined the Institute in 1976. As the Assistant Director/General Production manager of the institute, Pimenta played a pivotal role in the emergence of Mozambique. Born in the Central African Republic, Pimenta studied Economics in Portugal, and taught in Maputo before joining the National Film Institute.

32. Pedro Pimenta, fax to Rod Stoneman, Channel Four Television, London, February 1991.

33. The subsequent success of Mitterrand's government initiatives: CIDC, CIPROFILMS, INAFEC, is moot.

34. Françoise Pfaff, *Twenty-Five Black African Filmmakers* (New York: Greenwood Press, 1988) 174.

35. Gaston Kaboré, "Journees Internationales de Partenariat audiovisuel (JIPA)," *FESPACO NEWS,* 4 (February 26, 1991) 1.

36. Pedro Pimenta, phone conversation with author, August 1992.

37. Some films were caught in the transition from the Institute to the free market sector. Two young Mozambican filmmakers, Joao Ribeiro and Jose Passe, were finishing their film training in Cuba during the upheavals at the Institute. Ribeiro's *Fogata* (1992, 20 min., 16mm) is a drama based on a novel by Mozambican writer Mia Couto, where a peasant couple struggles with assuring each other's

proper burial. Passe's *Solidão* (1991, 30 min., 16mm), is a drama set on the eve of independence, and revolves around the despair of a white Portuguese settler about his marriage to a black woman and the subsequent inevitable changes coming with independence.

38. Jean-Pierre Garcia and Caroline Helburg, "Cinema et Television au Mozambique," *Le Film Africain* (Novembre 1992) 11–12.

39. Jean-Pierre Garcia, "Le Cinéma au Cap-Vert," *Le Film Africain* (Mai 1993) 24–25; Daniel Spencer Brito, personal interview, July 1993.

New Developments in Black African Cinema

N. Frank Ukadike

> Today, we can summon to memory the languages of our ancestors. What is important, though, is the rediscovery of the power of words of our people. Metropolitan French, English, Spanish—all languages of colonization to be colonized in turn.
>
> *Maryse Conde*

> I shall speak about Africa, with confidence both that some of what I have to say will work elsewhere in the so-called Third World and that it will not work at all in some places.
>
> *Kwame Anthony Appiah*

Contours of an Emerging Trend: Toward a New Cinema?

The above statements vividly address problems of cultural production within the African diaspora and Third World societies. With production processes becoming increasingly validated in terms of new modes of film practice, this endeavor, utilizing indigenous systems of thought, is creating aesthetic practices that illuminate developmental struggles. Even though positive results have emerged from this collective cultural experience, the urge to give more specific character to cultural products means widening the indigenous boundaries for the investigation of modes of representation to provoke discussion. We can see this in the reading of "nativist texts," when interrogation of production processes stimulates the selection, arrangement, and reception of cultural products. African cinema, as already established in previous chapters, is shaped by the producer's avowed determination to move ahead, unencumbered by the legacy of the dominant paradigms.

In recent years innovative works of historical, cultural, political, and aesthetic significance have emerged from "new breed" African filmmakers as well as from the pioneers, enabling us to speak about "interrogation of origins" (that is, transformation of negative conventions into a positive source of signification), "analytical complexities" (exploration of African alterity via sociocultural dynamics), and "aesthetic diversion" (nationalistic practices versus aesthetic internationalism). It is now possible to say that African cinema is at last penetrating the world market with major works of indigenous cultures that explore and adapt their own oral and literary traditions in the articulation of a new film language. Yet, the relationship between the perception of African cinema and the realities it tries to depict is a paradox marked by the filmmakers' indefatigable struggle.

On the one hand, experimental trends inherent in recent developments in African cinema exemplify a characteristic emphasis on film as cultural and political practice. As the concern for economic viability deepens, African cinema can also be understood from the perspective of thematic and aesthetic pluralism. On the other hand, the aesthetic orientation seen at first as hybrid was gainfully used to critique the aesthetics and ideology of alien film practices, later developing as a renovated film language. Also, the employment of oral tradition and other cultural codes that served as the base for indigenous aesthetics also helped to repostulate canonized Western paradigms (modes of production and spectatorship). However, while oral narrative illuminates the paradigmatic exigency of black African film practice, the engendering configurations for the new aesthetic are in danger of being misread as an alienating convention. It is my contention that this alienating posturing is inescapable—involving both the combative didactive practices of nationalistic concerns (Sembene, Hondo, Gerima, Kaboré, Cissé, and Ansah exemplify this category) and the politically compromised aesthetic internationalism of economic concerns (exemplified by the works of Ouedraogo, Sissoko, and Duparc).

In the first instance we have categories of films put in the service of political consciousness, and in the second, films that are thematically audacious and innovative but whose sociopolitical allegory is diffused by pandering to the imitation of alien conventions and commercialization. If it is true within the confines of political consciousness and cultural affirmation that the latter works would be deemed as abandoning the collective consciousness engendered in the pioneering works, it is the recourse to aesthetic internationalism—the fusion of traditional codes with canonized Western forms—with which the films aspire to break through national and international alienation.

Politically conscious films constituted the mainstream, at least through the early eighties, especially in the francophone and lusophone regions. In the late eighties, the new African films created by the new-comers launched the opposition to the dominant didactic formulas. Seeing themselves as no longer in the periphery, the new filmmakers discarded heavy-handed didacticism for narrative conventions that stressed entertainment over instruction. Although the films are still modeled after oral tradition, the majority, as we shall see, cannot be said to be revitalizing film language. There seems to be a movement away from the political use of the film medium, which addresses and relates to authentic cultures and histories, toward a concern with film as an object of anthropological interest. It has been argued that most films employ the same Western ethnographic conventions that have historically worked to limit the understanding of Africa's sociocultural formations.[1] Is African cinema in general lacking the dynamism and experimentalism of the sixties and seventies? How has its film practice developed operational tactics that promote (inter)nationalization of film form and ideological precepts? Has the diversity of textual expressions created new attitudes toward black African film culture? This chapter will provide a matrix of convergent and divergent perspectives by examining recent films, illuminating their historical, sociopolitical, and cultural affirmations, while at the same time foregrounding the approaches and structuring complexities of the emerging trends.

African cinematic practice of the 1980s, if nothing else, has done an enormous job in terms of restating the authentic values of African society and, given the popularity of some of the films, enhancing the position of this cinema within the stylistic repertoire of world cinema. With the protean notion of Africa as "cinematographic desert" obsolete, the question is no longer whether Africans can make films. The excitement and expectation offered by recent films compel one to accept that there can be a prosperous African film practice given the right incentives. The unprecedented success of some recent African films, both in critical acclaim as well as in box-office receipts (in Africa, Europe, and the United States), promulgates speculation regarding the emerging trend of new African cinema. Does the current trend constitute what might be considered the quintessence of African cinematographic art and industry? It is with this question in mind that I shall examine *Yeelen* (Brightness, 1987) by Souleymane Cissé, *Yaaba* (1989) and *Tilai* (1990), both by

Idrissa Ouedraogo, and the following social-realist films (also of cultural ramifications): *Finzan* (1989) by Cheick Oumar Sissoko, *Bal poussière* (Dancing in the dust, 1988) by Henri Duparc, and *Zan Boko* (1988) by Gaston Kaboré. I shall also address this study to the historical affirmations in Med Hondo's *Sarraounia* (1987), Kwaw Ansah's *Heritage . . . Africa* (1988), and Ousmane Sembene and Thierno Faty Sow's *Camp de Thiaroye* (The camp at Thiaroye, 1987) and shall examine their historical propensities and methods of construction.

The new African films have not strayed too far from the issues depicted in earlier films, yet the impact of the popular films by the "new breeds" are more strongly felt than that of their predecessors. I will argue that the impact of the popular African films of the eighties is a result of filmic subterfuge: passion for ethnographic information (thematic construction); profilmic and extrafilmic organization (immutable landscape); and aesthetic reconciliation (the incorporation of oral art).

In black African film practice, Ntongela Masilela identifies the historical and cultural meaning of African cinema as relating to the "dialogism" between Frantz Fanon's philosophy of culture and Amilcar Cabral's philosophy of history, noting that the former is a deliberate attempt to restructure Africa's political systems while the latter strives to redraw the map of the "social geography of African history."[2] Applied to the new African cinema of the late eighties, it is possible for one film to embrace both philosophies, as in the cases of *Camp de Thiaroye, Zan Boko, Heritage . . . Africa,* and *Sarraounia,* whose structures exhibit affinities with the social and historical philosophies of Cabral as well as the cultural and political philosophies of Fanon. Although films such as *Yeelen, Tilai, Bal poussière,* and *Finzan* openly exhibit an exquisite affinity with culture, they lack the political candor of Fanon's ideas. Exception, however, can be made for *Yeelen,* whose analytical complexity and (arguably) biased political undercurrents are hidden beneath the umbrella of traditional culture.[3]

The films as indigenous cultural expressions are situated within the social context of multicultural Africa. Hence, the symptomatic trend characterizes a wide range of representation and interpretation from works inspired by African and Third World resentment of Western processes of conversion (*Camp de Thiaroye, Heritage,* and *Sarraounia*) to culture-based representational structure (*Yaaba, Yeelen, Finzan,* and *Zan Boko*). It is this resistance to assimilation that unites black African cinema with Africanists' theories on African culture. Although the films display their "unity on a continental plane . . . they equally differentiate themselves from each other by simultaneously articulating national cultural patterns, national ideological conflicts and national class confrontations,"[4] and, we might also add, national economic patterns. For example, *Bal poussière* could be read as a political film, although its few political jabs get lost in the comedic structure. Yet, like the film's environment, some narrative codes (its theme and music) are typically African. Although the focus of each film is on geographically divergent zones, the contents and significations indicate similarity of purpose and goals. Ideologically, there is the persistent struggle to develop genuine film practice; politically, there is an attempt to use the film medium as a speaking voice of the people; and aesthetically, there is relentless experimentation with film form aimed at achieving indigenous film culture, although it has not been possible to reach this goal. Similarly, there is a profound urge to satisfy the tastes of both African and foreign audiences. Reflecting the economic realities in Africa today, the targeting of foreign audiences connects philosophically with the notion of economic viability, in which effective export of goods enhances the gross national product.

Narration, Transgression, and the Centrality of Culture

Black African film practice rejects all vestiges of colonialism and acculturation. Because cultural assimilation has involved the protracted colonial process of stripping Africans of their individuality, the various steps taken to counter this process argue for the linkage of self-affirmation with universal consciousness, for example, attachment of the individual to society, and society's place in the centrality and the collectivity of world culture. This process of inclusion and retrievability, apropos to what John A. A. Ayoade terms "reculturation and re-Africanization,"[5] echoes the need for the restitution of social institutions and relations and beliefs and practices congenial to indigenous traditions expressed in nationalist agendas. In his essay "Cultural Restitution and Independent Black Cinema," Tony Gittens identifies some essential characteristics of indigenous beliefs and practices. He states that "they are usually ritualized and celebrated in a process which helps to engender a sense of national, secular, regional, or ethnic pride, dependent on the type of cohesive force and social priorities binding a particular group together."[6] This explains why cultural productions are viewed critically and, regarding African films considered by Africanists as the epitome of decolonization, there is continuous demand for African dignity to be forcefully depicted on screen. "Dignity," in the words of the late Captain Thomas Sankara, the ex-president of Burkina Faso,

> has not been presented enough [in African films]. The cry from the heart, justice, too, the nobility and the necessity for struggle in Africa, that has not been shown enough. Sometimes one has the impression that Africans are striving in vain in a world of evil men. What we have experienced, what we have suffered, what we are

> now experiencing, what we are still suffering—this has not been publicized enough and we also know that the media elsewhere in the world are efficacious in preventing people in other countries from understanding the struggle which we are waging here.[7]

African films such as *Yaaba, Tilai, Zan Boko, Yeelen,* and *Finzan* offer meticulous anthropological renditions of African cultures. The films are true to life and do not attempt to rearrange natural settings or modernize them to look foreign. In their cultural manifestations, certain rituals, music, dance, and song, once considered primitive, are frequently adopted motifs in these films. It has not been easy to achieve a way of reconfiguring the images created, of transforming them into counterhegemonic images capable of confronting dominant paradigms in order to engender a sense of African pride and the sort of identity envisaged by Sankara. Sensibility and sensitivity as the mainstay of African film practice, and the need for indigenous characteristics, simply require a formula that presents the environment and characters without running into the Hollywood situation in which non-Western images connote the exotic "Other."

While sympathizing with the pioneering impulse that created African cinema, the "new breed" African filmmaker seeks to advance cinematic strategy beyond the ideologies that have defined the contours of didactic sociopolitical films. In the first place, there has been no inclination toward cultivating the services of professional actors. Second, in the absence of a sound infrastructure to pursue glamorous projects, the filmmakers feel there is an overabundance of local themes and an inexhaustible number of sociocultural motifs with which to sustain production. Thus, as long as village scenery and pastoral landscapes continue to provide the desired spectacle and backgrounds that urban landscapes lack (for example, skyscrapers that are similar to the ones in Paris, London, and New York), filmmakers capitalize on this simple arithmetic of economics.

The cultural dynamics of African cinema must also be examined within the larger significations of African cultural discourse. Africanist thinking has long recognized the dialectical link between one factor of development and another, in this case, the coalition of cultural and economic enterprises. For example, on the dynamics of culture. Sékou Touré of Guinea notes that "every people must struggle to exist by creating the material means of its existence."[8] For him, culture is "a social process, an infrastructure" whose characteristics are determined by "the level of development of productive forces and the nature of the means of production as determined by the historical and social context."[9] Bringing this point closer within the environs of the dominated people's struggle in Africa in particular, Touré goes on to emphasize that "the cultural level of a people . . . its means of conquering knowledge, its manner of explaining phenomena, will depend on the . . . degree of objectivity and abstraction attained in the heat of action to gain mastery over ever more perfect techniques."[10]

In the African film industry, the tripartite relationship amongst the cultural, economic, and historical forces responsible for the development of African cinema reflects the shifting conditions of African culture and its economy. If the responsibility of the African film industry is to cultivate audiences for African film, then the accomplishments gained in production—the instinct for self-preservation (cultural restitution) and fulfillment (aesthetic development)—must embrace societal and infrastructural requirements (audience reception). As it is now, African cinema has presented African culture both realistically and impressively on screen. But African cinema has not been able to exploit the continent's economic potential. An attempt to pull African film practice out of its current form of marginality can only be achieved if genuine development of the audience and reception patterns are refocused. And, if there is to be a common ground in this film practice, it will undoubtedly be found in the aspect of production to be nurtured rather than patronized in Africa, not perceived as a metaphor for otherness, isolation, and difference but as a continent central to world power institutions, no longer bound to the periphery.

Generally, while African film practice has succeeded in renovating its film language, it can also be argued that it has failed to produce a new African audience capable of exploring different patterns of signification while pondering this "new" African film language. Toward this goal, the most important endeavor would be the making of films that challenge the reception patterns of the African audience. While some African films succeed in attaining this goal, some films actually alienate their audiences, precluding the building of a strong African audience for African films.

Yeelen, Yaaba, Tilai, Zan Boko, and *Finzan* are African films made in the 1980s that have emphasized African culture specifically and, in the case of *Zan Boko,* have contrasted precolonial tranquillity with the turbulence of contemporary life. This is not the first time that African culture has been meticulously studied in African cinema. If compared with earlier black African films, these new films show shifting aesthetic and formal concerns. Here we find that the sociocultural dynamics that influence the cinematic language of explication can be both an asset and anathema. In this sense, structural configuration poses methodological questions not only about what the artists are saying with the images they create but also about how they say it. This surging interest in these recent African films stems from the curious anthropological images they proffer, some critics reading them as innovative, others as repugnant. Agreeably, there is vivid concern to stretch inquiries about Africa beyond cultural and anthropological limitations. The crux of the matter is that, while the majority of the films are packed with realistic images of everyday village life, overemphasis on "lowlife" details reinforces the fascination with the exotic African images in Western ethnographic films. Rather than destroying and deconstructing canonized codes of spectatorial fascination through rejuvenated film form and modes of production, some of the films perpetuate marginality and also invite reductionist reading of the films mainly within the confines of ethnographic film discourse.

From another perspective, some critics have argued that the images and verisimilitude present in African films indicate structures that manifest African notions of ethnography, and that by depicting the African peoples the way they are, their cultures and histories, these images explore value systems contrary to Western ethnocentric understanding (in which non-Western images are evaluated by Western norms). In this sense the films are also read as reversing the institutionalized condescension and master-race narcissism offered in Western ethnographic films.[11] It is in this light that we focus on the varying trends of the African films of the eighties by examining their thematic, aesthetic, and textual configurations.

Souleymane Cissé's *Yeelen* (prize winner at the 1987 Cannes Film Festival) eptiomizes the daring stance shared by African cinema pioneers. In every respect the filmmaker's quest for an indigenous film structure parallels the film's exploratory tactics and is realized in a complex cinematic style. The film narrates an initiation journey that puts father (Somo Diarra) and son (Nianankoro) on a collision course as they struggle for supremacy in the possession of a special knowledge—the secrets of nature—that is exclusive to the Bambara people of Mali. The father, seeking to prevent his son from acquiring this special knowledge, plans to kill him. But Nianankoro's mother intervenes by sending her son away (into an initiation journey that introduces him to the oral cultures and traditions common to the Sahel). This pilgrimage takes him across the semiarid regions spanning the Bambara to Dogon and Fulani lands, enabling him to acquire the ultimate knowledge that establishes his own supernatural powers.

Cissé documents that this film was accomplished through mutual understanding and respect, by placing the significations of form and content within social, cultural, and historical contexts and in a transgressive mode of address far apart from dogmatic codes and conventions. For example, *Yeelen* could either be seen as promulgating a strict reflectionist or deterministic notion of cultural and social connotations or, more daringly, it could be read as an antitraditionalist allegory that introduces "the correlate" and "the resilient" as dichotomous canons for the critique of tradition and change.[12] The very depiction of the "Komo" secret society, in which the camera lens witnesses the actual initiation, rituals, and ceremonies, would ordinarily have been regarded as sacrilegious and intrusive. In fact, traditional Malians have long been suspicious of what they term "diabolic images" of the magic machine of the devils' creation named *Tiyatra.*[13] But Cissé holds the middle ground between the Western ethnographic conception of the moving image, which seeks out the misery of the Third World, and the falsehood of repressing it on the grounds of cultural intrusion. Thus *Yeelen* is significant for a number of reasons: the film creates a dialectic of old and new; as a critique of culture, it displays new ideas; and as a critique of convention, the filmic strategy suggests a resolute trend in creativity, fresh and focused on African film language. Out of this practice, however, emerge certain questions of perspicacity and contradictions of ideological ramification. I attempt to expedite this study through an examination of Cissé's concern for authenticity and respectful treatment of the subject and the cinematic strategies contiguous to the film's realization.

Structured around mythological patterns, *Yeelen's* complex narrative allows one to explore the articulation of the precolonial and the traditional. The film's binary structure foregrounds the distinctive dynamic of traditional culture, also revealing its contradictions, such as the antecedent practices indicative of what Kwame Anthony Appiah, in another context, terms "exclusivity of insight."[14] By calling attention to the opposition and rejection that dichotomize traditional systems, Cissé's projection of the inevitability thus centers around apostasy, which works as a revisionist code that, though not on par with traditional systems, enables him to construct the dialectics of culture and morality. By having Diarra refuse to transfer generations of secrets to his son, the viewer is forced to sympathize with Nianankoro when he attempts the retrieval of this knowledge by force. In this respect, the father's action nullifies one of Africa's most cherished traditions—the quest for knowledge—while the son's rebellion is just another case of nonchalance typifying negation of the tradition of morality that is emblematic of respect and loyalty to the elders. The film does not show us, however, that Diarra is not obligated to reveal such secrets to his son since he is too young to be entrusted with the secrets of the Komo, which only members who have attained the rank of "Kore" can possess. According to Kate Ezra, it takes a neophyte "seven-year cycles through six grades of progressively more arcane knowledge" to reach, at old age, the Kore stage.[15] Because the Komo is highly restrictive, and because the powers wielded by the members are all-encompassing, like other secret societies all over the world, it is prone to criticism and rejection. But considering that Cissé's background is in the Soninke, early Islamized marabouts,[16] it is possible that the director is speaking on behalf of his own clan, which may be deprived of admission to this impregnable Bambara knowledge.

The nation of Mali is 90 percent Islamic, and Cissé's view, while its focus is mainly on contemporary Muslim urban and rural society, does not necessarily indicate that traditional systems should be discounted. Rather, he reflects on what has plagued African sensibility for many years. He ponders the potentials of traditional culture—for example, the practice of witchcraft—which is not put at the service of advancement, in what amounts to an Africanist injunction to garner such culture for posterity. Put another way, this implies that beyond its (witchcraft's) present scope, Cissé is advocating a new kind of thinking, a refinement in which such powers are no longer misused but are transformed from destruction to invention, as have been the inspirations that have shaped Western scientific and technological prowess.

It is interesting to note that he has the ability to tell a legendary tale against the backdrop of a cultural motif without resorting to reductionist conventions. His fabulous attention to detail is not simply exploitive, illustrative, or didactic; rather, he develops a way of introducing new cultural perspectives—new ways of looking, discovering, and identifying. He does not select his characters and locations in order to simplify or exoticize them; his composition in depth, camera movement, mise-en-scène, and sound coalesce into what might be termed a discerning vision. Together with the natural movement and mannerisms of his nonprofessional actors, the distinctions between society and culture and fiction and reality are interrogated as fictional characters are juxtaposed against ritual and the ritualistic to evoke specific sociocultural dynamics. This prerogative, charged with the inclination for authenticity, is attributable to the philosophy emerging from the "mastery of content," as he says, which forces one "to select forms that are appropriate."[17]

In *Yeelen* the characters must go through a series of encounters as part of an initiation process that would enable them to acquire more power and knowledge. In the end, both father and son perish. For Niananköro to attain the level of maturity necessary to confront his father's prowess means the acquisition of occult powers. He must recover a specter—a long piece of carved wood shaped like a wing which symbolizes knowledge and power—which is the only thing capable of destroying the Komo. Already he has retrieved parts of the missing wing from his mother, and she now reveals to him that the remaining part of the Kore is in the possession of his father's identical twin brother, Djigui, a blind man who fled the Bambara land. Later we learn that Djigui himself lost his sight trying to acquire the Komo secret. A series of encounters which attempts to cast doubt on the relevance of the whole process of this mystical acquisition and the negative connotations of the cult are set in motion: Nianankoro kills Baafing, one of his other uncles who tries to persuade him to give up his revengeful mission. In the land of the Fulanis (Peul), he seduces the king's wife, Attou, proclaiming, "My penis betrayed me." Later he marries her and she bears him a son.

On the other side of the conflict, members of the Komo cult are meeting secretly and deciding what kind of punishment to inflict upon Nianankoro. He must be stopped before he acquires the power to destroy them, and this means destroying him. It is not surprising that his father, armed with the pestle of the Komo, embarks on the dreadful journey of tracking him down. When they finally meet, Cissé unleashes one of cinema's most dramatic moments. Nianankoro, armed with the wing of the Kore, and his father, with the pestle of Komo, stand gazing at each other. In this confrontational stance there is a powerful and unpretentious composition; the extreme close-up of both men's faces compels us to witness a wide range of moving emotions. Few words are exchanged. We see both men perspiring feverishly, and when the tears in Nianankoro's eyes begin to trickle down, both Kore and the pestle emit magical rays. The screen turns translucent, an indication that the competition is over, as both men are destroyed and the earth scorched.

By allowing Attou and her son to survive, Cissé's prophecy is fulfilled. This was first implied when Nianankoro is seen deliberating with Djigui on the parched ground of the Dogon, when we learn that the Komo cult, if unchecked, will continue to use its powers to subjugate the inhabitants. In the last scene, Nianankoro's son is seen walking over a large sand dune through a beautiful desert. He uncovers two "orbs," gives one to his mother, who in turn gives him Nianankoro's *grand boubou* (robe). If an accurate reflection of a particular society is the main point of *Yeelen's* narrative structure, the film can be read as criticizing redundant aspects of the Komo cult and at the same time presenting it as an authentic culture specifically valorized to project its cultural significance. The film's utilization of the myth of origin with which it embellishes its structure is indicative of the credo of returning to one's roots as a paradigm of reaffirmation, politically as well as aesthetically.

Amie Williams's observation that *Yeelen* and some other African films "dip into the ancient past not to escape the present or lapse into nostalgia but to extract knowledge and history relevant to the present condition of its viewers as dispersed postcolonial subjects"[18] echoes Cissé's contentious stance of transforming spectator intractability (African or Western) into docility, a transformation involving active processes of "looking," "interpreting," and "discovering."

Nianankoro's son's position not only exemplifies hope for Africa but also ratifies, once again, what many Africans have wished for in the postcolonial era: eradication of all forms of dictatorship and the sweeping away of all dictators, tyrants, and corruption, making way for a new generation to emerge. In this sense, the dramatic closure reminiscent of most African films, à la Sembene, withholds judgment. Cissé's *Yeelen* imposes the filmmaker's vision.

Yeelen displays Cissé's powers of directorial invention in its combination of forms. As a poetic celebration of African culture it is dynamic; in the sense of authenticity and cultural iconography and in the sense of indigenous African art form, it is a classical epic reminiscent of the tradition of Sanjata, weaving aspects of repressed cultural motifs—rituals, folklore, and symbols—into a historical tapestry of ritual values. As a cinematic summa encompassing the indigenous and the mainstream, it specifically alludes to styles ranging from Sembene's candid didacticism to Hondo's eruptive style; in it we find the fusion of ethnographic and documentary modes, fiction and reality, cinéma vérité and oral literature.

Although *Yeelen* is not exactly a film that fits well into the tradition of social realism, its pattern of construction reflects a synthesis of ideology and aesthetics that is reminiscent of early Soviet filmmaking. For instance, Vertov's radical documentary practice was distinguished by artistic and innovative editing and camera technique. So too is *Yeelen* in its formalist criteria of mise-en-scène, lighting, photography, and directing, all fused with the cultural components of African oral tradition to produce a distinctive narrative structure and rhythmic patterning that make it brilliantly evocative. It is here that Cissé succeeds where many have failed. However, our concern is not to appraise the film as the "most beautiful film ever to have emerged out of Africa,"[19] as it would be naïve to subscribe to that view, but rather to see it as a dynamic construction whose beautiful photography demystifies the notion of the photogenic. This film compels the viewer to feel as if, in Alice Walker's phrase, "one has been in Africa during several centuries."[20] It is this that is central to our discourse.

Yeelen's inventiveness is also notable in its subversion of linear narrative structure. For example, favoring short vignette over lengthy narrative development, it could be argued, is instrumental to the film's complex structure. Many scenes and sequences are particularly illuminating for various innovative reasons. But this structure emanates from the complex paradigms of the cultures presented in each episode. The purpose is to foreground the use of cultural codes to force the viewer to articulate the complexity of the relationship of precolonial structures to the present. For instance, the film's flashbacks draw attention to the interplay between past and present, and, since the film is replete with so many allusions, it also illuminates the boundaries of fact and fantasy. This means that the language of indigenous modes of inquiry is presented in a dichotomous art that bridges the distinction between the African mode of address (precolonial past) and the technologically inspired mode of representation (neocolonial present). I will expand upon this with specific examples to illustrate that there is a dialectical connection between traditional art forms and the mainstream, based around reconstitution, and I will show how this duality affirms cultural existence and identity in a nonconvoluted fashion.

Nature and scenery have constituted the settings of the majority of African films, although not as a focal point of interest as is the case in Cissé's *Yeelen.* Most of the events take place in the bush, such as the initiations, or on the parched grounds of the Sahel, but these settings are neither romanticized nor reduced to exotic decor in the Hollywood fashion that would deny Africans their culture. Universal significance is accorded nature's gifts: for example, ancestral soil, water, and trees are assigned symbolic values. In the scene in which Nianankoro and Attou are seen bathing, water symbolizes purity and fecundity, and the respect attached to the shooting of this sequence clearly indicates Africa's notion of morality maintaining the human body as a sacred entity. Although seminudity is visible, it is presented in a wholesome, nonsensational way. Similarly, colors are significant, as in the black pigmentation of Attou, Nianankoro, and the close-up shot of Nianankoro's mother as her fingers mix the water in the bowl before she uses it to purify her son. Consider also the shot of the stream water and blue sky, together with the natural shadows cast by the sunlight; these are captured in their purest naturalism. This attention to visual simplicity compels one to think about the close-up shots of the cliff leading to the stream as Cissé's panoramic camera pans slowly around to highlight natural beauty at its best.

In Africa the tree is universally symbolic, and like any other of nature's creations, accorded multiple interpretations. In the West, what is called "woods" may just be as thick as what Hollywood depicts in Africa as "forest" or "jungle." There is what one might call the cosmic tree that Africans believe eternalizes the link between the earth and the cosmic universe, an intermediary between the natural and the supernatural. Trees are also valued for providing basic necessities of life: food, medicine, and knowledge. Residents of the Sahel attach special importance to the baobab tree; it plays an important role in initiations and ceremonies.[21] In *Yeelen* rituals are performed under a baobab tree, which serves as a meeting place for cult members. (Cissé was particularly concerned with appropriate representation of the rituals to instill their significance and meanings in order to transcend the images of Africa offered in ethnographic films.) The tree is also an instrument of power and has mystical dimensions; the wing of Kore and the pestle of Komo (symbols of power and knowledge) are both carved from the tree of wisdom. As cultural "eyes," they are deemed to have seen ancestors come and go. In many African films of the Sahel region there is hardly any one film without the baobab tree.

The level of experimentation, solid as it is here, ratifies Cissé's contention that "knowledge is built and consolidated by one generation [precolonial], it is destroyed by another [colonialism], and recreated by a new generation [indigenous practices]."[22] In his discussion of the production and commodification of African art, Kwame Anthony Appiah distinguishes between the "traditional" and the "neotraditional."[23] "Traditional" African art—likely to be favored by minority African elites, "if they [buy] art"—dons the accoutrements of precolonial established "styles" and "methods," whereas the "neotraditional" is tourist art "produced for the West."[24] Applied to African cinema, I would argue that there is no precolonial African structure of the cinema, but rather that oral narrative gave African cinema precolonial traits that were integrated into the postcolonial cinematic paradigm. It is more plausible, therefore, to state that since African cinema originated after independence from two conventions—the mainstream and the indigenous—it exhibits affinities with the "neotraditional" in terms of inception but not in terms of commodification as in the "neotraditional genre" of African art.[25] However, since Africa is deeply immersed in a culture of poverty that worsened in the late eighties, some sections of African film practice transmogrified to what one might call "bitraditional" structures. This in turn, with respect to the popular films of the late 1980s, made them susceptible to exploitation, commercialization, and misinterpretation. Simply put, what is typical about this genre (bitraditional structure) is the choice of a universalist theme distinguished by hybrid conventions, which targets foreign spectators for materialistic ends. Does this mean that the films are incapable of cultivating African film audiences, or confronting and challenging the dominant conventions that Hollywood uses to construct denigrated images of the "other"?

Gaston Kaboré, Indrissa Ouedraogo, and Cheick Oumar Sissoko, Pfaff notes, "seem to have a predilection for the filming of rural lifestyles," but their films are "not made with specific ethnographic intentions."[26] But of the monotonous images of women pounding millet or corn, or families cooking or eating, as is the case with almost every film, one cannot help but caution that Africans already know the process of pounding, weaving, and tilling the soil. While these images unquestionably connote vignettes of life as actually lived in the villages, they can only appeal to non-Africans at the expense of alienating African audiences at whom these films are aimed.

We must also acknowledge that in using these images of rural lifestyles, some filmmakers have been able to construct culturally, politically, and aesthetically inventive films. One such case is the director Gaston Kaboré. His film *Zan Boko,* like his other feature *Wend Kuuni* (1982), depicts the "confrontation between two humanities," as he puts it, and critically examines the disrupted lives of people forced out of their ancestral village as a result of urban expansion. *Zan Boko's* main focus is on social justice, corruption, arbitrary misuse of power, acculturation, and freedom of the media to deal with the problems of modernization and urbanization.

Tinga Yerbanga, the film's protagonist, is forced to abandon his ancestral land to urban expansion partly because his wealthy neighbor wants to use the land for a swimming pool. A fearless journalist is thrown into the center of this social injustice when he aspires to reveal Tinga's predicament. His live television current events show mounts a forum on the problem of forced urbanization which Tinga attends. Because it implicates government officials, the show is suddenly cut off the air when the minister of information orders an end to the broadcast.

Zan Boko is a courageous film and ironic in the sense that the very establishment it lampoons, the Ministry of Information, helped to finance the film's production. It thus echoes the audacity and the strategy of denunciation in Cissé's *Finye* and Arthur Si Bita' *Les coopérants,* which also critique the governments that financed them. Kaboré, however, makes it perfectly clear that his film was started before the regime of the late Thomas Sankara was installed.

The film starts at a leisurely pace as ethnographic realism dealing with images of everyday village life. But it uses this meticulous exploration of village culture to inform the audience about the politically charged atmosphere of its second half. Kaboré skillfully transforms what would have been an innocuous and romantic patina of exoticized life into an extremely combative sociopolitical critique. John H. Weakland refers to fiction film as story film that presents "an interpretation of some segment of life by selection, structuring and ordering images of behavior."[27] Here in *Zan Boko*, this reference is recontextualized as a heuristic device rather than as an ethnographic study, and should not be misunderstood as such. When Kaboré's camera slowly takes its viewers to witness scenes of workers in the field, we are reminded of communal labor activities spanning time immemorial. Women carrying water gourds on their heads do not signify misery; rather, we are informed that it is way of life. A woman giving birth to a baby outside a modern clinic makes crystal clear the prowess of traditional medicine rather than the backwardness of "witchcraft." When men gather around fires each evening and women chatter, the relevance of oral culture is amplified. These are selections of images that accent sociocultural parameters of this work as reflected by its title, which Kaboré explains in his production notes:

> Amongst many black African cultures the birth of a child is accompanied by rituals that are designed to prepare the introduction and acceptance of a new member of the community. Among the rites practiced by the Mossi in West Africa, is the burial of the mother's placenta. This act consecrates the first bond between the newborn

> child and the nourishing earth. It is also the home of the ancestors and of the spirits which protect the family and social group. The place where the placenta is buried is called "Zan Boko." These two words are used by the Mossi when speaking of their native land, with a meaning that is at once religious, cultural, historic and emotional but also signifies a real relation with place. "Zan Boko" is an expression of the concepts of "roots" and identity."

In *Zan Boko* the special relationship of attachment to one's ancestral land is disrupted, turning that energizing cultural icon of existentialism, popularly tagged "son of the soil," into a figure of displacement and disillusion. This obliteration of tradition begins when the modernizing government surveyors arrive and start defacing the "pretty mud brick huts" with "white harsh numerals," in *Variety* reviewer Yung's phrases.[28] In this scene where natural sunlight casts interesting patterns on the wall, Kaboré eloquently dramatizes his concern for relationship and, above all, identity. In an unusual long take, the camera lingers on a nine-year-old boy as he moves gradually toward the numerals of his father's wall, which he laboriously rubs off with a piece of brick. In this sequence Kaboré knows when to alternate psychological time with cinematic time. The first shot establishes the boy's intentions as the spectator is positioned to witness him perform the act. For emotional impact, Kaboré then cuts to a medium close-up of the boy's face, which reveals consternation. Finally, the third shot (a close-up) shows the numbers almost completely erased as the boy's hand goes in and out of the frame—still rubbing. The sadness in his face suggests everlasting memory paralleling the scar of remembrance indelibly etched on the wall. Although the numerals are gone, the wall no longer retains its original look, just as the boy's experience is likely to remain with him as he grows.

Visual collage of village life comprises the tapestry of indigenous cultures which Kaboré exploits for intelligibility. In the second part of the film, when the focus shifts to the interrogation of the role of the media in society, the conflict between the traditional endogenous systems of communication and the technological, imported television culture is dramatized. This dramatization can be seen to represent an imposing dialectic between appropriating Western conventions and deviating from its norms. As an example, *Zan Boko*'s focus is on the problems of urbanization and modernization and their impact on the people and their relationships with the environment. But when the dauntless journalist, Yabre Tounsida, decides to have his current affairs program debate the issue live on television, television's role as a symbol of enlightenment is reversed and it becomes one of disinformation. He introduces his guests, the secretary general of the mayor's office, the general manager of national lands, the general manager of public works, and a sociologist, after which he announces that he was expecting a fifth guest. This guest, Tinga, is deliberately made to appear late in order to disguise the main focus of the show—the misappropriation of his land, nepotism, and governmental corruption.

The whole process of staging a live-action show of this type undercuts the subversive nature of television culture in developing countries such as Burkina Faso. As Tinga walks onto the stage, we can tell that he is in an alien environment. The only forum he knows is the traditional type in which the elders meet under the baobab tree or in the king's court where knowledge is transmitted and disseminated. Like this traditional forum, when Tinga walks into the studio, he offers his hand to the other invited guests only to be told by the host, "It is not necessary to shake hands. Have a seat." He heads toward the floor before Yabre points to an empty seat reserved for him. Momentarily, Tinga is carried away, mesmerized by a beautifully painted backdrop, artificial decor that he finds alienating and far removed from the kind of natural environment replete with fresh air and sunlight where traditional meetings are held as opposed to the air-conditioned room with hot lights pointed at his forehead. The program continues for a short period of time, and when the host acknowledges that Tinga's problem is scheduled to be discussed, the program is ordered off the air. Tinga's presence is symbolic because it is his story that causes rupture, forcing the minister of information to arbitrarily cut off the program.

During this period Kaboré's camera frames the participants in television news style, using medium shots and close-ups and eschewing the long takes and long shots that are characteristic of the first part of the film. A female announcer, dressed in a colorful African print with head tie to match, walks in a few seconds after the screen goes blank, to announce, "We ask every faithful viewer to please excuse us for this technical problem. But our program will continue with the third episode of the magnificent serial 'The Golden Dream,' which will take us to the enchanting Riviera." This statement mocks the very identity, the appearance that the announcer represents. In her statements, "faithful viewer" and "enchanting Riviera" metaphorically connote noncompliance and cultural erasure. Perhaps we would have accepted the naïve excuse for technical problems if the audience were taken to the splendor of Victoria Falls (at the border of Zambia and Zimbabwe) or to Maroko, in Lagos, Nigeria (to witness urban slums, chaos, and helplessness). That a program which aired to debate national problems is replaced by a foreign program suggests a myopic vision that is indicative of national disaster. Kaboré uses this irony to look critically at society, its cultural heritage, and diversity. While questioning, he also subverts and destabilizes anachronistic dichotomies rooted in the traditional and modern.

The pervasive codes of media communication in this circumstance pose a serious challenge to traditional oral communication patterns. The rich man's message is communicated to Tinga via word of mouth as opposed to the newsroom situation in which telephones function as an oppressive device. It is possible for the government dignitaries to connive with one another through the telephone to stop Yabre's broadcast. In both cases, television and telephone, as subversive communication devices, demarcate what we see, how we see it, and how we evaluate the images represented. The interesting thing is that it is through the omnipotence of oral literature, which modern technology here tries to subvert, that *Zan Boko* becomes indefatigable. Therefore, it is right to observe that the film's peculiar hybrid nature is capable of introducing a situation that makes it possible to negotiate between generic codes and cinematic forms both affirmatively and transgressively, creating a nationalist work as Kaboré does.

The film also shows how modern society subverts traditional culture through double standards. A clear example would be the kind of support Tinga's rich neighbor receives from high government functionaries who have sworn to uphold the law of the land and faithfully serve its citizens. It demonstrates how nepotism and personal alliances can easily be used to outmaneuver others in one's own society. The effectiveness of the film derives from the simplicity of Kaboré's style, his editing, camera placement, and mise-en-scène, recalling his concern for "two humanities." In effect, the first part is realized in minimalist editing as every shot, every long take, comes right out of the camera as if unmediated. Characters are observed in normal behavior as they go about their business in the realist traditional sense typical of Satyajit Ray (*Pather Panchali*) and Ousmane Sembene (*Mandabi*), but also incorporated are some elements of the here-is-the-villain-shoot-him structure that is reminiscent of Jorge Sanjine and other Latin American documentary filmmakers[29]—as the pace quickens in the second half. There is manifest interest in community and humanity (the good and bad) in the framing of a disdainful young boy as he aspires to reassert his parents' dignity by erasing the numerals painted on their wall, or an affluent couple's display of insensitivity and double standard as they cannot bear the smell of their neighbor's "soumbala," a condiment that Tinga's wife, Napoko, prepares and sells. In actuality, this condiment is an indispensable cooking staple for both the rich and poor, elite and illiterate.

To show how traditional culture is colliding with contemporary realities, *Zan Boko* reverses the order of traditional mores; confiscation of land goes against the tradition of Mossi culture, and to think of offering money for a piece of land is an act of betrayal. Pfaff notes that "land was collectively owned and assigned to individuals by heads of lineages. . . . Land could, traditionally, be borrowed or inherited but never sold."[30] The griot's words as he sings "The monster has triumphed" in a local outdoor bar endorses the feelings of Tinga, who represents the majority of the underclass who are likely to be trampled on. As he tells his wife, "What bothers me is being treated like outsiders on the land of our ancestors." The problem of illiteracy in the film recalls *Mandabi*'s theme. Because Deing, the hero of *Mandabi,* cannot read or write (he can only write his name, which works to his detriment), he signs his money order away to an unscrupulous relative. In *Zan Boko,* the lack of education seems to hinder Tinga's comprehension of the television debate regarding his predicament. Like Dieng, he is not stupid. Mirroring that prophetic statement at the end of *Mandabi* when Dieng says, "We shall change all these," in the final shot of *Zan Boko,* Tinga's final statement echoes the voices of the marginalized who make up 90 percent of the population. Looking at Yabre's agonizing face, he says, "I don't understand French but it is easy to see something serious has happened. I urge you to remain true to your convictions and to yourself as human being."

Finzan is the second feature-length film by Sissoko following his successful *Nyamanton* (Garbage boys), winner of a gold medal at the Mannheim Film Festival. *Finzan* is a story about women and their resistance to the social systems and traditions they consider oppressive. The story focuses on the plight of two women: Nanyuma, a mother who is now widowed, and her niece, Fili, a socially sophisticated young woman who has not undergone the traditional clitoridectomy. Nanyuma was married to an older man (possibly as old as her father), who had two other wives. He dies after eight years of their being together. Her predicament is not yet over since tradition demands that she remarry, to her late husband's next of kin, Bala, the village buffoon. This obligation is ratified by the village chief despite Nanyuma's indignation. Nanyuma, however, is in love with a younger bachelor about her own age whom her family rejected as a suitor before her marriage and who now once again is being denied the opportunity to marry her. She escapes to another village, taking refuge in her brother-in-law's house. This situation sets up a division amongst relatives as her brother-in-law tells her that she is obligated by custom to marry Bala, while Fili supports Nanyuma's position.

In another twist, the film shifts focus to a general emancipation theme designed to reinforce Sissoko's social critique. When a government dignitary arrives and demands that villagers surrender tons of millet to the government at a fixed price well below market value, the villagers' protests lead to the king's arrest. He is later freed when the women mobilize themselves to express their disapproval of his arrest. The king's freedom in turn empowers the women to protest their own marginalization—women's role in society and Nanyuma's plight—by threatening their husbands with a sexual boycott if their requests are not met.

Meanwhile, Nanyuma is caught and returned to her village, where she will be forced to marry Bala. Her young sons, who despise this treatment of their mother, retaliate by planning retribution against Bala. In subsequent scenes laced with comedy, Bala's water is adulterated with a poisonous concentration capable of inducing diarrhea and flatulence in order to make him believe that the gods demand that he let Nanyuma go. Other humiliating measures follow, which really make him look foolish, including using Koteba "ghost" costumes and voices to send threatening messages to frighten him. However, all these measures do not stop Bala's obsession with Nanyuma, who refuses to consummate her marriage with him. She becomes very friendly with Fili, who is soon to be accused of irreverence toward society for not having endured the required clitoridectomy. This is where the real tragedy lies. The villagers are divided and the women who support the king's position of letting tradition prevail (even though they know that there is a possibility of infection, which might ultimately cause her death) outnumber Fili's supporters. Subdued and helpless, she is forced to undergo the procedure after which she bleeds profusely and is rushed to a hospital by her father.

In the end Nanyuma leaves the village with her youngest son, surprisingly, with no challenge. This occurrence signifies a return to sanity—a progressive synthesis between an antiquated past and demystified present which the director uses to project a remodeled society in which women and men negotiate on an equal basis and solve their problems together.

Like his first feature, *Nyamanton,* which seriously depicts the tragedy of children in Mali's capital city of Bamako, *Finzan* examines traditional systems and culture, vehemently denouncing the contradictions inherent in them. From the beginning of the film one recognizes the handiwork of a professional seeking to give illumined definition to film form and cinematic art. In attempting a compelling narrative style, there is no doubt here that cinematic art reinforces the strategy of rendering explicit the cultural and the political. However, from the perspective of forging indigenous film culture respectful of African sensibility, one is forced to ask if what is presented is not some kind of structural asymmetry that tends to perpetuate the entire paradigm of cinematic representation of African cultures, as Western practices do through seeking the exotic.

The film opens with a graphic reflection on motherhood as a beautifully composed shot of two goats shows them giving birth and straining at their tethers. This shot is followed by a catchy text in French which announces that "a world profile on the condition of women reveals the striking effects of double oppression. Women are 50 percent of the world's population, do about two-thirds of its work, receive barely 10 percent of its income and own less than 1 percent of its property." This statement is not merely descriptive; it also reflects the populist cry for equal justice. For anybody who has been following feminist writings, it could easily be misconstrued as a film made by a woman, or as a United Nations' documentary seeking out Third World "barbaric" cultures for debate. Rather, *Finzan* is a moving account of desperation told with sincerity and boldness, a film that looks at women through an empathetic male's eyes.

Although the film deals with a number of social issues, the main focus revolves around women's emancipation. One focal point centers on the cultural practice of clitoridectomy which, still in effect today in over twenty African countries, by United Nations' estimate, affects from twenty million to seventy million women. This practice has traditional significance in Africa, for excision was thought to be a proper way of guarding virginity as well as discouraging extramarital intercourse and an insatiable sex drive. It is important to restate that the subject of "excision," "female circumcision," or "clitoridectomy," as it is variously called, is a controversial one. From the perspective of a Western cultural norm, the practice is usually denounced as sexual mutilation performed by "ethnic" cultures seen as "barbaric" or "anachronistic." The practice's defenders, on the other hand, see its criticism as nothing short of racist, anti-African diatribe. It is pertinent to emphasize at this point that our reading of *Finzan* is not based on dogmatic principles. Following the same usual pattern of decipherment applicable in all circumstances, ours is not a matter of whether circumcision is still relevant in our present time, or whether one believes in it or not; rather, the main focus is to examine Sissoko's film style and his overall method of analysis.

In *Finzan,* Sissoko presents a critique of indigenous culture in a manner recalling the liberationist injunction to fight for freedom, a call for rebellion. I will argue that while the filmmaker presents this issue with utmost concern, his camera fails to present a detailed examination of the culture from a logical perspective enabling us to understand how the ritual evolved. Rather, what we see is a farcical analysis which treats the subject of excision, in the words of Françoise Lionnet, "peremptorily, in an impassioned, reductionist and/or ethnographic mode which represents the peoples who practice it as backward, misogynistic, and generally lacking in humane and compassionate inclinations."[31]

In a scene preceding the one in which the female traditionalists mobilize to track Fili down, we are shown what appears to be callous attitudes on the part of the children who repeatedly tease Fili for not being circumcised. This parallels the brandishing of a razor by the woman assigned to perform the excision in a macabre manner reminiscent of Bunuel's *Un chien andalou,* where in one of cinema's most bizarre moments, an eyeball is caught in big close-up as it is slit open. The irony here, though, is that Bunuel is a surrealist interested in the bizarre, incongruity, shock, and rejection of causality while Sissoko is a realist who prides himself in representing African reality for the purpose of enlightenment and entertainment. He succeeds in the latter (but not necessarily in *Finzan*), his narrative drawing from the popular Koteba theater combining slapstick comedy and satire with raucous scatological humor. Some of these devices are tried in *Nyamanton* and they work well, but in *Finzan* they function to distance the African spectator from the narrative, particularly in the bush scenes when the little children play tricks on Bala, or when a snake is made to crawl on his body and into his jacket before he smashes its head. Following this, Bala dances, glorifying his prowess for having killed a snake.

Regarding film and entertainment, if *Finzan* is strong in the entertainment sector, it serves to satisfy non-African spectators; it is unilluminating since the film's structure adheres to "the new pressure to seek pleasure and fascination in the exotic"; nor is there sustained cinematic inventiveness to suggest overt insurgency or challenge to the subjugative dominant structures.[32] Throughout the film, Bala's acting cannot go unnoticed. Although he overacts at most times, he instinctually commands professionalism, but his performance alone is not enough to salvage the film from its amateurish style of representation.

Indeed, the question of inventiveness is so often raised by critics and filmmakers who want to emphasize either the equivalence between education and entertainment or cultural productions reflective of genuine indigenous art and aspirations that we must examine its premises. For example, through constructive criticism the relevance of cinematic structure can be tied to its ability to inform the viewer of important sociocultural issues such as excision, which is prone to disdain and misinterpretation. A cautious approach, rather than haphazard assemblage of reality, in my opinion, might best illuminate Africa's "fitful process of shifting from one set of rules to the other" (from colonial and neocolonial influences to cultural critique and changes) when "loyalties are stretched between commandments of the Bible" (missionary influences) and "obligations to the ancestors" (traditionalism).[33]

As Abdou Diouf, the president of Senegal stated,

> Female mutilation is a subject that is taboo. . . . But let us not rush into the error of condemning [genital mutilations] as uncivilized and sanguinary practices. One must beware of describing what is merely an aspect of difference in culture as barbarous. In traditional Africa, sexual mutilations evolved out of a coherent system, with its own values, beliefs, cultural and ritual conduct. They were a necessary

> ordeal in life because they completed the process incorporating the child in society. These practices, however, raise a problem today because our societies are in a process of major transformation and are coming up against new sociocultural dynamic forces in which such practices have no place or appear to be relics of the past. What is therefore needed are measures to quicken their demise. The main part of this struggle will be waged by education rather than by anathema and from the inside rather than from the outside. I hope that this struggle will make women free and "disalienated," personifying respect for the eminent dignity of life.[34]

There is ample room from this policy statement for us to agree or agree to disagree. It is interesting how Diouf challenges us with the task of handling the issues and becoming informed as never before about African social systems, beliefs, and thoughts. For Sissoko, *Finzan* "deals with excision as an oppressive practice," but he "does not want it to be known as a film about excision," saying that he "made it generally as a film about women's right and struggle for freedom."[35] Yet it is the "mutilation" sequence, which lasts for less than fifteen minutes, that is the most powerful and talked about because of the way it was shot. In Sissoko's filming of the brutality accompanying the capture and tying together of Nanyuma's hands and feet with a rope that looks to be as fat and strong as that which anchors ships to the docks, the long take, long shot (of hands and feet) and the close-up (of the rope) lingers for an unsympathetic time.

It is true that filmmakers have to contrive an adventure, and Sissoko's automatically finds a center of attraction—the bush and the bizarre. It could be argued that most of the scenes in the bush could have been filmed at a better location or left out entirely and the film would make better sense. To substantiate this claim, let us look at a particular sequence in which Jean Rouch's[36] influence is most profound and retrogressively appropriated. In this sequence, after the king rules in support of Nanyuma's marriage to Bala, she tries to escape from the village by hiding in the bush. This triggers a night search by the villagers, a well-orchestrated sequence comparable to the search sequence in Euzhan Palcy's *Sugar Cane Alley* (1983) (when the whole village, led by José, played by the talented child actor Gary Cadenat, mounts a formidable search for Medouze). Villagers are seen here holding lit torches scrambling to find Nanyuma. No definition of body contour is visible except when certain forms need to be highlighted, as when Nanyuma stops running to oil her body. The only other thing that Sissoko draws our attention to is a close-up chiaroscuro effect of a ferocious lion growling as it charges Nanyuma or the villagers. The lion's menacing presence is shown three times: in the first and third it is visible and in the second we only hear it growling as the villagers run for their lives. This particular sequence is especially disturbing because of the deployment of retrogressive practices reminiscent of African images proffered in Tarzan jungle melodrama films. Fashioning his cinematic structure after this practice, a strange misguided judgment that subverts the director's attempt to liberate himself from constrictive norms, canons, and myths the mainstream media has used for a long time to undermine Africa, indicates a failure to deal with the burning question of indigenous mode of African film language—something that has preoccupied the minds of African cineastes since the inception of African cinema. It also highlights the desperation that is in many hearts, African critics and audiences alike: How to completely decolonize the image?

If the goal of African film practice is to formulate strategies for the production of work relevant to pan-Africanism, our concern here is to build a community of support for cultural and creative endeavors and to explore possibilities for collaboration in pan-African spirit. This is emblematic of the position espoused by Kwame Anthony Appiah in my second epigraph to this chapter taken from his essay "Is the Post- in Postmodernism the Post- in Postcolonial?" regarding African art, its commodification, and interpretation. *Finzan* leads one to question what guidelines are necessary to construct a "positive image" of Africa. By "positive image" we mean the representation of what really matters, not that the image has to be favorable and conforming to the values of the norm (hence a critique of Islam and African systems, for instance, in *Ceddo*). That is, "positive" refers to the appropriateness of the method of selecting reality in a particular context. All to often, artists' affinities with specific cultures conjure immediate knowledge; sensitivity to that culture's configurations becomes instinctive. Unfortunately, this is not always the case, *Finzan* is beautifully filmed, as in the opening scenes, where composition-in-depth (deep focus in one case) is used to capture the village environment. The night sequence, while also beautifully shot, defies logic. This sequence, as in so many other scenes in this pockmarked film, is "pretty bad." And if sensible depiction of Africa means working to dismantle demeaning mainstream canons of representation, we might as well ask if *Finzan* has helped to reverse the popular assumptions perpetuated by existing forms of cinema and television. Commodification, as in the "bitraditionalism" discussed above, is not on a par here with African identity and sensitivity, especially the search sequence's "Tarzan" overtones. *Finzan* is a film that no African will watch and feel proud of, nor want to purchase for subsequent viewing. *Finzan* is extremely popular with Western audiences and widely distributed for classroom showings as, according to one critic, "an important new resource for studying rural sociology . . . [that] can bring to Women's Studies curricula a badly needed African perspective."[37]

There are recent African films by aspiring African filmmakers that do not celebrate contemptuous colonial and postcolonial messages which have worked historically to limit knowledge about Africa. Their structures may be steeped in the traditions of ethnographic vérité, but they do not promote the entrenchment of inferiority or the "development of underdevelopment," in Hondo's words, but rather, a development path to artistic maturity. For reaching a truer sense of African reality, they have been able to push art toward a celebratory role of a culture and of a life that is as warmly as it is urgently articulated. In this category are *Saaraba* (1988) by Senegal's Amadou Saalum Seck, which uses utopian motifs to demystify the illusions of independence in postcolonial Africa, and *Angano . . . Angano . . . Tales from Madagascar* (1988), coproduced by Marie-Clemencé and Cesar Paes (Madagascar/France), a highly innovative ethnographic film, brilliantly conceived and evocative, which places oral tradition specifically in its social context. As Donald Cosentino writes in California Newsreel's catalogue, "This film reveals the ever shifting, perhaps illusory, boundary between reality and myth." The others in this selection include Idrissa Ouedraogo's *Yaaba* and *Tilai* (Burkina Faso, 1989 and 1990), to which we will soon return.

Saaraba in Wolof stands for a mythical environment, equivalent to the Western idea of utopia, supposedly free from the troubles of everyday life. Like Kaboré, Seck has a penchant for neorealism, and like *Zan Boko, Saaraba* starts as an ethnographic document, changing its style to a more conventional narrative structure as it progresses. It is the ethnographic value of the first part, however, that gives the whole film its African feel. *Saaraba* recapitulates some of the themes that popularized early African films and their combative posturing. It denounces with unflinching temerity the corruption and lust of materialism and the excesses in the life-style of the older generation of Africans. In doing this, Seck foregrounds the disillusionment that has crippled Senegal's younger generation since independence—including the director's age group—who see preindependence promises melting away in the postcolonial period.

Seck's first feature functions as a political critique and speaks well of African film projects that scrutinize contemporary developments in traditional values of the first generation of African filmmakers. As we notice in *Saaraba* a craving for a personal style, we are also reminded of the future of African cinema, the new direction proposed by the African film practice of the 1980s.

Idrissa Ouedraogo, the dean of the "new wave," citing Burkina Faso as an example, contends that films made in Africa with Africans solely in mind "cannot generate revenue to defray production expenses. If we want them [films] to be truly profitable, we have to internationalize our methods of making films. I personally do not dream of industrializing my films; I try with the little means I have to deal with subjects concerning human beings."[38] It is from this perspective that Ouedraogo has succeeded where many others have not broken the bondage imposed by distribution and exhibition problems. His films have been extremely popular and well received. Africa and its cultures are the mainstay of his thematic convergence, in which the humanistic and the universalist interweave with the ancient and the present—probably deliberately avoiding political confrontations (and confrontations with the status quo).

In terms of universal acceptability and commercial viability, *Yaaba* is exceptional if not unprecedented in the history of Black African cinema. It is one of the few African films that has made it to international commercial screens, and remarkably so, in box-office terms. Following its critical acclaim at Cannes (it was selected as the opening night event of the Directors Fortnight), the film has played to large, enthusiastic audiences in Africa, Europe, Asia, and the U.S., thus shattering the exclusion of commercial theaters that has long plagued African cinema.

Ouedraogo's next film, *Tilai,* about "incest, revenge, crime and punishment," is the winner of the Special Jury Prize, Cannes, 1990, and also won the Grand Prize of the 12th FESPACO—the coveted Étalon de Yennega. Ouedraogo has since made another film, *A Karim na Sala* (1991), also shown at the latter festival. Bowing to international pressure to produce another miracle following previous successes, this work failed to meet the expectations of people to see a well-polished film like *Yaaba* or *Tilai. A Karim*'s postproduction was reportedly rushed for the 12th FESPACO's opening ceremony. It is a poorly edited film that may never be resurrected from FESPACO.

Yaaba is a deceptively simple parable, filmed in an unadorned, lucid style. It derives its plot from an African tale of the oral tradition. It is the story of two youngsters, Nopoko (the girl) and Bila (the boy), and an elderly woman. The woman is ostracized by the villagers, who accuse her of being a witch and therefore responsible for the village's misfortunes. However, the youngsters recognize that the elderly woman is often blamed for what she has not done and decide to befriend her. They nickname her "Yaaba," meaning grandmother.

Yaaba exemplifies a number of important trends in contemporary African cinema.[39] It is made in a style that could be called "elitist" and "individualistic"—combining various comedic modes with moral nuance in a fashion that reveals the contours of a revivified pan-African strategy. This strategy manifests itself in the shared concensus of African filmmakers' quest for authentic treatment of African issues in a style that not only inscribes African identity in African films but also renders the films competitive in the international market. The "new" African cinema interweaves elements of melodrama, satire, and comedy in a manner that attempts to satisfy the spectator's appetite for entertainment. *Yaaba* typifies this trend and its success and international acclaim exemplify the goals of the new crop of African filmmakers.

The appropriation of neorealist techniques by Third World filmmakers has been well documented, and Ouedraogo fuses the neorealist penchant for eliciting polished performance from nonprofessionals with the African narrative tradition of the griot (oral storyteller). Ordinarily, *Yaaba's* story, in terms of cinematic storytelling, would be dismissed as too amateurish. But, as in the oral tradition, a story's interest and attraction for an audience depend on how creatively a storyteller embellishes what he has heard or taken from his own experience. Although *Yaaba's* story line is basically simple, the film itself commands a universal appeal owing to the sophisticated ability of the director to unite all the essential elements operating in the film. Here, bits of humor, comedy, and satire coalesce into an idiosyncratic cinematic style that highlights societal mores and those who attempt to undermine them. In terms of mise-en-scène, every detail is conveyed by the extraordinary framing of a static camera. In a sequence in which the camera lingers on villagers playing a traditional West African game, participants proceed in and out of frame in a fashion that unobtrusively captures the flux of traditional life. Similarly, skillful editing promotes rhythmic progression of disparate episodes. Attention to visual detail is coupled with Francis Bebey's effective score, which makes use of indigenous string and reed instruments to heighten solemn moments (for example, Yaaba's burial) as well as carefree ones.

The film's emotional power is largely due to the choice of Yaaba as the village scapegoat. Her old age leaves her vulnerable to superstition, since elderly people are among those most suspected of being witches. This stigma does not, however, render her any less the quiet woman of wisdom she is. Anyone familiar with African culture knows the tremendous respect accorded elders—which is, of course, withheld from Yaaba. This respect begins in the family and is ordinarily unconditional. The aged, like the sick, for instance, are taken care of within the home, and members of the family, including the extended family, are obligated to contribute their quota. Contrary to what occurs in the West, where, for instance, government welfare programs are substituted for family care, in African society close-knit family relations eliminate the horrors of old age. Old people are indispensable, and above all, they are loved, no matter how rich or destitute they may be. In *Yaaba* the old woman is denied this important benefit of her life, ostracized, and driven to despair by the villagers. Yaaba is apparently childless; although she grew up in the sophisticated caring system of African communalism, she cannot reap the benefits of an extended family, having been dismissed with ignominy. This desertion causes other members of the village community to ridicule her; in one scene we see them throwing stones at her. She survives the resulting wound and it is Nopoko and, even more, Bila, who show sensitivity to her predicament. At one point, when the children call her "Yaaba," she is elated. "This is the first time that someone has called me Grandmother, and that makes me happy," she says, as she and Bila eat the chicken that Bila has stolen from the village to feed her. As a rule, stealing in the village, as in many other societies, contravenes ethical and moral codes. However, there is good reason for Bila to steal the chicken for Yaaba to eat; it is out of sympathy as well as a matter of subsistence, since Yaaba is unable to trap meat for herself.

Nopoko and Bila represent the values of modern Africa and serve as a bridge between the present and the past which is Yaaba's tumultuous environment. Here the film seems to suggest that there is need to abandon superstition and critically examine the causes and effects of all societal circumstances before crucial judgments can be arbitrarily imposed. Yaaba's ordeal reveals that what is associated with the past is not necessarily at odds with the present; her wisdom is reincarnated in the vision of Nopoko and Bila, who in this respect represent both future and hope. This quest for future and hope is strongly entrenched in the solidarity of Bila's prophetic utterance, when, after Yaaba's house is burned down by mean-spirited villagers, we hear him say to her, "I will build another house for you."

In another sequence when Bila is accosted by the village bullies (a group of youngsters about the same age as Bila and Nopoko), a fight ensues and Nopoko rushes to the aid of Bila, and she is cut by a rusty knife held by one of the assailants. The cut becomes infected and she falls ill, causing her mother to fear she has tetanus. After all medical attention fails to cure Nopoko, it is the medicine of Taryam, a native doctor who is connected with Yaaba, that provides a cure for her ailment. Once again, Yaaba's wisdom prevails. Bila serves as the go-between, the link between Yaaba and the native doctor, although Nopoko's mother must hide Taryam's medicine from her husband, who does not want anything to do with the witch or her native doctor. Nopoko's recovery synthesizes the film's attempt to reconcile past and future. Bila's endeavor to repudiate suspicion, superstition, and ignorance speaks of a desire for understanding between the young and the old, past and present. The temptation to disparage Yaaba's traditionalism and Taryam's values (the healer's medical roots originate from traditional culture) is offset by the fact that recourse to these ancient remedies prolongs Nopoko's and Bila's lives. Ouedraogo's respectful treatment of Taryam is part of the film's dialectic between traditional wisdom and the sometimes dubious "progress" wrought by modernity.

In all African societies, adultery is a sin and divorce is an offense, if not taboo. Within the rural milieu of *Yaaba,* adultery and divorce are condemned, since they threaten communal solidarity and family cohesiveness. In the film, Ouedraogo displays all three, not through commentary but as plot elements meant to entertain spectators, not instruct them. Thus, when he shows us an attractive young woman, Koudi, married to Noaga, an alcoholic, who is having an affair with a local charlatan, it is not surprising that he bypasses the implication of such a serious taboo subject by showing her comically upbraiding her husband for being impotent. Since Noaga is unable to sexually satisfy Koudi, a farcical transference of guilt occurs in which the promiscuous wife absolves herself of any responsibility. Yet, the beautiful woman cannot leave her husband because of the matrimonial bond consecrated in traditional doctrine. This narrative ploy is in sharp contrast to Sembene's *Xala,* a film in which impotence suggests a larger political metaphor, an allegorical strategy that is absent from the more prosaic plot twists of *Yaaba*.

Stealing, bickering, gossiping, eavesdropping, and juvenile mischief are portrayed, no doubt flaunting the strict moral code of a community deemed closely knit. But behavioral strictures are not the primary focus in *Yaaba,* even if the dilemma seems to extend far beyond the mere question of moral decay. Although much of the film's strength lies in its whimsical irony, Ouedraogo nonetheless does not seem to have a clear vision of the African future, and there is cause to wonder whether *Yabba's* parabolic structural oppositions truly illuminate societal conflicts. In one sequence, for example, a quarrel breaks out between a couple, but instead of showing them fighting, in the next shot a door opens and both of them end up in a romantic embrace—this is like saying, sarcastically, of course, "If this is how couples fight, no one will ever cry." In another instance, the mother of three delinquent youngsters who are constantly berating other members of the community is shown turning a deaf ear to the misbehavior of her own children.

In *Yaaba,* the director creates expressive rhythm through characterization and meticulous attention to detail. The marvelous performances of the nonprofessional actors increase the stark realism and simplicity of the narrative, which in turn focuses attention on the film's drama. In the end, the spectator is left with the feeling that the compassionate human drama as it unfolds in the village where love, evil, fear, superstition, and intolerance coalesce are revelations that only an insider can convey. For example, by making Yaaba a scapegoat, in this case accused of being a witch and therefore responsible for the evils afflicting the society, the film touches upon a subject that is crucially important for understanding African culture. For instance, in colonialist discourse witchcraft is denigrated as "superstition," demonic," and "primitive," endemic to the "dark continent." Although the subject of witchcraft is used to enhance the story line, it is not used tendentiously to either valorize African tradition or dismiss it as irrational. In this respect the film exemplifies an African sensibility in its effort to sensitively represent certain aspects of African culture.

The various episodes and segments that compose *Yaaba*'s structure are seamlessly constructed. Along with the utter simplicity of the narrative pattern and his careful attention to detail, Ouedraogo knows when to alternate long takes and fast cutting, silence and sound, light and dark. For example, the scene in which Bila takes food to Yaaba and discovers she has died progresses with rapid cutting—Bila running to inform the villagers and then returning, forward to the burial sequence where the action is slowed down deliberately by the employment of the long take. The scene in which Koudi is seen wanting to have sex with Noaga, her alcoholic husband (which comes to no avail), is showered with chiaroscuro lighting but also features an alternation between silence and sound—reminiscent of Fritz Lang's *M*—that expedites our anticipation. Thus a static camera watches the young beauty as she struggles to arouse her husband, who is snoring loudly. When she fails to awaken him, she resigns to melancholy, twisting and wiggling in silence. This silence is broken momentarily by whistling from outside. When Koudi walks to the window, the camera shows her submerged in light and dark looking out the window into the night until we are made to understand that it is her lover, Razougou, signaling for attention.

The dialogue spoken throughout the film is sparse and delivered in Mooré, one of Burkina Faso's mother tongues, with English subtitles, though most people who have seen the film agree it can be understood without them. The emphasis on image over heavy-handed dialogue is increasingly common in African cinema. The multiplicity of African languages demands the breakdown of language barriers, enabling films to cut across boundaries. The desire to internationalize black African cinema in order to gain larger audiences and reap greater financial benefits currently defines the structure of the "new" African film.

However, *Yaaba*'s innovative tendencies seem to obscure some of the flaws in this film: many sequences seem all too predictable owing to the deployment of stock characters and clichéd plot devices. *Yaaba*'s admittedly diverting vignettes do not have the power to suggest the historical resonances achieved by a film such as Hondo's *Sarraounia* or the nuanced depiction of African ritual explored in Cissé's *Yeelen*: but they are clearly antipodal with Sissoko's *Finzan.*

After *Yaaba,* two African films, *La vie est belle* (Life is rosy, 1985) by Ngangura Mweze and Benoit Lamy and *Bal poussière* (Dancing in the dust) by Henri Duparc are distinguished by their engagement with popular culture and Africa's polyphonic rhythms tempered with Hollywood enticements. But *La vie est belle* is not at all similar to a Hollywood film. It is an interesting film, an unusual African film, commercial but not trivial.

It was released as a coproduction involving Belgium, France, and Zaire, Ngangura Mweze wrote the script and was assigned a codirector by the authorities in the Belgium Ministry of Culture who provided two-thirds of the financing as a condition for cosponsorship. The outcome of this problematic collaboration is a contemporary comedy, the roots of which are tied to the Zairean popular culture and to commedia dell'arte—television programs of years ago where singing and depiction of the sexes were very popular with the audience.

In *La vie est belle,* Zairean popular theater and music are also used, allowing the audience to experience the Kinshasa boisterous night life. Here, a popular Zairean musician, Papa Wemba, a star and a crowd-pleaser, plays the protagonist, Kourou, a neophyte in the process of exploring the musical road to stardom. Although the film is a musical comedy, it is not strictly structured as an entertainment film using the African prototype of a modern city (Kinshasa) as mere decor. Rather, Kinshasa comes to exemplify many of the film's thematic preoccupations: the city as a fashion capital à la Paris in which we see men and women chicly dressed; the city as a boisterous music capital continually hosting the region's hundreds of musicians, many of them the world's best, in which we find the development of new tunes and dance steps presented as routine;[40] and a city that exhibits the problems of urbanity, the entrenchment of the survival of the fittest. *La vie est belle,* like *Kukurantumi* (also set in a big city, Accra, Ghana), deals with the theme of misplaced priorities, as with the show of affluence in the lust for expensive goods, such as Mercedes Benz cars, polygamy, class division, male-female relations, urban migration, and economic imbalance. Although the structure of the film is comedic, one can still recognize societal discrepancies as they are filtered out of joke-filled scenes that are sometimes humorous and sometimes ironic, as in the scene in which a dwarf who hawks "kamundele" (shish kebab, in Lingala) in front of a nightclub reminds us that "life is rosy" as he watches Kourou and his feisty girlfriend frolic on the doorstep.

Male impotence among the African bourgeoisie, the theme of Sembene's *Xala* (1974), is also explored. However, unlike *Xala,* in which the subject is used as a caustic satire metaphorically signifying the political impotence of Senegal's ruling oligarchs while at the same time valorizing the expediency of traditional medicine, *La vie est belle* uses it to induce laughter. One such moment is when Kourou's rich employer, Nvouandou, comically practices the dance steps prescribed by the traditional medicine man as a cure for his impotence. Ngangura Mweze makes the following observation on the search for a popular film form apropos of a paradigm that argues for a new way of seeing.

> The term "commercial cinema" is quite pejorative, let me call it "popular cinema." I like to make a film that many Africans come and see and feel good with it. If I want to say commercial cinema, it means concocting stories for money, reinforcing it with sex, violence, speeding sports cars . . . that is not what interests me. Rather to make films which are best popular with African culture as well as which will feature African problems and accomplishments at the same time. I don't feel like a teacher or a messenger of a particular idea. I am a filmmaker, as there are carpenters and bricklayers.[41]

Like *La vie est belle,* Henri Duparc's *Bal poussière* makes use of popular forms. The film is replete with entertainment features as in the emerging trends. His, though, is being pushed toward a more bizarre, vulgar tradition, as his new film *Sixième doigt* (Sixth finger, 1991) suggests, which unfortunately is not only nihilistic but also reproduces many of the racial and ethnic stereotypings that African films purport to attack. When these films were shown at the 12th FESPACO, African critics, who still associate "sexual modesty with African dignity and sexual exhibitionism with Western decadence," as Rayfield puts it, were very critical of the overblown sexual scenes. On the surface *Bal poussière* is a social comedy focusing specifically on polygamy, making references to corruption, contradictions of tradition and culture, but cannot be taken seriously as the issues are treated with only one thing in mind—amusement.

The focus of *Bal poussière* is a local rich man about fifty years old who is obsessed with power and calls himself Demi Dieu (Demigod). He already has five wives and wants a young school girl, Binta, played by T'Cheley Hanny of *Visages de femmes,* as his sixth wife. She is not happy to have been sent out of the city by her uncle. She meets up by chance with Demi Dieu, who falls in love at first sight and promises to marry her. He is wealthy and Binta's parents agree to Demi Dieu's proposal. As soon as she comes into the household, the rebellious Binta exposes the other more traditional wives to untraditional behavior, such as naked bathing, splitting the family into two factions: those wearing Western-style dresses and those wearing traditional attire. Soon, Demi Dieu, who used to think with his penis, believing, according to the production note, that "a sixth wife will only harmonize the week: a wife for every day of the week and Sunday for the one who behaves best," now knows that that is, after all, not a "supreme reward."

Bal poussière's photography is stunning, the acting exceptional by African film standards. The colors, composition of shots, and rhythm and pacing epitomize polished craftsmanship. It is highly entertaining and is so at the expense of the changing role of African women or women's liberation, although that is exactly what it supposedly argues for. On a few occasions, the film manages a self-conscious attempt at germane criticism. For instance, at one point a prostitute who attempts to seduce a bar client is asked, "And what about AIDS?" Similarly, the film concludes that polygamy is somehow retrogressive, since the overbearing character Demi Dieu mishandles it as he does. Binta is allowed to leave him for the man she loves—in the spirit of a liberated woman—but only when she has proven that she could not be a traditional housewife like Demi Dieu's other wives.

Duparc's structure and his method of construction become questionable in view of his cinematic voyeurism. Consider, for example, his camera's preoccupation with the conspicuous, leering at Binta's features and contours, to expose them in close-up—her bouncing breasts, and in full-screen close-up, her buttocks. For Western viewers, these features may be more gripping and profoundly alluring, but for African critics, including feminists, Duparc's experimental interludes are simply jaundiced, misguided, and stereotypical.

Within the domain of the new African film practice, therefore, nonconformism, more than a flight from established traditions, is rather an inevitable transgressive mode of practice if only to accomplish the Herculean task of getting a good film distributed widely in important theaters. This has led to competition and aesthetic proliferation of inquiry into diverse filmic applications. I argue that some of the conventions used to attain this aspiration have, however, been misappropriated.

Notes

1. For the motivations and affectations of this ethnographic paradigm, see my article, "Framing FESPACO: Pan-African Film in Context," *Afterimage* 19, no. 4 (November 1991):6–9.

2. Ntongela Masilela, "Interconnections: The African and Afro American Cinemas," *The Independent* 11 (January–February 1988):14–17.

3. For example, *Yeelen* can be read as a critique of traditionalist culture (the Komo cult) since Mali is 90 percent Islamic and Cissé, the filmmaker, is Muslim. However, skilled in obsfucation, the film evades this primary source of influence.

4. Masilela, "Interconnections," 14–17.

5. John A. A. Ayoade, "The Culture Debate in Africa," *The Black Scholar* 20 (Summer–Fall 1989):2–7, illuminates and synthesizes recent discourses on the subject.

6. Tony Gitten's article is published in *Black Cinema Aesthetics: Issues in Independent Black Filmmaking,* ed. Gladstone Yearwood (Athens, Ohio: Center for Afro-American Studies, 1982), 115–120.

7. Captain Thomas Sankara, cited in J. R. Rayfield, "FESPACO 1987: African Cinema and Cultural Identity," paper presented at meetings of Canadian Association for African Studies, Edmonton, May 1987, 11.

8. Sékou Touré, "A Dialectical Approach to Culture," *The Black Scholar* 1 (November 1969):13.

9. Touré, "A Dialectical Approach to Culture," 23.

10. Ibid., 13.

11. See, for example, Françoise Pfaff's article "The Films of Gaston Kaboré and Idrissa Ouedraogo as Anthropological Sources," in *The Society for Visual Anthropology Review* (Spring 1990):50–59. Other references to this article are from the author's original copy, which she kindly made available to me before it was published; therefore, page numbers are omitted in subsequent citations.

12. See note 3.

13. Malian dialect specific to cinema, see Amadou Hampaté Bâ, "Le dit du cinèma Africain," in *UNESCO Catalogue: Films ethnographiques sur l'Afrique noire*. Paris: UNESCO, 1967.

14. Kwame Anthony Appiah, "Is Post- in Postmodernism the Post- in Postcolonial?" *Critical Inquiry* 17 (Winter 1991):342.

15. Kate Ezra, *Human Ideal in African Art: Bamana Figurative Sculpture* (Washington, D. C.: National Museum of African Art, 1986):15. Quoted in Rachel Hoffman, review of *Yeelen* in *African Arts* 22 (February 1989):100.

16. Cited in Hoffman, "Yeleen," 100.

17. See Cissé's interview, "Souleymane Cissé's Light on Africa," *Black Film Review* 4, no. 4 (Fall 1988):12.

18. Amie Williams, review of *Zan Boko, African Arts* 23, no. 2 (April 1990):93.

19. See, for example, Gilbert Adair, "The Artificial Eye: *Yeelen,*" *Sight and Sound* 57, no. 4 (Autumn 1988):284.

20. In the catalogue "Library of Africa Cinema," a project of California Newsreel, distributor of *Yeelen.*

21. Abdoul Dragoss's article in the 1991 FESPACO catalogue, "Environment: An issue which needs to be addressed by African Cinema," draws attention to the significance of the 'cosmic tree.' "

22. Cissé's interview in *Black Film Review,* 15.

23. Appiah, "Is Post- in Postmodernism the Post- in Postcolonial?" *Critical Inquiry* 17 (Winter 1991):346.

24. Ibid.

25. The problem of exhibition is still a persistent one, and the sector is still foreign-dominated. Since films produced in the francophone region during the pioneering years were financed by the French government, France also controlled the rights to distribute them. Some of the films were considered too political to be promoted in their home countries.

26. Pfaff, "The Films of Gaston Kaboré."

27. "Feature Films as Cultural Document," in *Principles of Visual Anthropology,* ed. Paul Hockings (The Hague and Paris: Mouton, 1975), 231–251, as quoted in Pfaff, "The Films of Gaston Kaboré."

28. D. Yung, *Variety,* 16 November 1988, 23.

29. As in the process of using still photographs on screen to identify dictators or people who were responsible for unjust punishments on the society, *Courage of the People, Hour of the Furnaces,* and *Battle of Chile* are some of the films that employ this strategy.

30. Pfaff, "The Films of Gaston Kaboré."

31. Françoise Lionnet, "Dissymmetry Embodied: Feminism, Universalism and the Practice of Excision," *Passages Issue* 1, no. 1 (1991):2.

32. N. Frank Ukadike, "Framing FESPACO," 6–9.

33. Blaine Harden, *Africa: Dispatches from a Fragile Continent* (New York: W. W. Norton and Company, 1990), 18.

34. Abdou Diouf, in Olayinde Koso-Thomas, *The Circumcision of Women: A Strategy of Eradication* (London: Zed, 1987), 27. Also quoted in Lionnet, "Dissymmetry Embodied," 3.

35. See Pat Aufderheide's interview with Cheick Oumar Sissoko in *Black Film Review* 6, no. 2 (Winter 1991):4.

36. Rouch contributed to the training of African filmmakers, including Sissoko. Rouch's numerous films about Africa are menacingly degrading.

37. *Finzan* is distributed by California Newsreel. This description is contained in their catalogue *Library of African Cinema,* 9.

38. Françoise Pfaff, "Africa Through African Eyes: An Interview with Idrissa Ouedraogo," *Black Film Review* 4 (Winter 1987–88): 11–12, 15.

39. Some of the information here was presented in my review of *Yaaba,* coauthored with Richard Porton in *Film Quarterly* 44, no. 3 (Spring 1991): 54–57.

40. William Fisher provides valuable information on Kinshasa's boisterous musical life, stating that Claud Cadiou's French Côte d'Ivoire coproduction *La vie platinée* (1984), a highly successful film "showcasing guitarist Zanzibar and his group 'Les Têtes Brûlées,'" initiated the musical comedy genre. See "Ouagadougou: A Beacon for African Culture," *Sight and Sound* 58 (Summer 1989):172.

41. From an interview with the author in New York City in 1986.

Caribbean & South America

Shape and Shaping of Caribbean Cinema

Mbye Cham

With the exception of Cuba and perhaps Puerto Rico and Venezuela, film-making in the Caribbean by Caribbean people is primarily a phenomenon of the 1980s and beyond, even though the decade of the 1970s witnessed the appearance of a handful of films from Jamaica, Haiti, and Guadeloupe. The very small number of films that managed to emerge in this period did so in the absence of an infrastructure of production, distribution, and exhibition, not unlike the experience of many African and other Third World films. Caribbean cinema is in its infancy at the moment, the most recent "arrivant" (to borrow a term from Edward Braithwaite) in the domain of Black world film practice, hence its designation as "un cinéma au rez-de-chaussée des nègres," ("a cinema at the basement of cinema by Blacks") and "un cinéma 'pou zot' " ("a cinema for others").[1] If cinema by Caribbean people is a new phenomenon in the Caribbean, it is not the case with the phenomenon of cinema itself. Like many countries in the Third World, many islands in the Caribbean region were exposed to cinema at a very early period following the invention of the form by the Lumière Brothers; for example, in the case of Haiti, the exposure came only four years after the invention. If, as argued by Michaëlle Lafontant-Médard in Chapter 3 of *Ex-Iles,* Haiti, "like Italy, France, England, Germany, Russia, and the United States, was indeed one of the very first countries to be part of the civilization of cinema," its initial position in this civilization was that of a receiver/consumer of film products from these other western countries, the producers/transmitters. The Caribbean

From Mbye Cham, ed., *Ex-Iles: Essays on Caribbean Cinema* (Trenton, N.J. Africa World Press, 1992), pp. 1–43. Reprinted by permission of the publisher.

had a long acquaintance with cinema, but only as a resource for foreign productions which exploit(ed) the natural/physical endowment of the tropical islands and invented other endowments to manufacture an image of the Caribbean radically at odds with the reality of the people of the Caribbean. Some of these images manufactured in Euro-American image factories with "raw materials" drawn from the Caribbean found their way back to the Caribbean to be consumed by Caribbean people. Haiti, like the majority of the islands in the Caribbean, is still in this position, subjected to a dominant regime of foreign film culture, itself a derivative of the history of plantation slavery and western imperial and colonial exploitation and of the resultant socio-political, economic, and cultural relations of dependency of the Caribbean on the West. If the Caribbean has historically been positioned as a receiver/consumer of and a resource for Euro-American productions (in terms of its [mis]use as location) in the world of film, it is only in the 1970s and, more significantly, the 1980s that one begins to witness the beginnings of a shift, no matter how feeble, toward the position of producer/transmitter.

Rarely, in the history of cinema, do such shifts occur in a vacuum. Mutually interacting currents in the local and international socio-political, economic, and cultural context always come into play and invariably leave their imprint on and/or receive the imprint of cultural productions. In the case of the Caribbean, a number of developments, transformations, and challenges in these areas in the 1970s and 1980s influenced the small corpus of indigenous Caribbean films that emerged under very difficult, political and economic conditions; these films that are different, in some cases, from the normal imported/imposed fare, thus marking what was figured to be the start of a sustained move toward a repositioning of the Caribbean in the civilization of cinema.

Politically, we note, for example, events in Manley-Seaga-Manley Jamaica, in Maurice Bishop's Grenada with the New Jewel Movement and American military intervention, in Guyana with Walter Rodney and the WPA and the brutal elimination of Rodney, in Haiti with the Duvalier reign of terror and plunder and the post-Duvalier aftershocks, in Guadeloupe and Martinique with budding militant movements advocating independence from France. These moments, and many others, in Caribbean politics in the 1970s and 1980s were instrumental in creating a climate for the emergence of new currents of political awareness among different strata of society, for the intensification of competition, at times violent, among/between different strata and tendencies, and for the intensification of an already existing gamut of radical militant activist practices tending toward a radical transformation of society.

Economically, the Caribbean was not spared the general Third World experience of a wave of generalized decline, the intensified hegemonic control of western capitalism in the form of World Bank/IMF-imposed Structural Adjustment Programs and U.S.-inspired Caribbean Basin Initiative, and increased emigration toward the west. Deteriorating infrastructure, declining quality of life indices, declining terms of trade, vagaries of heavy reliance on single commodity/sector economy (tourism, in many cases), and foreign exchange problems all combined to impose severe limitations on many Caribbean nations and individuals.

Culturally, new articulations of a distinct Caribbean identity began to emerge on the foundations of the thought and practice of literary/political/intellectual figures such as Frantz Fanon, Aimé Césaire, Joseph Zobel, and Edouard Glissant of Martinique; Jean Price-Mars, Jacques Roumain, and Jacques Stephen Alexis of Haiti; C. L. R. James of Trinidad; and Roger Mais, V. S. Reid and Louise Bennett of Jamaica; as well as the Caribbean heritage of oral traditions. Particularly noteworthy in these new articulations is the project of a group from Martinique, Jean Bernabé, Patrick Chamoiseau, and Raphaël Confiant, whose "manifesto," *Éloge à la Créolité*[2] [*In Praise of Créolité*], as well as their more general cultural agenda take as their point of reference the seminal thought and practice of their mentor and compatriot, Edouard Glissant, one of the most articulate advocates of a Caribbean identity based on the idea of Créolité.

Underlying their project is a strategy to rediscover and appropriate the diversity that is the Caribbean in order to construct a new, more authentic sense of identity. In their terms, Créolité, which is mosaic, disordered, open, and unpredictable, is referred to as a testimony: "And this testimony deals with the mental framework which today enables us to write, to carry out our research projects, to claim the right to exist for our Creole language and culture, to demand the political sovereignty of our Nation, and to think the world."[3] Elaborating further, Chamoiseau states that Créolité is, in turn, i) "a literary project" designed to keep alive the Creole language, ii) "the designation of a geo-cultural solidarity with people with a Creole language and culture, and a geo-political solidarity with people of the Caribbean Basin with a view to constructing what Glissant refers to as Antillanité," iii) the designation of "a major contemporary fact" in so far as human identity the world over "is in the process of modifying itself in order to be placed at the confluence of several histories, several languages and several cultures, what we call in our text 'the state of Créolité' " ["l'état de créolité"], and iv) "a desire to figure the necessary linking ["mise en relation"] of peoples, individuals and their culture, taking care to preserve their original diversity and even their obscurity—this is what we try to capture in proposing the term 'Diversalité' " [in place of 'Universalité'].[4] Along with other cognate articulations such as "marronisme moderne"[5] ("modern maroonism"), Créolité constitutes a continuation of as well as a shift away from a history of intellectual, cultural, and political practice in the Caribbean ranging from Negrismo in Cuba, Indigènisme in Haiti, Negritude in Martinique and Guadeloupe, and Pan Africanism.

Caribbean music like reggae, kaiso, soca, kadans, and zouk, and musical stars and groups like Bob Marley from Jamaica; Dr. Slinger Francisco, more popularly known as The Mighty Sparrow, from Trinidad and Tobago; Kassave from Guadeloupe; and Tabu Combo from Haiti experienced a phenomenal increase in local and international attention and popularity in the 1970s and 1980s. In concert with their entertainment aspects, calypso and reggae became more and more identified with satirical and radical political and social practices, indicting injustice and inequality, exhorting the masses to resist injustice and fight for justice and equality, making fun of human foibles, and extolling the virtues of love, peace, and pride in the Black heritage.[6] In particular, Bob Marley, the Rastafarian reggae star, emerged as the embodiment of a form of radical difference, canonized, commodified, and appropriated in different ways in different contexts both within and outside of the Caribbean.

These political, economic, and cultural currents of the 1970s and 1980s mark the context within which the fledgling practice of film production in the Caribbean emerged and is struggling to prosper, and, looking at the nature of the extant corpus, the impact and influence of these political, economic, and cultural currents become evident. The films engage aspects of these currents as subject matter, and, in some cases, aspects of these currents manifest themselves in matters of form, style, and orientation.

Prior to the appearance of *Rue Cases-Nègres (Sugarcane Alley)* by Euzhan Palcy of Martinique in 1983, awareness of indigenous film practice by the people who inhabit the various islands of the Caribbean tended to be limited to one film, Perry Henzel's *The Harder They Come* from Jamaica, which relies for most of its construction and orientation on reggae music which the film also helped popularize. Even more limited was/is awareness of the 1970s film work of pioneers like Gabriel Glissant and Christian Lara from the French possession of Guadeloupe and Raphaël Stines, Bob Lemoine, and Rassoul Labuchin (Yves Médard) from Haiti. Although known to a few people in the respective countries as well as to restricted film circles within the metropolitan capitals of Europe, France and Britain, in particular, Canada, and the United States, the work of these and many other new filmmakers is just beginning to command attention on a wider scale. The phenomenal local and international success of Palcy's *Sugarcane Alley* has created a new awareness of the importance and potential of indigenous film practice in Caribbean society and culture, and a number of new works have emerged recently from Martinique, Guadeloupe, Haiti, Jamaica, and, for the first time, from the Dutch Antillean island of Curaçao and the Dominican Republic.

Comparatively small in terms of quantity but technically well accomplished in many instances, the major portion of this corpus of indigenous Caribbean films comes from rather unlikely quarters of the Caribbean. Guadeloupe and Martinique, two of the smallest islands that are still overseas colonial possessions of France (Departments d'Outre-Mer—DOM), Haiti, economically the poorest nation in the western hemisphere and ravaged by brutal dictators and violence, and Jamaica, which also has its share of economic and political turbulence, account for a good part of the feature films made in the Caribbean by Caribbean people. Paradoxically, outside of the highly-acclaimed work of Horace Ove[7] who works mostly in Britain, there is little or no activity in the area of feature film work in Trinidad, one of the more economically well-off islands in the Caribbean. This is also the case with Barbados and Guyana. This fledgling corpus of indigenous Caribbean films is also marked by a diversity of forms, styles, subject matter, themes, and ideological orientations, but underlying this diversity are both a desire to engage the full range of Caribbean experiences from Caribbean perspectives as well as a determination to rescue the Caribbean from its usual misuse as exotic background to Euro-American romantic narratives and spectacles. Placing the Caribbean differently in the foreground, these films construct a film image of it different from the reigning limited stereotypic "islands of enchantment," or, as Alain Ménil puts it in *"Rue Cases-Nègres* or the Antilles from the Inside," "a journalistic synthesis of the three 'S's' (*sea, sex,* and *sun*)."

Like many other Third World film practices, Caribbean cinema also has significant segments of its production that take place outside the physical geographical space of the Caribbean undertaken by people born in the Caribbean but living outside of the Caribbean and by people with Caribbean parents born and raised outside of the Caribbean. History, economics, politics, socio-cultural, and natural factors have played key roles in creating a Caribbean diaspora in Britain, France, the Netherlands, Canada, and the United States, among other places. Emigration from the Caribbean to these places has become an integral part of the Caribbean experience, and a fair number of films have been made in these contexts, speaking to the experience of life outside of the Caribbean. This experience has also raised new questions regarding identity and belonging, questions that frame, in particular, much of the film discourse of third and fourth generation people of Caribbean parentage born and raised in Britain and France. These questions also form the focus of much debate within the Caribbean itself. The film work of London-based collectives such as Ceddo, Black Audio, and Sankofa; of individuals such as Menelik Shabbaz, Henry Martin, and Horace Ove in Britain; of Willy Rameau, Julius Amede Laou, Jacques Ferly, Benjamin Jules-Rosette, and Elsie

Haas in France (only part of the work of the last two); of Raoul Peck in the United States and Germany; and of Guilly Koster and Gloria Lowe in Holland forms a significant part of film practice by people of Caribbean descent in the Caribbean diaspora. The films of these individuals are shot for the most part in these countries, and they speak mostly to various experiences of Caribbean and other Black people in these countries, some of them linking these experiences to related ones back in the Caribbean.

There is also the case of Sarah Maldoror who was born and raised in France, one of whose parents is from Guadeloupe. The majority of her film work has thus far focused on revolutionary struggles in Africa and the experience of immigration in France. The only film about a Caribbean subject is her 1977 documentary on Aimé Césaire, *Un Homme, Une Terre.* Does she belong to African cinema, to French cinema, or to Caribbean cinema? Euzhan Palcy's 1990 Hollywood-produced anti-apartheid film, *A Dry White Season,* shot on location in Zimbabwe, has occasioned a reemergence of a similar line of questioning.

Besides these two categories of Caribbean film practice, i.e., films made by Caribbean filmmakers inside the Caribbean and films made by Caribbean filmmakers outside the Caribbean, there is a third category which consists primarily of films about the Caribbean, but made by people—both Black and white—who are not Caribbean in any way. Some of these films perpetuate dominant industry practices and stereotypes while others inscribe themselves within a tradition of independent and Third Cinema film practice that differs radically from dominant Euro-American industry productions. Films in this latter group, the more interesting one, generally tend to be concerned with the same range of issues and challenges that constitute the subject of films in the first two categories mentioned earlier, and many of them attempt to adopt/approximate a perspective that places them in positions of solidarity with those films. A prominent example is *West Indies: Les Nègres Marrons de la Liberté* (1979) by Med Hondo from Mauritania, who adapted this film, a radical reconstruction of Caribbean history, from a play by the Martiniquan playwright Daniel Boukman. There is also the 1987 film *Hayti Même Bagay,* shot in Haiti by African American filmmaker Ronald Wayne Boone; *Bim* (1975) shot in Trinidad by African American actor/director, Hugh Robertson; *One Hand Don't Clap* (1988) by Kavery Dutta on various Calysonians; and a work in progress—*Nunu*—on resistance to slavery by Ethiopian filmmaker, Haile Gerima, who shot on location in Jamaica in the early part of 1990. Others include left-leaning French productions such as *On N'Enterre Pas le Dimanche* (1959) by Michel Drach, *Toutes les Joséphines Ne Sont Pas Impératrices* (1977) and *En l'Autre Bord* (1978) by Jérome Kanapa, *Le Sang du Flamboyant* (1981) by François Migeat, *Les Enfants de la Guadeloupe* (1984) by Olivier Landau, and

Souvenance (1990) by the German Thomas Harlan. Generally, films in this category are not considered Caribbean in the way that *Anita* and *The Harder They Come,* for example, are considered Caribbean.

In the last few years, a couple of semi-mainstream Canadian film adaptations of Caribbean literary works have emerged, adding new dimensions to the question of what constitutes Caribbean cinema. With their screenplays usually written with the collaboration of Caribbean writers, these films are directed by white filmmakers and shot on location in Canada, and they usually address a mainstream spectatorship. Two films in this category are Rebecca Yate's and Glen Salzman's *Milk and Honey* (1988),[8] written in part by Trinidadian playwright Trevor Rhone, and Jacques Benoit's *Comment Faire l'Amour avec un Nègre sans se Fatiguer* (1990) (*How to Make Love to a Negro without Getting Tired*),[9] adapted from the novel of the same title by exiled Haitian journalist/writer Dany Laferrière, who also cowrote the screenplay. Both of these films deal with the experience of Caribbean people in exile in Canada.

On the whole, then, Caribbean cinema seems to be a composite of two discrete but related entities: productions by Caribbean people inside the Caribbean, on the one hand, and those outside of the Caribbean, on the other. It is a cinema "exîles"—from the islands/of people from the islands—and it is also a cinema of "exiles"—by people from the islands living in exile. However, the issue is not as clear-cut and unproblematic as it may appear from this general categorization. The nature of many islands-based productions and the relationship between islands-based productions and those productions that situate themselves outside of the islands are the subject of some debate, not unlike that which has been rehearsed in other areas of Third World cultural practices revolving around issues of home/emigration/exile/diaspora relationships. The essays by Keith Warner, Stuart Hall, and Ed Guerrero in *Ex-Iles* examine these very difficult and complex issues in Caribbean cultural practice. The first and second editions of the Images Caraïbes film festival in 1988 and 1990, respectively, also provided an impetus and forum for questions like "who is a Caribbean filmmaker?" and "what qualifies as a Caribbean film?" to be raised and debated, sometimes with passion.

In a 1982 interview, the Guadeloupian filmmaker, Christian Lara, stipulated five requisite conditions for a film to qualify as Antillean or Caribbean: "the director should be from the Caribbean, the subject matter should be a Caribbean story, the lead actor/actress should be from the Caribbean, Creole should be used, the production unit should be Caribbean." He stressed the importance of the last point from a legal standpoint, because, for him, "each film takes the nationality of its producer." He also pointed out that ". . . seldom do all five conditions obtain together in any one case."[10] By this criteria, he was compelled to disqualify his own first film, *Une Glace avec Deux Boules,* which was shot in Carcassone, France. Although criticized as overly simple and general in many respects, the thrust of Lara's stipulation, nevertheless, hints at the complexity and difficulty of the question of what films and which filmmakers belong properly to Caribbean film practice.

This question of belonging was given a new twist by Guy Cabort-Masson of Martinique following the first Images Caraïbes film festival in Fort-de-France in 1988. In an article entitled "Images Caraïbes ou Festival des Paradoxes,"[11] Cabort-Masson laments the dominance at this festival of what he labels "des films négropolitains," that is, "films which have nothing, absolutely nothing to do with what is called 'Caribbean Images.' " Included in this category are films of the diaspora, such as *Lien de Parenté* by Willy Rameau, for example, who was born and raised in France. However, his parents are from Martinique, and his film deals with the experience of a young Black man of French and Caribbean parents living with his white grandfather in a remote village in Southern France. Cabort-Masson states,

> When one looks at these films of France, one has the impression of looking at a French film with trivial French problems with emigrés. I clearly asked Willy Rameau this question: "in their writing and themes, do these films of the diaspora belong to French film history or are they part of the history of the people of the Caribbean Basin?" They are psychological dramas, they are vaudevillian with little or no humor. . . . We are not castigating these directors who live abroad, it is just that they do not belong to a festival of Caribbean Images, particularly when these filmmakers essentially claim the label "Emigré" in order to play on two fields.[12]

He praises the steadfastness of Sarah Maldoror, an "emigre" herself and member of the 1988 festival jury, in insisting that awards be given only to strictly Caribbean films. In this latter category, though, he mentions only one example, *Dérive ou la Femme Jardin* (1977) by Jean-Paul Césaire, a film shot in Martinique that is highly critical of the political and socio-moral status quo in the Antilles. It can be safely inferred from this that what is meant here by a strictly Caribbean film is one made by a Caribbean filmmaker in the Caribbean, criteria apparently fulfilled by Césaire's film. But what about Alain Ménil's and Daniel Boukman's criticism[13] of Christian Lara's films, shot by a Caribbean filmmaker in the Caribbean, but which they consider to be severely limited in their Caribbeanness? Interestingly, Cabort-Masson also makes mention of a Canadian film which documents the capitalist exploitation and miserable conditions of Haitian sugarcane workers in the Dominican Republic and argues that this film, apparently because of its focus on island realities, and never mind the fact that it is made by Canadians, belongs next to the work of Jean-Paul Césaire and to a festival of Caribbean images.

Amon Saba Saakana, a Jamaican writer and filmmaker (*Stolen Image* [1983]) in attendance at the second Images Caraïbes festival in 1990, also makes some observations which have implications for the issue of identity in Caribbean film practice. In his review of the festival, he draws attention to what he calls "historical inaccuracies that were allowed to filter through,"[14] for, in his view, it is ironic to have a white person represent Trinidad and Black people represent Britain at the festival:

> *Crossing Over* by Chris Laird of Trinidad was a documentary about two musicians, one from Africa and the other from Trinidad, who visited each other's countries and experienced the similarities in culture.
>
> The sensitivity of the film dramatized the point that a white Trinidadian could make an artistic and insightful film about black people. The company, Banyan, also showed the ironies in a white British man, Bruce Paddington, representing Trinidad and the Caribbean, and committed to Caribbean cinema outside the Hollywood or mainstream industry. It was equally ironic to witness blacks representing Britain, although considered from the African diaspora.[15]

Perry Henzel is also a white Jamaican ("a conscious white Jamaican," according to Mike Thelwell) who made *The Harder They Come,* which has been generally accepted as a Jamaican film; the film also inspired Mike Thelwell's highly acclaimed novel of the same title.

The debate on the question of belonging continues, and film practice in the Caribbean continues to struggle in order to flourish, the absence of vital resources and infrastructure notwithstanding. In a few areas, local public intervention has helped create structures which have enabled filmmaking inside the Caribbean. The case of Martinique stands out. Here, through the foresight and initiative of Aimé Césaire, President of the Regional Council, the Service Municipal d'Action Culturelle (SERMAC) was created in 1976 with the aim of promoting cultural productions of all kinds in Martinique. SERMAC has provided training in filmmaking for a number of aspiring cineastes, and it was under its auspices that Jean-Paul Césaire was able to make his 1977 feature film, *Dérive ou la Femme Jardin,* an adaptation of "Alléluia pour une Femme Jardin," a short story by the Haitian writer, René Depestre. This institution has also enabled the production of another feature film in 1978, *Hors des Jours Étrangers,* as well as many short films. Euzhan Palcy talks about the importance and role of this institution in Martinique in her interview with June Givanni in *Ex-Iles.*

It was also through the initiative, persistence, and sacrifice of a Martiniquan woman, Suzy Landau, that the most crucial and important film related event, the biennial Images Caraïbes film festival, was instituted in Fort-de-France, Martinique, in 1988. In October 1985, Landau, in conjunction with another Martiniquan woman, Viviane Duvigneau, founded the Association for the Promotion and Development of Caribbean Cinema (APDCC) with the general aim of encouraging and promoting the development of a distinct and original Caribbean cinema by Caribbean people. The goals of the APDCC include the following:

a) make known and promote Caribbean cinema, especially by organizing film festivals with focus on the Caribbean.
b) contribute to the establishment of regular exchanges and meetings between Caribbean film professionals and others of African descent.
c) privilege and encourage the production and development of original Caribbean films.
d) encourage and aid the training of young Caribbean people in filmmaking.
e) participate in changing mentalities and foster friendship among people through audiovisual means.
f) organize an audiovisual communication network to help people within the Caribbean.
g) establish a Caribbean cinema resource center.[16]

Thus far, the APDCC has been successful in its exhibition efforts, especially, with four successful events already to its credit. In May 1987, in collaboration with the Office Municipal Culturel du Marin, on the occasion of the celebration of the abolition of slavery, the association presented a film program, "Regards sur l'Emigration," which featured three films, *La Vieille Quimboiseuse et le Majordome, Solitaire à Micro Ouvert,* and *Mélodie de Brumes à Paris,* by the Parisian-born Martiniquan filmmaker, Julius Amede Laou. In August of that same year, the APDCC took part in the Festival Marin-Village with a week-long program of films by independent African American filmmakers William Greaves, Larry Clark, and Haile Gerima, all of whom were present at the festival. It was, however, in June 1988 that the first fully fledged region-wide and diaspora-wide festival of Caribbean films took place in Fort-de-France, with more than sixty films and videos—fiction as well as documentary—from the Caribbean, Britain, France, and Canada taking part. The second edition of Images Caraïbes took place 1–8 June 1990 with a vastly expanded program of more than ninety films and videos from many more countries participating.

The program of this second festival included the following categories:

I) MOTION PICTURES
  1) —panorama of Caribbean films
     Competition: Short and full-length fiction, documentary, animation.
     Prize: —Short film and full-length production
       —Special Jury Prize
       —Best actor; Best actress
  2) "Banzil Creol" in images
     (films, videos of the Creole world, the Indian Ocean, Louisiana, Caribbean)
  3) a look at plantation cinema from Brazil to Louisiana
  4) "Premières Images": siléma antan lontan ("First Images": movies of the good old days)
     Cinemathèque: invited guest—Puerto Rico
  5) "Images d'Afrique" ("Images of Africa")
  6) Outsider perspectives on Caribbean reality
  7) Tribute to film actors:
     Norman Beaton (UK/Guyana)
     Robert Liensol (Guadeloupe/France)

II) VIDEO SECTION
  1) panorama of Caribbean videos
     Competition: TV programs
     Prize: —best video
       —best commercial
       —best clip
  2) a look at Caribbean television

III) DISCOVERING THE ISLANDS SECTION: HAITI
  Spotlight on Haiti throughout the festival (art, culture, literature, culinary arts etc)

IV) ENCOUNTERS/DEBATES
  1) "Banzil Creol"
  2) "Cinema de plantation"
  3) "Division de Educacion de la Communidad Puerto Rico"
  4) "Fespaco"
  5) "Actors"
  6) "Broadcast distribution—production with countries of the North and between countries of the South"

V) COLLOQUIUM
  Theme: Produce, Distribute, Exhibit (Caribbean, Europe, Africa, the Americas)

VI) MARKET: BUY/SELL/TRADE
  —Television programs
  —Films and videos[17]

In spite of the enormous resource and logistical limitations of the 1990 festival, as well as its predecessor,[18] the program reflects the efforts of the APDCC to solidify its position as an outlet for Caribbean productions and, more significantly, as a forum to bring together all Caribbean filmmakers to organize and to advocate measures that would promote and enable more production, distribution, and exhibition within the Caribbean itself. One of the products of the 1990 colloquium was a decision to form an organization, Federation of Caribbean Audiovisual Professionals (FeCAViP). The manifesto reads:

> We, producers, filmmakers, screenwriters, technicians and actors of the second Images Caraïbes Festival 1990, being aware of the need to further develop the space within the Caribbean, for professional workers in film and video, reflecting our special needs, and after having made a deep analysis of our reality, acknowledging the importance of film, TV, and video, decided to give ourselves the means in order to obtain the conditions necessary for the realization of the expression of the professionals working in film and video.
>
> So together, we have to:
>
> 1. Create, produce, distribute, and broadcast the works of our young Caribbean artists
> 2. Contribute to the training of our young artists and technicians
> 3. Collect, record, archive, and preserve our cultural heritage
> 4. Overcome the existing linguistic, legal, technical, and commercial barriers
> 5. Promote Caribbean cinema, video, and TV productions
> 6. Develop the exchange of information between Caribbean professionals
> 7. Establish relationships between all the associations and audiovisual events of the Caribbean and its diaspora
> 8. Create new contacts with countries facing similar problems (in Africa, South America, for example)
>
> In order to achieve our goals, it has been decided a Federation be created which name will be Federation of Caribbean Audiovisual Professionals (FeCAViP). Fort-de-France, June 8, 1990.[19]

Supported by almost all of the filmmakers and others in attendance at the festival, this manifesto, together with plans already underway to implement its provisions, especially in light of the significance of 1992 for the Caribbean, represents a historic moment in Caribbean cinema. It reflects a sharp awareness on the part of Caribbean filmmakers of the enormous challenges confronting them and the measures required for negotiating these challenges. As pointed out by Victoria Marshall in the specific case of Jamaica, which is the situation with the Caribbean, in general, there is no film industry in the Caribbean (Cuba excepted). There are no formal region-wide associations of filmmakers, and those that exist within individual islands, such as Sistren in Jamaica[20] and Cosmic Illusion

Productions, formerly in Curaçao and now in Holland, have no formal channels for region-wide networking. (The situation is a little better with television where, as Bruce Paddington observed, some networking via programs like "Caribvision" and "Caribscope" takes place.) In the few islands where rare local ownership of film theaters exists, such as Circuit Elizé in Martinique, Caribbean films are seldom programmed on a regular basis. Prior to Images Caraïbes, the only regular festivals in the Caribbean were (still are) the Martinique-based Festival Antillais du Film Fantastique[21] (Horror Film Festival! The film *Evil Dead* won the best film prize in the 1983 festival.) and the Festival de la Francophonie (dubbed "sotisme francophone" [Francophone foolishness] by Guy Cabort-Masson) which allows films from the francophone areas only. A viable indigenous Caribbean film will have much difficulty thriving under these conditions, but the efforts initiated in Martinique and elsewhere have the potential to go a long way in changing these conditions for the better, especially, in terms of creating an awareness of the importance and need for local investment in all aspects of Caribbean cinema.

The Shape of Recent Production

Many of the essays in *Ex-Iles* describe, analyze, and make mention of a few fairly well-known Caribbean films as well as their context of production. There are, however, many more films and videos, some of them relatively recent, that are not that well known outside of the circles identified earlier. The brief selected island by island overview that follows highlights some of these works.[22]

Guadeloupe and Martinique

Generally referred to as the French Antilles, Guadeloupe and Martinique account for a fairly large portion of the corpus of films made by Caribbean people within the Caribbean, and the Caribbean filmmaker with the highest name recognition, Euzhan Palcy, is from Martinique. Filmmaking by people from the "French" Antilles started not in the islands but in France in the 1970s.[23] A short twenty-minute 16mm black-and-white film, *Le Retour,* made in 1971 by a young Guadeloupean graduate of the Institut de Formation Cinématographique in Paris, is usually credited as the first film by a filmmaker from the Antilles. This was followed in 1972 by another twenty-minute 16mm color film entitled *Le Pion* by Gabriel Glissant, also from Guadeloupe and a graduate of IDHEC (Institut des Hautes Etudes Cinématographiques) in Paris. Both of these pioneer films engage the experience of exile in the Metropole, a theme taken up again in many subsequent films of the Antillean diaspora.

The mid-1970s saw the appearance of a few more films by Antilleans, some of them shot in the Antilles and addressing a popular Antillean audience. Glissant followed his 1972 effort with another film in 1975 exploring economic realities of the Antilles. Entitled *La Machette et le Marteau,* this seventy-minute film documents a series of strikes by agricultural workers in Guadeloupe in 1974. In pursuit of his belief in creating a cinema that appeals to a popular mass audience, Glissant accomplishes in 1976 a Creole dubbed kung fu film, *Chiba Ti Mal-Là,* a feat that Catherine Ruelle describes as a "fantastique détournement."[24] Jacques Ferly is another Antillean filmmaker in the period of the 1970s who also had a vision of a popularly oriented Antillean fiction film, but who was able to make only six short films: *Liberté Coupée* (1975), *Ça Ne M'A Pas Plu* (1975), *Tambour au Loin, Joli Son* (1976), *La Charpente de Marine* (1981), and *Chronique du Coeur* (1984). In 1975, Euzhan Palcy came out with her first piece of audiovisual work in the form of *La Messagère,* a fifty-two-minute black-and-white television drama, and in 1981 she made her first short fiction film, *L'Atelier du Diable.* Palcy talks about these works in her interview with June Givanni. In 1977 Sarah Maldoror came out with a fine tribute to Aimé Césaire, *Un Homme, Une Terre.*

The importance of SERMAC in Martinique has already been discussed, as well as its enabling role in the making of two Antillean films in the 1970s—Jean-Paul Césaire's *Derive ou la Femme Jardin* (1977), depicting the mental trauma of an Antillean woman who returns home to Martinique after twenty-five years in France and has to adjust to the pressures of being caught between two cultures, and *Hors des Jours Etrangers* (1978), dealing with fraud in legislative elections and the disastrous consequences of the existing colonial status of the Antilles as overseas departments of France. Along with SERMAC, another institution reemerged in the 1970s to further encourage film-related activities. This institution is the Centre Martiniquais d'Action Culturelle (CMAC), created in 1952 by Anca Bertrand. In addition to its screening sessions, CMAC underwrote the making of eight medium- and full-length films between 1974 and 1985.[25]

Perhaps the most prolific filmmaker in terms of number of films made, Christian Lara from Guadeloupe occupies an important, if controversial, position in Caribbean cinema. In addition to his first effort, *Une Glace avec Deux Boules,* which he dismissed as a non-Caribbean film, as pointed out earlier, he has made a total of six feature films to date. His first feature is *Chap'la* (1977). It is described as a tropical detective story, a comedy about the commercial bourgeoisie of the islands.

Lara's next film, *Coco la Fleur Candidat,* was finished in 1978, and it is one of the few Antillean feature films, in addition to Glissant's, to use Creole first. The film deals with the political situation in Guadeloupe through a narrative about electoral politics, but critics have emphasized the failure of Lara to engage adequately some of the most crucial political issues and activities of the moment, such as the pro-independence nationalist and workers agitations in Guadeloupe. *Coco la Fleur Candidat* was followed in 1980 by *Mamito,* a film considered by some to be Lara's best work. *Mamito* narrates the process by which the eponymous heroine comes to an awareness of the socio-economic and political deprivation of and challenges for the people of Karukera, the traditional Caribe name for Guadeloupe, through her association with a militant trade union activist who also advocates independence from France. Lara's next film, *Vivre Libre ou Mourir,* (*Live Free or Die*) was made in 1980. Drawing on research undertaken by his grandfather, Oruno Lara,[26] Christian Lara reconstructs in *Vivre Libre ou Mourir* two centuries of the history of Guadeloupe, indicting European and French slavery from the time of Columbus to that of Victor Schoelcher and rehabilitating the memory and image of Ignace, a nationalist hero who resisted French domination. These three films, *Coco la Fleur Candidat, Mamito,* and *Vivre Libre ou Mourir,* constitute what Lara dubs a political trilogy.

In 1981, Lara came out with *Adieu Foulards,* a film about the problems and challenges of being Antillean in the world of French music and show business, and his latest feature to date is a sixty-minute light-hearted adventure story, entitled *Black,* made in 1987, about a "tribe" of Black actors from Africa, America, and Europe who arrive in an African capital for an artistic performance. Most recently, Lara has found himself mired in controversy over a television series, *Yoka,* which he is in the process of producing.[27]

His work has been hailed as pioneering, but it is mostly criticized as imitative and unoriginal with a tendency to reinforce the exotica of Euro-American productions about the Caribbean and to unconsciously subvert its avowed militant pro-Antillean pretentions. Whatever the artistic and ideological limitations of his work, Lara is a force to reckon with in Antillean cinema. He has the savvy of an entrepreneur able to raise resources required to make films and he has been instrumental in introducing young Caribbean talent to cinema and aiding in their development. Three well-known Antillean actors, Robert Liensol, Greg Germain, and France Zobda were first put on screen by Lara, and he is also one of the first to feature the music of a group of young people, Michel Alibo, Claude Vamur, and Jean-Claude Naimro, who were later to make it big as part of the musical group Kassave which popularized zouk music locally and internationally.[28]

The decade of the 1980s in the "French" Antilles was undoubtedly dominated by Euzhan Palcy's *Rue Cases-Nègres* (1983) which eclipsed the few other works that appeared during this period. It was a decade that also witnessed the increasing use of video by Antillean artists to dramatize and document various aspects of individual, social, and cultural experience. In 1982, Benjamin Jules-Rosette of Martinique came out with

Bourg La Folie,[29] about the life of a community of fishermen and women on the island, adapted from the novel *Martinique des Cendres* by Roland Brival. This was followed by *Bon Die Bon,* a touristic "socio-detective" (Jules-Rosette's term) film shot in the island of Reunion in 1986.[30] He is also the director of a recent film, *Errance,* shot in France about the mental trauma of a young woman exiled in and alienated from France but at the same time strongly attracted to her adopted country. *Et Survint la Vipère* is a 1983 psychological drama by José Egouy, a Martiniquan living in Guadeloupe, who narrates, in this first film, the mental torture of a woman twice at the receiving end of misfortune at very crucial moments of her life.[31] Guy Deslauriers, also a Martiniquan who worked as intern in the production of Palcy's *Rue Cases-Nègres,* is the director of *Les Oubliés de la Liberté,* a thirty-eight-minute narrative on events and movements in Saint Pierre, Martinique, in August 1789 moments after news of the French Revolution reached the island. A 1989 video documentary from Guadeloupe entitled *Noirs et Blancs en 1789* also deals with the repercussions of the French Revolution in Guadeloupe, Martinique, and Haiti.

A handful of short films also came out in the latter part of the 1980s to join the earlier work of Jacques Ferly. Some of these include a group of what could be referred to as "dream movies," in that they all invoke elements of popular beliefs in the unconscious, dealing with individuals in a state of dream, who hear voices and act or want to act on what these voices say to them. Deslauriers's *Quiproquo,* Harmel Sbraire's *Frayeur au 6ème Ciel,* Henry Vigana's *Premonition,* Michel G. Traore's *Les Fruits de la Passion* (1989), and *La Nuit de la Saint-Sylvestre* by Patrick Baucelin all from Martinique, deal with some form of dream, hallucination, and freak accident. Traore is also the director of an earlier short narrative film, *Mizik Rez de Chaussée Neg,* shot in Paris in 1981 and dealing with the problems and challenges of Antillean musicians in Paris, a theme treated by Lara in his *Adieu Foulards* of the same year.

Production by Antilleans in France in the 1980s include Guadeloupean Constant Gros Dubois's *O Madiana* (1981) about the life of an Antillean couple living in Paris, Willy Rameau's *Lien de Parenté* (1986), and the films and drama works of Julius Amede Laou, *Solitaire à Micro Ouvert* (1983), *Melodie de Brumes à Paris* (1984), *Sonate en Solitudes Majeures* (1986), and *La Vielle Quimboiseuse et le Majordome* (1987).

Haiti

In *Ex-Iles,* Michaëlle Lafontant-Médard provides a detailed history of the Haitian experience with cinema and a thorough thematic analysis of a landmark Haitian film, *Anita* (1982), by Rassoul Labuchin (Yves Médard), a poet, actor, critic, screenwriter, and director who has shaped and influenced a great deal of the nature and direction of Haitian cultural history and politics. Chapter 6 by John Stewart is also devoted to an exploration of Haitian history and culture in relation to their filmic [mis/under]representation.[32] In addition to the works and filmmakers cited and discussed in three chapters of *Ex-Iles* there has emerged in recent years two major figures in Haitian film practice whose work has elevated Haitian cinema to new levels of excellence and maturity. These two filmmakers are Elsie Haas, who lives in Paris and has made films both in and out of Haiti, and Raoul Peck, who studied film in Germany and now lives there. Peck's first major film, *Haitian Corner* was shot in New York in 1988. Their films privilege Creole, and they deal with the human costs of oppression and dictatorship for the Haitian in Haiti and in exile, the pains and challenges of exile, the trauma of the pull of two different cultures, and the strength and potential of Haitian culture.

Haas is the director of seven documentary, short and medium-length narrative films: *La Seconde Manche* (1979), *Zatrap* (1980), *Des Saints et des Anges* (1984), *La Seconde Manche II* (1985), *La Ronde des Tap-Taps* (1986), *La Ronde de Vodu* (1987), and *No Comment* (1988). The two *Seconde Manche* films chronicle the relationship between a Martiniquan couple in Paris and their respective attitudes towards their native land. In the first part, a letter arrives from Martinique inquiring if the man wants to sell a piece of land he owns in Martinique to an interested buyer. Because he wants the cash to buy a car in Paris, the man argues for selling, but the woman opposes him on the grounds that she has a desire to go back to Martinique in the future to settle on this land. The man prevails temporarily, but, before he implements the plan, the woman is able to convince him to see her point of view and appreciate the invaluable nature of land as opposed to the short life and temporary usefulness of a car. The sequel deals with the victory of the woman over the man.

Zatrap (Creole for "the trap") and *Des Saints et des Anges (Saints and Angels)* engage the question of identity for Martiniquans at home and abroad. In *Zatrap,* a mother who works as a domestic helper for a bourgeois family passes on the responsibility of caring for her own children in her absence to her own eldest daughter. But the latter dreams of leaving Martinique for France, just like all her peers. It is a film that looks at life in Martinique through a series of anecdotes which blend together to convey graphically the ways in which some Martiniquans feel trapped between the desire to be themselves and the necessity to be French. Hence the significance of the title.

Des Saints et des Anges explores this sense of entrapment in the context of exile in Paris. Here, the setting is an African beauty salon in Paris where the quest for the appropriation of the image of the other is undertaken through the medium of hair and hair style. European hair is seen as a symbol of civilization by the mother and her daughter who come to the salon to get a do. Haas' latest work, a short television piece called *No Comment,* shows a Black woman preparing herself for a date and agonizing over what she should wear for dress and make-up and how to fix her hair and other worries. These films speak to other similarly oriented films such as Maureen Blackwood's *Perfect Image?* (1988) and Ayoka Chenzira's *Hairpiece* (1986).

La Ronde des Tap-Taps and *La ronde de Vodu* are two films Haas shot in Haiti in the wake of the demise of formal Duvalierism. The first chronicles the nature of urban living and public transportation in Haiti by focusing on the hordes of minivans, converted jeeps and trucks called "tap-taps" (the equivalent of the "jeepneys" of Manila, the Philippines, and the ubiquitous "mammy wagons" of urban West Africa) that form the core of the means by which the majority of Haitians move around. The second is by far Haas's most acclaimed work, and it deals with the place of religion, Vodu, in particular, in the history, politics, and culture of Haiti, as well as its place in the lives of Haitian individuals. Well balanced and critical, the film differs radically from previous accounts of Vodu, such as those by Maya Deren. By intercutting interview shots of a broad cross section of the population with shots of diverse religious rituals/performances and everyday life moments and activities in different parts of the country and by skillfully integrating song, poetry, dance and painting into the fabric of the narrative, *La Ronde de Vodu* links this old and highly influential aspect of Haitian culture to other areas of Haitian life, history, economics, politics, and culture in ways never done before. It accomplishes a fairly comprehensive and moving portrait of a society and people struggling to negotiate a legacy of oppression and denial, a people with resilience and the wisdom and will to question aspects of their cultural heritage in order to renew it and make it more responsive to the demands of today's economic, political, and spiritual life.

Haas is currently working on her first major feature length film, *Boni,* scheduled to be shot on location in Surinam. *Exile II* is a work in progress for television, and it deals with Black intellectuals and artists in exile in the west and the meanings of the west's refusal to communicate with them in any significant ways. The title plays on the name of a Haitian musical group, Exile I. She is also doing research for a film project on Blacks in Russia with focus on Alexander Pushkin.

Raoul Peck's *Haitian Corner* has garnered all manner of acclaim from a broad variety of spectatorship both in Haiti and abroad. In what is perhaps the most meaningful homage for Raoul Peck, Rassoul Labuchin, the filmmaker generally credited with inaugurating a new film discourse in Haiti with *Anita* and a former political prisoner himself under Duvalier, labels *Haitian Corner* "the best ever made by a Haitian," and expresses his gratitude to Peck, "the best Haitian filmmaker for his film *Haitian Corner* which made me relive my season in hell and which brought me to believe that the future will be enchanting."[33] Shot in Brooklyn and featuring a fine performance by newcomer Patrick Rameau and seasoned Toto Bissainthe, *Haitian Corner* is about memory and the immediate past, it deals with power and how the misuse of power can return to haunt abusers; it deals with the human capacity to cope with and transcend feelings of anger and revenge; it is the triumph of humanity over base animal instincts. More significantly, *Haitian Corner,* like Elsie Haas's *La Ronde de Vodu,* casts a jaundiced look at the legacy of Duvalierism and the urgent need to go beyond this sad moment in the history of Haiti.

The film narrates the experience of Joseph Bossuet, a poet from Haiti now living in exile in the Haitian community in Brooklyn, New York. A factory worker who devotes many moments of his free time composing poetry, dealing with a troubled romantic relationship, hanging out at the "Haitian Corner" with fellow Haitian exile-victims of dictatorship and economic hardship, and reflecting on his past experience of torture while a political prisoner for seven years in Haiti, Bossuet lives in a constant flux of alienation and confrontation. The feeling of alienation from his new environment derives from his inability to erase the memory of his past which constantly haunts him in a series of flashbacks that Peck skillfully sneaks in at key moments of the narrative. Unable to blend in with the other exiles who have accepted the realities of their new adopted environment, Bossuet chooses silence and turns inward.

One day he spots a fellow working as a kitchen hand in a restaurant and recognizes him as Theodor, one of his former prison torturers now hiding in New York. After tracking him down, Bossuet corners this former tonton macoute who had also fled Haiti. At this moment the film flashes back to Bossuet's time in prison with Theodor the macoute torturer in his face. When the film returns to the present, with Theodor practically on his knees, weeping and pleading for mercy, the situation is reversed, and Bossuet is in a position to take his revenge. However, instead of killing him, as his initial instincts and posture dictated, Bossuet deliberately lets him go, thus refusing to succumb to the base animal instincts that characterize the macoute torturers. From this moment on, Bossuet is able to forge ahead with a new sense of place and direction.

A Haitian newcomer to the field is Jean-Claude Fayolle who recently made his first video fiction feature in Montreal entitled *Tilom à Letranje* in Creole (*Tilom Abroad*). It is a story about the trials and challenges of Tilom, a man who can neither read nor write and who leaves his native land for the city of Montreal.

The "Netherlands" Antilles

The dominant filmmaker from this region of the Caribbean is undoubtedly Felix de Rooy from Curaçao who is also fast becoming one of the Caribbean region's most creative filmmakers. Based in Holland where he runs a production company (Cosmic Illusion Productions) along with Norman de Palm from Aruba, Felix de Rooy is the director of twelve films of various length and format, including two highly creative and widely acclaimed feature films, *Almacita di Desolato* (1986) and *Ava & Gabriel* (1990), both of which were shot on location in Curaçao. Like many of the recent films that use Creole, de Rooy's films privilege Papiamentu, the language used by the majority of the people of Curaçao.

Based on old legends compiled by Elis Juliana and Father Paul Brenneker in the Sambumbu volumes "Folktales from Curaçao, Aruba and Bonnaire," *Almacita* draws heavily from the heritage of oral traditions and myth of Curaçao to narrate a story about the struggle between creative and destructive forces in a fictional agricultural community at the turn of the century. This is a community of Blacks with their own forms of worship and lifestyle far from the influence of white slave traders and landowners know as "shons." Following is a detailed synopsis of *Almacita* supplied by the director in the film's press packet:

> The story takes place at the turn of the century in Curaçao in an isolated agricultural community of former slaves. In the village [of] "Desolato" traditional lifestyle continues far away from the white "shons," the landowners.
>
> Alma Sola, the symbol of evil, the patriarch of the "shons," is ever present on the arid plains of "Tera Kora," the red earth surrounding the village [of] Desolato. Alma Sola transforms into male, female or animal and always strikes . . . when the vigilance of Desolato weakens.
>
> Solem, the priestess, has the duty to guard the spiritual well being of the village and to maintain contact with the world of spirits. Her knowledge of medicinal herbs allows her to ease illness and pain. Her fertility is sacrificed for the welfare of the community. As a consequence, she is not allowed to have a relationship with a man. Solem's longing for the mystery of physical love provides the Alma Sola with an opportunity to lead her astray.
>
> While searching for herbs and food, accompanied by Lucio, a 13 year old boy, she encounters a young man who is wounded. She decides to take care of him and hide him in a cave far from Desolato. In doing so, she violates the rules of Desolato. The young man manages to seduce Solem and disappears shortly thereafter.
>
> When Solem's pregnancy becomes apparent she is forced to leave Desolato, taking with her the evil forces she afflicted on the village.

> She leaves with her new born baby, Almacita, and Lucio in search of "Matriz di Piedat," the place where the spirits of the ancestors dwell. She hopes that the light of the spirit world will cleanse her and Almacita from evil.

Almacita is steeped in metaphors and myths, many of which invite parallels with Labuchin's *Anita* (1982) and ancient fertility myths, and the moralistic, even Biblical, undertones of the story are inscribed in the very person of Almacita, "the little soul," born of the forced union between Alma Sola (force of evil) and Solem (force of goodness) and figured to be the one to unite these conflicting opposites.

Felix de Rooy's latest film is *Ava & Gabriel* (1990). Based on an original story by de Rooy and a screenplay by producer Norman de Palm, *Ava & Gabriel* is also set in Curaçao, but the time period in this instance is more recent, the late 1940s. In the words of the filmmaker, the story goes like this:

> Upon request of Father Fidelius, parish priest of St. Anna's, the Surinam painter Gabriel Goedbloed arrives from Holland to paint a mural of the Virgin Mary in the St. Anna's church.
>
> The drama unfolds from different angles.
>
> First, the clergy and the locals are confused by the fact that the painter is black, originating from Surinam, but resettled in The Hague, where he received a Fine Art education. The close knit Antillean society did not easily welcome strangers who would not conform to their colonial way of life in those days.
>
> Contributing [complicating?] factors arise when he [Gabriel] chooses a young teacher, Miss Ava Recordina, who is of mixed origin, to be his model for the painting of the Virgin Mary. Ava is engaged to the white police major, Carlos Zarius, who is not too happy with his fiancee posing for the painter. The fact that the Dutch Governor's wife, Louise van Hansschot, is interested in Gabriel also fuels the tension.
>
> In the end, Gabriel Goedbloed falls victim to the controversies, hypocrisies and intrigues that have arisen around his person and his paintings.

Ava & Gabriel echoes the mythic tones and moral subtext of *Almacita* as well as the professional background of de Rooy who is also a painter, and, even though the historical context changes from the era of slavery in *Almacita* to that of formal colonialism in *Ava & Gabriel,* there is a common set of political, racial, and moral issues and metaphors and religious allusions that run through both films. In fact, de Rooy sees *Almacita* and *Ava & Gabriel* as the first two parts of a trilogy concerned partly with a reinterpretation of religion, Christianity, in particular.[34]

Surinam (sometimes called Dutch Guyana) is the other country in the "Netherlands" Antilles with filmmakers, but many of them work out of Holland. Among the ones known here are Pim de la Parra whose best known works are *Wan Pipel (One People)* (1976), *Aruba Affair* (1981), *Odyssée d'Amour* (1987), and his latest film set in Amsterdam, *Two People: Analysis of a Seduction* (1990); Guilly Koster who made a short film in 1987 entitled *Self Portrait,* about Koster himself and his migration from Surinam to Holland; and Gloria Lowe whose work has so far been in video. She is the director of a forty-minute production on the issue of racism in Holland entitled *We're Doing It for the Children/Santa Claus* (1982). Pim de la Parra's *Wan Pipel,* which uses three languages (Sranang Tongo, Surinamese Hindustani, and Dutch), tells the story of Roy, a young Surinamese student in Holland who returns to Surinam to attend to his mother who is ill. As a "been to," i.e., someone who has been to Holland, his image and position in the community are enhanced, but this fact amounts to nothing when it comes to relations between races in the island. Roy's love for Rubia, a Hindu woman, falls victim to entrenched prejudice on both sides. Pim de la Parra uses this often humorous story to provoke reflection on the meaning and implications of racial and cultural diversity and possible ways of coping with and transcending its negative modes. As for his other film, *Odyssée d'Amour,* done in English and Papiamentu, the focus is on male bonding at the mid-life stage. It is the story of the relationship between three men, Paul, Ramon, and Nicholas, on reaching age forty.

The small corpus of films that constitutes Caribbean cinema at the moment was formed over the last two decades, the 1970s and 1980s, a period marked by momentous socio-political, economic, and cultural shifts and transformations which influenced the shape and direction of these films. This corpus exhibits the marks of first arrivals and beginnings. As the latest "arrivant" in Black world independent film practice as well as in Caribbean cultural practice, Caribbean cinema is struggling to assert a presence within spaces dominated largely by foreign films and by Caribbean literature and music, to which it sometimes turns for inspiration and models in its project to construct Caribbean notions of self, of collectivity, of place, of culture, of history, and of the future.

Notes

For all references to *Ex-Iles*, see Mbye Cham, ed., *Ex-Iles: Essays on Caribbean Cinema* (Trenton, NJ: Africa World Press, 1992).

1. Alfonso Sadi, "Un Cinéma au Rez-de-Chaussée des Nègres," *Regards Africains,* no 12 (November 1989): 28–29. Special issue on "Afro-ciné, vous connaissez?"

2. Jean Bernabé, Patrick Chamoiseau, and Raphaël Confiant, *Éloge à la Creolité* (Paris: Gallimard, 1989). I should emphasize that the term "manifesto" used to refer to this text does not signify a rigid programmatic primer which functions to police, direct, and regulate cultural production and practice. The authors themselves refuse any such dogmatism and sectarianism for their text. See interview with Patrick Chamoiseau by Henri Pied, "Eloge d'une Identité," *Antilla,* no. 346 (août 1989): 4–8. An English translation by Mohamed B. Taleb Khyar can be found in *Callaloo,* no. 13 (1990): 886–909.

3. Chamoiseau, "Éloge d'une Identité," 5.

4. Ibid., 5–6. Chamoiseau, Bernabé, and Confiant also use the term "Divers" to replace "Univers" (Universe) to further anchor their concept of Créolité.

5. See Réné Louise, "Le Marronisme Moderne," *Antilla,* no. 51 (28 avril–5 mai 1983): 37.

6. See Keith Warner, *KAISO!: The Trinidad Calypso* (Washington, D.C.: Three Continents Press, 1982) and Stephen Davis and Peter Simon, *Reggae Bloodlines* (New York: Anchor Books, 1979).

7. Ove's most recent works include a 1989 television mini-series entitled *The Orchid House* about the life of Jean Rhys, the Dominican novelist and storyteller (*Wide Sargasso Sea* and others) who left the Caribbean for residency in England in 1910. This series was shot in Dominica. Ove has also produced a short one hour television drama titled *When Love Dies.* He is currently working on the Caribbean segments of a BBC project, *After Columbus,* scheduled for broadcast in 1992. (I thank June Givanni for this piece of information.)

Trinidad has a well developed video production infrastructure, and one of the more well-known production companies here is Banyan, whose works include *Crossing Over* about the meeting between a Trinidad calypsonian and a Ghanaian musician, and *Gayelle at Images Caraïbes* (1988), a documentary on the first Images Caraïbes festival in Martinique in 1988 and the coverage given it by *Gayelle,* a popular magazine based in Trinidad and Tobago. Banyan's latest work is a thirty minute documentary, *FilmCarib* (1990), on Caribbean cinema, featuring interviews with Horace Ove, Elsie Haas, Euzhan Palcy, Felix de Rooy, and others.

8. For a critical review of this film, see Roger McTair's review in *Cinema Canada,* no. 159 (January 1989): 20.

9. See Hal Weaver's review in *Cinema Canada,* no. 162 (April 1989): 19; and Jacquie Jones, "How To Make Trouble," *Black Film Review* 6, no. 2 (1990): 14–15.

10. See Brigitte Tallon, "L'An II du Cinéma Antillais" [Interview with Sarah Maldoror and Christian Lara] *Autrement* 49 (1983): 242. Special issue entitled "Blacks."

11. Guy Cabort-Masson, "Images Caraïbes ou Festival des Paradoxes," *Antilla,* no. 294 (1988): 30–31.

12. Ibid., 31.

13. See Daniel Boukman, *"Vivre Libre Ou Mourir," Antilla,* No. 8, 1981, 32, and "L'Humour Bête et Con de Christian Lara: *Adieu Foulards," Antilla,* No. 68, 16–23 Septembre, 1983, 35–6.

14. Amon Saba Sakaana, "Diaspora Film Festival Hosted in Martinique," *Caribbean Times,* Tuesday, 19 June, 1990: 24.

15. Ibid.

16. *Premier Festival Images Caraïbes* [Festival brochure] (Fort-de-France, 1988), 19.

17. *Deuxième Festival Images Caraïbes: Cinema Television* [Festival brochure] (Fort-de-France), 1990.

18. For details on the problems faced by Suzy Landau in putting together the Images Caraïbes festivals, see the following interviews: G. C. M, "Suzy Landau: Une Vision de l'Ame Caraïbe," *Antilla,* no. 283 (1988): 10; Tony Delsham, "Suzie Landau: 'Au Niveau de la Caraïbe la Martinique et la Guadeloupe Sont Très Peu Connues," *Antilla,* no. 293 (1988): 16–17; J. G. Cauver, "Un Entretien avec Suzy Landau," *Antilla,* no. 294 (1988): 34–37.

19. The manifesto is distributed by the APDCC/Images Caraïbes, Fort-de-France, 1990.

20. Sistren is a Jamaican popular theater collective composed mostly of women. It has so far made one documentary video entitled *Miss Amy and Miss May* dealing with the life stories of two Jamaican women activists and their struggles for women's causes and for social justice. For a discussion of Sistren's theatrical and cultural work, see Ketu Katrak, "The Sistren Collective: Women Dramatize Their Lives," *African Commentary,* August 1990: 27–29.

21. See "*Evil Dead* Jugé Meilleur Film par le Jury et le Public," *Antilla,* no. 60 (30 juin–7 juillet, 1983): 6.

22. This overview passes over Trinidad and Tobago and Jamaica because recent works from here are dealt with by Keith Warner, Victoria Marshall, Ed Guerrero, and Bruce Paddington in their essays in *Ex-Iles.*

23. For a general overview of film production in the Antilles, see Catherine Ruelle, "Le Contexte et l'Histoire du Cinéma Antillais," *Racines Noires 1985* [Festival brochure] (Paris: Association pour la Promotion des Cultures du Monde Noir, 1985): 38–41. This festival of Black Arts and Culture took place in Paris in 1985 and was sponsored by the Paris-based Association for the Promotion of Black World Cultures. See also Daniel Boukman, "Antilles," *CinémAction/Tricontinental,* Numéro Special, "Le Tiers Monde en Films" (1982): 95–97. This section on filmmaking in the Antilles draws from Ruelle's and Boukman's overview.

24. *Racines Noires 1985,* 38.

25. Ibid., 39. See also, Roland Suvelor, "Martinique et le Cinéma: Éléments d'une Découverte," in *Premier Festival Images Caraïbes.* Festival Catalogue.

26. *Racines Noires 1985,* 42.

27. This controversy revolves around the [mis]use of money and actors/actresses and broken promises, etc. See Bruce Drayeur, "C. Lara et l'Argent de Nos Impots," *Antilla,* no. 356 (2–8 Novembre 1989): 10–11. For Lara's response to Drayeur, see Christian Lara, "Informer en Déformant," *Antilla,* no. 362 (14–20 Decembre 1989): 12.

[At the time of this writing, Lara was editing a new film in Paris. As a matter of practice, Lara does not talk about projects in progress, so repeated attempts to get him to talk about this new film were met with silence. Ed.]

28. For more details on Lara's work, see Boukman "Antilles"; see also

"L'Humour bête et con de Christian Lara: *Adieu Foulards,*" *Antilla,* no. 68 (16–23 Septembre 1983): 35–36.

29. For a detailed review of this film, see Lambert-Félix Prudent, "—Rouge le Bourg,—Morne la Folie,—Tristes Cendres du Cinema Martiniquais," *Antilla,* no. 24 (15 Août 1982): 32–33.

30. See Marine, "Bon Dieu Bon: La Réunion à l'Affiche," *Réunion Magazine,* no. 476 (27 Août, 1986), and Jules-Rosette's interview with D. Guadarama in *Le Quotidien* (Réunion), Vendredi 17 Juillet, 1987.

The island of Reunion does not have much in terms of productions by Reunionese. One of the rare local filmmakers is Madeline Beauséjour who made a 1988 thirty-minute short feature called *Koman I le La Sours* about a young mother of one child whose house is the popular hangout of the neighborhood.

31. See José Egouy's interview with Raphaël Confiant in *Antilla,* no. 69 (23–30 Septembre 1983), continued in *Antilla,* no. 70 (30 Septembre–7 Octobre 1983).

32. For other accounts of cinema in Haiti, see Arnold Antonin, "Problèmes Politiques du Cinéma Haïtien," *Antilla*, no. 75 (4–11 Novembre, 1983): 33–34, and Antonin's "Cinéma Haïtien, in *Cinémas d'Amérique Latine*, ed. Guy Hennebelle et Gomucio Dagron (Paris: Editions Lerminier, 1982).

33. Rassoul Labuchin, "*Haitian Corner* de Raoul Peck: Le Meilleur Jamais Réalisé par un Haïtien," *Le Nouvelliste,* (Jeudi, 26 Janvier 1989). See also in the same issue an interview with Raoul Peck by Pradel Charles, "*Haitian Corner:* Un Nouveau Pas du Cinéma Haïtien," and Joubert Satyre, "*Haytian Corner* de Raoul Peck: Entre la Fiction et la Réalité," *L'Union,* (13 février 1989). In addition to analyzing various aspects of the film, these articles also provide good accounts of how the film was received by Haitian spectators when it premiered in Haiti in January 1989.

34. Interview with Felix de Rooy contained in the press packet of *Ava & Gabriel.* The full quote is:

> It is a re-interpretation of Christianity. Gabriel the archangel who brings the joyful tidings to the world and impregnates the virgin. After he has completed his task, he returns to the spiritual world through death. It is a parable.
>
> It is very dangerous to present a retrospect on spirituality and Christianity in the nineties, since people will meet it with quite a lot of cynicism.
>
> That is why the story is multi-layered, to avoid the danger of it becoming a sermon. This spiritual way of thinking has its origins in the Messiah myth.
>
> The young girl with the iguana at the beginning and at the end of the movie symbolizes the announcement of the Messiah.
>
> If I were to make a new movie in the present, then I would place the Messiah in the present. He has been mythologically announced in *Almacita,* has been shaped in *Ava* and is now ready to manifest Himself in the present. I am currently working on a trilogy of which *Ava & Gabriel* is the second part.

This was in answer to a question regarding the essence of the film *Ava & Gabriel.*

On Adapting a West Indian Classic to the Cinema

The Implications of Success

Keith Q. Warner

It is rather significant, for the purpose of the discussion that follows, to note that the Martiniquian writer Joseph Zobel included a very detailed description of a West Indian cinema audience of the period between the world wars in his novel *La Rue Cases-Nègrès,* which has itself now been transposed to the silver screen under the same title for French-speaking audiences, and *Sugar Cane Alley* for English-speaking ones (at least in North America).[1] At the time of writing, Zobel most probably did not envisage that one day his work would be interpreted for a cinema audience—one which, though possibly more experienced, would exhibit some of the same gut reactions as those of his fictive audience:

> The lights would go out one by one and everybody would scramble to the chairs to sit down.
>
> At the first images on the screen, the cinema would become relatively silent. For all that, in the darkness, conversations and comments continued, attracting anonymous replies that clashed, exploding into violent discussions laced with jeers and threats. In the long run, however, the atmosphere turned out to be inoffensive and even pleasant—simply foreign.[2]

Zobel's description highlights two aspects of the cinema audience of the period that bear emphasizing. First, it shows the impassioned reactions of one section of the patrons as they become more and more involved in the events unfolding before them on the screen. This state of affairs that

From Renee Tajima, ed., *Journey Across Three Continents* (New York: Third World Newsreel, 1985), pp. 41–45. Reprinted by permission of the publisher.

has hardly disappeared completely, especially among those in the stall or "pit" section (the lowest priced seats that are situated closest to the screen), who, according to Michael Lieber in his book *Street Life: Afro-American Culture in Urban Trinidad,* "tend to interact with the film, much to the annoyance of those in other sections," are "raucous and continually on the lookout for double-entendres."[3]

Second, the description further shows that, when all was said and done, what remained was an atmosphere that was "simple foreign." In other words, films, like written literature, remained almost the exclusive domain of the outsider, with the West Indian having no input at any stage: none in the original writing, none in the interpretation, none in the eventual production. Now, whereas in the field of literature there has been considerable "localization" over the past two decades, the same cannot be said for the world of films. With very few exceptions, all the films we see in the West Indies come to us from abroad, there being nothing even close to what one could call an indigenous cinema. What, then are the implications of an indigenous written work finally brought to the screen?

Zobel's is the first major West Indian work in English or in French to be made into a film and given wide distribution and press coverage ever since the 1967 film *To Sir, With Love,* the commercially-popular adaptation of E. R. Braithwaite's novel starring Sidney Poitier. The adaptation of Jacques Roumain's *Gouverneurs de la Rossee (Masters of the Dew),* filmed in Haiti with Haitian actors in 1975, was produced for television in France, but did not go beyond that medium. The very successful *The Harder They Come* from Jamaica, was originally a film, and only later was it novelized. Other indigenous film productions, like Trinidad's *Bim,* were not adaptations of well-known works, though credit must still be given to the writers of those original scripts.

The West Indian public has apparently grown so accustomed to regarding commercially-oriented feature films as strictly the province of the foreigner, that once indigenous scenes and locations began appearing, it found difficulty relating to them from an artistic point of view. Frequently, in the exuberance of self-recognition, the cinema audience shifts its focus from the actor in character to the actor as person—one recognized, say, as a neighbor or a local personality—from a location used as a backdrop to its private relationship with the place. A constant distraction from the screen often results, as members of the audience point out, either to their friends or to no one in particular, their personal experience with the person or place being shown. It would be difficult, in such an atmosphere to develop and maintain the necessary critical impartiality, if indeed this is even deemed necessary. The problems that beset early West Indian literary works, namely the public's unwillingness to accept the local and familiar as genuine literature, also beset the emerging West Indian cinema.

In view of the obvious interaction between film and audience, it is reasonable to assume that in the West Indies the cinema would have great impact in its portrayal of themes that purport to make a particular statement. Once an audience gets over the initial "exuberance of self-recognition," the familiarity and self-identification with people and places on the screen would create a degree of relevance to the lives of West Indian viewers that a foreign film with foreign themes could never achieve. A director seeking to do this, as opposed to one merely seeking box-office success, would, as a result, find the task relatively easy. An example is Euzhan Palcy's adaptation of Zobel's novel, which continued the work that the novelist started with this semi-autobiographical look at Martinique in the years between the first and second World Wars. Zobel's success has clearly been due to the fidelity and sensitivity of his portrayal of life on the sugar cane plantations and among the lower-class blacks in the urban Fort-de-France, in the shadow of the wealthy (and white) property owners. Palcy's task, then was to translate this fidelity and sensitivity to the screen—that is, to decide what to emphasize, what to omit, and what to change. According to Gerald Mast in his essay "Literature and Film:" "A film adaptation of an important literary work has an obligation to be faithful to the spirit (or, even, the letter) of the original text and, at the same time, to be a cogent and unified work in its own terms."[4]

We are told[5] that Palcy discovered the novel when she was 14 years old (around 1970), and that she wrote her first script based on it at the age of 17 when she went to work for the Radio Télévision Française office in Fort-de-France. Palcy is quoted as saying: "In my mind, it was urgent to make a movie of this story . . . Zobel's book was a great revelation and shock because all our books were about France. It was the first time I read a book written by a black man of our country about the fruits of our country." By her own admission, then, the director approached her task with a sense of urgency. One, it can be safely assumed, that was aimed more at the overall education of her fellow Martiniquians and others, than at the unpredictability of box-office success. This brings us back to the question of film as statement, or, more explicitly, as an expression of artistic commitment.

On the subject of the treatment of Africans, for example, Palcy has stated that when she first went to Paris she was shocked by the state of their relations with the Parisians. "After slavery was abolished in the West Indies, the government broke all relations with Africa as if they wanted to forget they were from African slaves . . . I know Africa only from films that Aimé Césaire arranged to have shown in Martinique," said Palcy. Her sympathetic portrayal of the elder African Médouze, like Zobel's, reflects the director's attempt to reach out to present-day Africans, and to bridge the gap. "If Médouze exists in the film," she has said, "that was my way to say 'Africa, hello! Africa, we didn't forget you.' "

By and large, Palcy has remained faithful to the spirit of Zobel's original, even to the point of reproducing snatches of dialogue that sound like verbatim liftings from the novel. Her scenes of the children's mischief while their parents and guardians are off working in the cane fields; with old Médouze teaching the boy José about riddles and life across the seas in Africa; with the plantation workers disputing their abysmally low wages in return for the toiling and moiling they had to do in the cane fields; and with the children being drilled in their catechism before First Communion. All of these scenes, and many others have an air of Zobel-like authenticity—the director complementing the novelist.

Palcy captures to perfection the importance of education in the lives of these plantation youngsters. It is their only way out of a seemingly hopeless situation. So we share M'Man Tine's disappointment when her grandson José is only awarded a partial scholarship to attend school in Fort-de-France, but applaud her resolve when she decides to do whatever it takes to ensure that he goes. In one scene Palcy's camera dwells on the teacher's enthralled expression as José displays his skill at word usage, and we share the pride filling the teacher's heart. Zobel had made the quest for education the focal point of most of the novel, and Palcy, by highlighting this focus, does the novelist full justice.

Naturally, the director displayed her own creativity through the artistic liberties taken with the text of the novel. In some cases, it appears, the innovations had more to do with practical demands than with a search for artistic fidelity. Primary of these is the "killing off" of José's mother, Délia, who in the novel, works as a maid in Fort-de-France. When José's friends break M'Man Tine's favorite bowl in their search for the sugar that they believe to be hidden somewhere in the shack—no sugar on a sugar cane plantation?—we learn that this deed is all the more reprehensible because the bowl represented a close link between the grandmother and her dead daughter. Palcy thus economizes on one potentially major character, and as a result was the grandmother, outstandingly played by Darling Légitimus, to take over Délia's role in Fort-de-France. The integrity of the adaptation is not severely compromised. Since Délia is not really missed, and M'Man Tine comes across all the stronger for this extension of her role.

Economy, plus a feeling for those events or episodes that were of most significance to her audience, occasioned Palcy's introduction of a character such as Léopold, José's mulatto friend, a composite of several Zobel characters. In one episode in the novel, told almost *en passant* by José and M'Man Tine's friend Carmen to explain why he was sleeping with so many of the *béké*'s (locally-born Caucasians) women, an unnamed youth is being refused the name of his white father. Even on his deathbed the man proclaims: "It is not a name for mulattoes." For the film Palcy embroiders on this episode, shaping it into something larger by making Léopold this unfortunate mulatto—and allowing her camera to do the rest. She thereby puts into sharper perspective the deep racial division so prevalent in West Indian societies, one which they are ever-willing to pretend does not exist.

Palcy also economizes in the film by restricting the time span of the action. Whereas the novel follows José over a period of some ten years, the film revolves around the boy at the age of eleven (thus saving in the process the trouble of having to "age" the young Gary Cadenat, who plays José with a promising degree of professionalism. In the final analysis, these changes matter little, since the spirit of Zobel's novel remains. For instance, Carmen's dreams of Hollywood, a Palcy innovation, is not too unlikely a scenario. Since it provides an even stronger motivation for Carmen to allow the younger José to tutor him; keeping close to Zobel's own powerful attraction to the cinema as shown in the novel.

Another one of Palcy's refreshing innovations is the liberal use of *créole,* whereas none was used in the novel. It is understandable that Zobel even wrote his dialogues in standard French. But, as I have stated in the introduction to the novel's English translation: "Most French West Indians have another language beside standard French—*créole.* In fact, for many this is the only language. *Créole* is widespread precisely among the class of people with which Zobel deals."[6] It is clearly better to opt for the authenticity afforded by the characters speaking the language that comes most spontaneously, than to force them to "dress up" their language, and come across as stilted or false. Instead, *Sugar Cane Alley* was subtitled in standard French—a technique used quite effectively in *The Harder They Come,* where Jamaican *creole* was subtitled in standard English. In this regard, then, Palcy went one step further than the novelist, completing what he probably felt precluded from doing during that particular period of socio-political history.

Palcy achieves an overall effect of simplicity and charm. Many reviewers, quite justifiably, used words like "touching" and "moving" to describe the film. The two recognized professional actors, Darling Légitimus, and Douta Seck, Césaire's original Christophe in the play *La Tragédie du Roi Christophe (The Tragedy of King Christopher)* as Médouze, join with newcomer Gary Cadenat and a huge cast of islanders to produce a film that captures the essence of the plantation lifestyle, without wallowing in the overly exotic. One North American critic, clearly caught up in the tourist-oriented view of the "islands," found Palcy's use of sepia tones "a mistake for a movie trying to capture the life of a Caribbean island, where strong, bright colors are bound to be part of anyone's experience and memory." This critic no doubt missed, or disregarded the connection with the snapshots of the period that opens the film. On the other hand, some have praised this technique for the air of retrospect it provides, thereby ensuring its charm, as well as its effect on countering the stereotyped image that many have of the West Indies.

Sugar Cane Alley has been successful in many places, playing to receptive audiences and critical acclaim in major cities of France, Canada, and the United States, and to enthusiastic and record-breaking crowds in Martinique. Palcy won the Silver Lion at the Venice Film Festival and a *Caesar* in France for best first film, and Darling Légitimus received the award for best actress in Venice. All of this should suffice to show that even with a relatively limited budget—approximately $US 800,000, we are told—to adapt our classics to the screen in a manner that does both the written work and the film justice. One wonders, though, how a West Indian film festival, with West Indian judges, would have responded to *Sugar Cane Alley.* One suspects that non-West Indian audiences react to the film for reasons that are not quite the same as those motivating their counterparts in Martinique. Whereas the non-West Indian may look for an image of exoticism, the West Indian looks for realistic portrayals. Palcy must have thought about this duality of pitch. In other words, for which audience was she filming? To further complicate matters, the film received international recognition and blessing. Alain Ménil has commented on what may have been a change in reaction due to this success: ". . . the perception of the film would of necessity be changed by this official recognition. The Metropolitan viewer goes to see a film that is now authenticated by international acclaim, one that has made its mark. In the end, this viewer is no longer part of a homogenous community, to which the film had been destined but is necessarily taken up with different preoccupations: as a West Indian, he goes to see an image of a time and a place he has probably never known; as a Frenchman, he simply discovers a universe unknown in all respects, and foreign to his being. One cannot erase this duality from one's analysis of the film."[7]

Obviously, outside of the West Indies—except among West Indian exiles and their descendants—the element of identification is missing. However, the film struck a very responsive chord among Martiniquans precisely because so many people recognized themselves in the portrayals. One elderly man, Palcy relates, felt he was seeing his life on film and pronounced himself ready to "die happy." If such a reaction seems exaggerated, it must perforce be seen in the light of the empassioned relation between film image and viewer that we first spoke of in this paper—occasioned by what is perceived by West Indians as the lack of opportunity to see oneself as a serious participant in the entire cinematic process.

Does self-recognition guarantee success in the West Indies? Apparently not, if we are to judge by the reported failure of Christine Lara's *Adieu Foulards,* another West Indian film that was released shortly after *Sugar Cane Alley.* According to the Martiniquian scholar Alain Ménil, this failure stemmed from the fact that the self-recognition was nothing more than a continuation of the same foreign-based view of the West Indies that had caused so many problems in the first place.[8] The accolades showered on *Sugar Cane Alley* have so far been heavily foreign-based—that is, from critics outside of Martinique and the West Indies. In the West Indies we have been led to believe that success before a supposedly unbiased non-West Indian audience confirms the overall artistic appeal and basic soundness of a work. On the other hand, success before a local audience—while confirming the fact that the artist has made a sufficient impact—comes with the nagging feeling and suspicion that the response is based more on audience emotion and familiarity with parties and events than with the attendant critical scrutiny. We West Indians therefore have to call into action all our intellectual honesty to fight this untenable situation.

One can only hope that a film such as *Sugar Cane Alley,* after its initial success, does not pass into the realm of the brilliant but forgotten, and that it serves as a catalyst for the production of others based, loosely or otherwise, on the outstanding written works our authors have produced. We can dream, perhaps, of seeing a film version of *A House for Mr. Biswas* or of *La Tragédie du Roi Christophe* as part of what could some day be called the West Indian cinema. Such a cinema, ideally, would seek to correct the state of affairs which foists upon impressionable audiences films that, according to the Gambian literature and film scholar Mbye Cham, "tend to be of the most alienating type: colonial propaganda productions that extol, justify and glorify the virtues and magnanimity of the so-called European 'civilizing mission' "; American "B" movies concerning cops and robbers, cowboy and Indian Westerns and war extravaganzas; escapist James Bond spectacles; "Spaghetti Westerns"; Indian romance fantasies and, lately, Bruce Lee karate and other Kung Fu packages from Hong Kong, and so-called blaxpolitation films."[9] West Indian cinema must become as real as West Indian literature. With unflinching tenacity, it must translate and interpret what our artists are saying to the West Indian society in particular, and the rest of the world in general. The audience is already there, willing and receptive, if not yet very critical. The technology and the various filmmaking talents, as Euzhan Palcy demonstrates so creditably, are also available. What, one wonders, are we waiting for?

Notes

1. I am aware that the film was presented at the 1983 Bombay Film Festival as *Black Shack Alley,* under which title appeared my translation of the novel in 1980 (Washington: Three Continents Press, and London: Heinemann Educational Books). The English subtitles on the film also use the phrase "Black Shack Alley" whenever the narrator talks of "rue cases-nègres." This leads me to believe that the American distributor sought a more catchy or marketable title.

2. Zobel, Joseph, *Black Shack Alley,* p. 168.

3. Lieber, Michael, *Street Life: Afro-American Culture in Urban Trinidad,* Boston: G. K. Hall/Cambridge: Schenkman Publishing Co., 1981, p. 95.

4. Mast, Gerald, "Literature and Film," in Barricelli and Gibaldi, eds., *Interrelations of Literature,* New York: MLA, 1982, p. 280.

5. I have culled much of this information from the many interviews Palcy gave to the press as a result of the success of *Sugar Cane Alley* in the United States.

6. Zobel, *Black Shack Alley,* p. xxi.

7. Ménil, Alain, " 'Rue Cases-Nègres' ou les Antilles de l'Intérieur," *Présence Africaine* 129, 1st Quarter, 1984:97.

8. Ibid., p. 100.

9. Mbye Cham, "Art and Ideology in The Work of Sembene Ousmane and Haile Gerima," *Présence Africaine* 129, 1st Quarter, 1984:89.

Sergio Giral on Filmmaking in Cuba

An Interview

Ana M. Lopez and Nicholas Peter Humy

Born in Havana in 1937, Sergio Giral lived for a few years in the United States before returning to Cuba in time to witness the triumph of the Revolution in 1959. After making a number of documentaries for the Cuban Film Institute (the ICAIC, or Instituto Cubano de Arte e Industria Cinematograficas) in the 1960s, Giral produced his first feature film, *El Otro Francisco* (The Other Francisco) in 1975. This was followed by two other feature films that, like Francisco, explored the history of slavery in Cuba: *El Rancheador* (The Bounty Hunter in 1976) and *Maluala* (1979). These three films are known as his "slavery" trilogy. In this interview, Giral discusses his two most recent films—*Techo de Vidrio* (1982) *(The Glass Ceiling)* and *Placido* (1986)—as well as his experiences at the ICAIC.

Giral was interviewed at his Havana home by Ana M. Lopez and Nicholas Peter Humy last August. The interview was later translated by Lopez and edited by Humy.

BFR: You have been with the ICAIC almost from the very beginning, haven't you?
Giral: Yes, since 1961. I returned to Cuba from the U.S. in 1959. I did the opposite of what a lot of other Cubans were doing then. I came to Cuba and I found this incredible phenomenon of the Revolution. At that time I was barely surviving in the U.S. I was 22 years old and I had no answers. When I returned to Cuba, little by little the Revolution began to

Black Film Review, vol. 3, no. 1, 1986/87, pp. 4–6. Reprinted by permission of the publisher.

give me answers, with many contradictions, but with contradictions that were sufficiently rich for me to dedicate myself to them. Time passed, and suddenly I realized that I was staying.

When I arrived in Cuba, I tried to be useful and began to study agricultural engineering at the University. At that time we had a tremendous need to develop our agricultural system. But that profession had nothing to do with my real interests. So I decided I was better off trying to help out in areas that were at least within the realm of my real abilities and possibilities. That's how I became part of the ICAIC—I looked for a way to be useful. There was this new institute and I figured out ways to become affiliated with it.

At that time in Cuba, there were few people skilled in the cinema. The few cinema technicians we had, had either already left or were packing their bags. This was around the time when the U.S. companies left the country and took their employees with them by guaranteeing them jobs in the U.S. The fledgling industry we had—a micro-industry of publicity and advertising shorts—disappeared. The Cuban cinema emerged from the residue of small efforts and isolated resources left to us.

My apprenticeship, like that of the great majority of ICAIC filmmakers, was autodidactic; it happened while we worked. We began making documentaries right away in order to learn how to make films. And we watched a lot of films. I am part of what is known as the second generation of Cuban filmmakers, a generation that first has been spectators. I grew up in New York—my parents moved to the U.S. in 1953—but I always loved the movies and used to spend every weekend in those Times Square theaters that back then showed great movies all day long for only 35 cents or so.

BFR: You are of the second generation of ICAIC filmmakers. How did the third generation emerge?

Giral: Once our industry became more developed, it had other needs. So now, for example, one of the prerequisites for entry into the ICAIC, at any level, is to be a graduate of the university in some field that seems somehow related. Before, when I came on board, this wasn't the case. There weren't too many of us with university degrees then. This makes a great difference. We went around the entire operation, doing all kinds of different jobs until we found the ones that we liked and that we were most suited for. Now the young kids come in directly as analysts or as assistant directors, and they stay in those positions for a long time learning and training before they begin to work on their first solo documentary films.

BFR: What is it like to work for the ICAIC today?

Giral: Our situation as filmmakers is superprivileged. Of course, we do not profit materially from our work; we do not get rich. But I do my work—which is my central interest, passion, and obsession—under the best possible conditions.

First of all, I do not have to worry about money. We are paid a monthly salary for our work. For example, we get a year and a half with salary to write a script, to write whatever you want wherever you want. This is a year and a half of financial and social security. Now, I have a year and a half to produce a script, but if I finish it sooner, say in six months, and I go into the preproduction phase with the script, then I also begin to collect a salary from production budgets. This is a great stimulus; it encourages us to work faster. But it is also a great luxury.

Another thing is that we have tremendous artistic freedom. From the outside, people always miss the point here. The issue of artistic freedom is somehow always wielded against us when people start talking about freedom of expression (which should always be placed in quotation marks in any part of the world anyway). But I will tell you that here I have complete artistic freedom of expression.

BFR: Cuba is culturally very different and very interesting because it is precisely in between most other Latin American nations. It has a unique situation. It doesn't have, for example, the large Indian population of the Andean countries nor does it have the bourgeois superstructures of, say, Argentina.

Giral: Yes, it's a middle ground. And in addition, there's something else very interesting that has an impact on the cinema. Here the filmmaker—and, of course, this also depends on his or her relative experience—is not obligated by material needs and circumstances to always make films for the mass public. You can express yourself, if you so desire, in the most hermetic or most personal way, and if that is your way of expressing yourself, then it is valid and acceptable. Or you can make more popular films, say comedies, if you so desire. You are not always trying to reach the same public, the same subcultural group.

BFR: But I imagine that in the planning of the ICAIC, that has to be mediated. A more popular film here and a more difficult film there, no?

Giral: Yes, it is all a matter of equilibrium, of course. We now make some 12 feature-length films a year and approximately 40 documentaries, so there's actually a lot of room in the system.

BFR: The most recent films of yours that we know in the U.S. are the films of the slavery trilogy: *The Other Francisco, The Bounty Hunter,* and *Maluala.* What about your new films?

Giral: My last film is *Placido.* It was inspired by the life of a poet of the last century named Gabriel de la Concepcion Placido, a poet from Matanzas who had a very interesting history. The most important thing about his life is that he was executed by the Spanish during the war because he was accused of being a conspirator and a traitor. The thing is that no one knows for sure whether he was truly involved with the revolutionaries or whether he was manipulated by the colonial order and made to serve as a scapegoat. This has never been historically clarified. And what his case has allowed me to do is to analyze the position of the artist at this particular historical moment, his compromise, his contradictions. What I seek to do in this film is to underline, to prove, the importance of commitment, of always being committed because ultimately, whether you are committed or not, you will pay. As we say, *"los justos por los pecadores"* [the innocents always pay for the sins of the sinners]. And that's sad, no?

BFR: In 1982 you made a film that we had not heard about, *Techo de Vidrio* [Glass Ceiling].

Giral: Yes. Unfortunately, I think it is a failed film. I tried to make a film about contemporary topics, in a more or less critical tone, but I wasn't pleased with the results. It is very difficult for us to broach and elaborate on contemporary topics. Our reality changes so much and so quickly. After all, this is a revolution that is never finished, that never rests.

First of all, I don't like the picture aesthetically. I am very interested in creativity, in the possibility of creating an aesthetic universe in my films. The developed cinemas—the U.S. cinema, for example—have reached that magical stage where they are able to deal with actual problems and with contemporary life within an aesthetic system of its own which we all accept as appropriate and which we all like. But our approach to contemporary life, to our contemporary cultural realities, has not pleased me so far. Neither mine nor that of other *compañero* filmmakers who have made films on contemporary topics. We are not aesthetically settled yet; we still have not found the aesthetic key for the representation of our contemporary reality.

Glass Ceiling deals with the need to make demands on individuals not only at the level of individual consciousness, but also in terms of social responsibility and in terms of the need to respect social (rather than private) property which belongs to everyone. I actually think that I will take on this topic again, that I will make the film over some day when I find the aesthetic answers that I am looking for. And I will find them.

BFR: What is the difference between the producer, the director, and the screenwriter here in the ICAIC? In the U.S., as you know, the producer is primarily in charge of finding funds.

Giral: The difference is that the producer doesn't have to look for the funds, although he is responsible for them. He is the administrator of all the resources. He answers to the company, represents its interests. But here, something else interesting happens. And that is that the director has complete authority during the production process. That is, I can spend my resources in any way I choose. If I have $3,000 to do three scenes and I instead choose to invest them all in one, then the producer does not have the authority to stop me.

BFR: Have you completely given up on documentary filmmaking? Are you totally devoted to fiction films?

Giral: I had almost forgotten to mention this. I am planning to make a documentary of still photographs, a short film based on the photographs of a friend of mine who has taken some beautiful shots of old Chicago blues men. I love his pictures and am thinking of doing a documentary using only the stills.

Besides this project, I am basically doing only fiction films now. I want to develop an idea—another contemporary problem—for a film on how romantic relationships all over the world, and also here in Cuba, are affected by material, economic concerns and needs. For example, the problem of the professional woman who has to go to the provinces and leave her husband behind in the city. That was inconceivable before, but now it happens all the time, and it changes the nature of romantic relationships. Another topic that interests me is more amusing. A film that would take place in the 1950s, in a cabaret, in the world of the *espectaculo* of the 1950s, but in relation to the mafioso types who controlled that entire world of drugs and prostitution: in other words, the Mafia in the context of Cuba in the 1950s.

BFR: Have you seen *The Cotton Club?*

Giral: Yes, and I didn't like it. The film is very beautiful, but it doesn't have any narrative strength. Coppola's films fascinate me though—for example, *The Conversation* and *Rumblefish*—although this last one is very strange.

BFR: How would you describe the representation of blacks and black issues in the Cuban cinema and your own position as a black filmmaker in the ICAIC?

Giral: It is very difficult for me to see things from the perspective that is implied in statements such as "black issues" or "the position of blacks" in the Cuban cinema because our social reality is one that does not make such a phenomenon possible. Such issues do not come up because the problems underlying those issues do not exist here. That I am black and a film director does not differentiate me in the least from other directors.

Perhaps what I may be able to address more directly are the topics that I have addressed in my films, because here there is a direct relationship between my personal experiences as a black man and filmmaker and the films that I have made. First of all, I have to indicate that this relationship has been made possible because of the nature of our social life in Cuba. It is the nature of our reality that has permitted me to elaborate, investigate, approach, re-create, and transform my personal experiences and concerns as a black man into a cinematographic experience. What I have tackled are phenomena that before the revolution were taboo not only for me but also for large segments of the general population.

BFR: In the U.S., black filmmakers who have tried to produce a black cinema have encountered the problem of audiences. The black audience is used to the Hollywood product; filmmakers want to begin to educate the audience to relate to cinema in different terms. Given that a somewhat parallel situation existed in Cuba at the time of the revolution, could you tell us some of the things that were done to educate this audience?

Giral: This is the problem of cultural colonialism. I know the problems of black filmmakers in the U.S.—I have met many of them and they have also traveled here. And their problems are very, very difficult. The biggest stumbling block is that they are not in power; they don't have power. We have to start from that. It would be an injustice to apply the same principles that we followed here 25 years ago because our situation was completely different.

In any case, there are some things that can be said. What did we do? First of all, I must underline that we never rejected any method, approach, or influence *a priori.* That would have been a terrible mistake. One of the greatest strengths of the Cuban cinema has been its ability to take the best of everything—that is, the best of that culture that colonized us—because we cannot negate it. But we must remain aware that the cinematographic tradition that formed us had, at the same time, another function, and that's where things must change. What function are you going to give your new form of expression? What are you going to be telling the spectator? What are your objectives? You can use the tradition that has subjugated you to new ends.

BFR: Here you are talking about the kind of work you did in *The Other Francisco,* where you took a classic Cuban novel, the classic anti-slavery novel, and turned it upside down?

Giral: Yes, you have to take "their" stories and make them "yours." . . .

BFR: But I'm not sure you have answered our question. You mention the function of film—the cinema is designed, is principally used, to entertain according to those "manipulative" molds. Didn't you have a lot of work to do to develop an audience that would expect and accept something other than that form of entertainment, the Hollywood pleasure machine? Do you have any advice for American independents on how they can go about developing their audience?

Giral: Giving advice to American independents is very difficult, and I feel somewhat presumptuous about doing so. Our situations, our realities are so different. In the beginning, we had a great need to create a national film industry. After the American companies left, we had theaters but nothing to show in them. So we had to bring in films from wherever we could get them.

What could I tell these black filmmakers? I have seen a kind of black filmmaking that was popular in the 1970s, what was called the black capitalist movement, that was a kind of black telephone film.

BFR: You are referring to the "white telephone" films produced in Italy in the '30s?

Giral: Yes, many of these films of the '70s remind me of them, and I was astounded because their splendorous vision of blacks was just terrible. I have a friend in New York who used to work for a magazine that shared in this same kind of vision, which has nothing to do with racism, with real social problems, or with the ghetto, but which of course did have a lot to do with the automarginalization of the black self, with self-racism. This is all false, a paralysis.

What I am trying to say is that what is needed is to make a cinema of social approximation, of analysis, and to develop specific aspects of black society, of the sociology of the blacks, of the problems of the blacks.

I saw *Purple Rain* and I thought it was an interesting film. I don't know whether the director was black or white, though I assumed he was white. But it actually does not matter. Sometimes the film seemed very attractive to me, seemed to offer an interesting analysis of the problems of the black family. But finally the film falls apart because it traces the origins of its hero's problems to that family and limits those problems to the individual, instead of locating them in social reality.

In a black family in the U.S., you might have a family situation where the mother is an alcoholic and the father is a bum, but where are you going to locate the origin of those problems? In the individual psyche? In the individual's lack of moral fiber? That's ridiculous. You have to look outside the family in order to investigate the nexus of social relations that gives rise to these problems. Otherwise you are lost. These are not psychosexual problems; they are social problems, economic problems, problems of the system—and not only of the political system—but of specific economic systems. Racism, remember, is also an economic system.

Samba, Candomblé, Quilombo

Black Performance and Brazilian Cinema

Robert Stam

> Blackness is like a vibrant and luminous combustible which supplies a kind of telluric energy for a humanity more and more in need of it.
>
> *Gilberto Gil*

For many Americans, the phrase "Brazilian Cinema" instantly elicits the memory of what was in fact a French film—Marcel Camus' *Black Orpheus.* More than any other film, *Black Orpheus* created in the international consciousness a powerful association between three related concepts: Brazilianness, Blackness and carnival. North American critics raved about "intoxicating samba music, frenzied dancing and violent costumes" (Bosley Crowther, *New York Times*) and of "garishly costumed natives doing the samba with hypnotic fervor to the insistent pulse of multiple brass bands" *(The New Yorker).* In his filmic adaptation of *Orfeu de Conçeiçao,* the Vinicius de Moraes source play, the French director combined actual carnival footage from the 1958 Rio carnival with staged footage in which thousands of Brazilians, generally without pay, played at carnival for the cameras.[1] While in many ways offering a French touristic view of carnival, *Black Orpheus* did at least have the virtue of foregrouding Black talent and Black performance. With the exception of Marpessa Dawn as Eurydice, all of the major roles were played by

The Journal of Ethnic Studies, vol. 13, no. 3, 1985, pp. 55–84. Reprinted by permission of the publisher.

Afro-Brazilian actors and actresses. Although a *New Yorker* reviewer attributed the "seduction" of the film to "its naive quality, emanating from its untrained negro actors . . ."[2] in fact most of the leading actors and actresses were professional performers, many of them from Abdias de Nascimento's Teatro Experimental do Negro (Experimental Black Theatre).

Black Orpheus called attention to a highly Africanized cultural phenomenon—carnival—combining ecstatic polyrhythmic percussion with the elaborate folk opera of the samba school pageants. The problem with the seductive imagery, when seen out of context, is the danger of a synecdochic reduction, of taking the festive part for the social whole. In this sense, *Black Orpheus* advances a romantic and mystified vision of carnival and Brazil. The Brazilian characters play out the archetypal patterns provided by a European myth, all against the backdrop of Rio's photogenic beauty and the contagious energy of carnival. The film suggests a primitive capacity on the part of the happy Black Brazilians to enjoy life no matter how devastating the conditions. As a kind of Brazilian *West Side Story, Black Orpheus* prettifies the *favelas* of Rio and isolates carnival from its social context. Authentic carnival offers a dialectical critique of the injustices of everyday life, not a metaphysical transcendence against a postcard backdrop. *Black Orpheus* enlists all the elements of carnival—dance, rhythm, music, color, laughter—but ultimately in the service of a stereotypical vision.

A Sociological Preamble

Because of the danger that a celebration of the Afro-Brazilian cultural contribution to Brazilian cinema might appear to imply an endorsement of Brazil's putative "racial democracy," it is important to provide a modicum of historical and sociological background. Brazil is the largest Black nation outside of Africa, with at least 70 million citizens who descend, to one degree or another, from the slaves brought from Africa. Despite the argument, advanced by Frank Tannebaum, Stanley Elkins and others, that Brazilian slavery was "humanized" and "tempered" by religious institutions which recognized the moral and spiritual personality of the slave, the fact remains that slavery was an unspeakably cruel commodification of human beings, and that whatever the Church's presumed respect for the slave's soul, the slave's body was treated with terrible harshness.[3] The abolition of slavery in 1888, meanwhile, was in many respects a purely formal measure intimately linked to the needs of the British industrial revolution and its preference for "free labor." An emancipation realized by and for Whites, abolition allowed the masters to liberate themselves from the responsibility of caring for the slaves. Thrown ill-prepared onto the free labor market, Brazilian Blacks soon had to face the competition of European immigrant labor invited to Brazil by Brazilian entrepreneurs.

Without developing a thorough-going analysis of the racial situation in Brazil, we can posit some initial points as a backdrop for our discussion. First, Blacks and mulattoes represent the marginalized *majority* of Brazilian citizens. In this sense, it is a eurocentric misnomer, as Abdias de Nascimento points out, to call Brazil a "Latin" Country; it might more accurately be called an Amerindian-African-Iberian country.[4] Second, Brazil is, in structural terms, a racist society. Although Brazilian history is not marked by the virulent racism of the Ku Klux Klan or by racial ghettos and segregation, Afro-Brazilians remain economically, politically, and socially oppressed. The causal relation between oppression and skin color is somewhat obscured by the lack of apparent racial tension and by the fact that Blacks and mulattoes often share similar living conditions with many lower-class Whites and near-White mestices. This shared position within the social structure results in a wide range of social intercourse between these racial groups—evidenced in countless Brazilian films–ranging from friendly contacts at work to more intimate friendships and marriages. But the fact that many Whites share living conditions with Blacks and mulattoes does not mean that the oppression of Afro-Brazilians is not *also* racial. Race, in this sense, is both a kind of salt rubbed into the wounds of class, and a wound in itself. The essential point is that the structural mechanisms of the Brazilian social formation systematically deprive Blacks, and to a lesser extent, mulattoes, of economic and political power. Blacks and mulattoes are excluded from positions of influence; they are virtually non-existent in the government and the diplomatic corps, and under-represented in the universities, while they are over-represented in the favelas, in the prisons, and in the ranks of the under-employed. Brazilian racism is subtle, camouflaged, diffuse (consisting in an aggregate of quietly discriminatory practices), less a matter of virulent hatred than of paternalist condescension. It consists not in a binary White-over-Black but rather in the superimposition of an official integrationist ideology ("racial democracy") on a reality pervaded by a subtle prejudice that White is better.[5]

One striking feature that separates Brazilian from mainstream North American culture is the extent to which the former is highly and consciously Africanized (I emphasize "consciously" because White North America has yet to become fully aware of the ways in which it too has been changed and enriched by the cultural contribution of Afro-Americans). When one speaks of Black performance in Brazil, one is also talking about Brazilian performance in general, in the sense that the Brazilian manner, the *jeito,* the way of walking and talking and touching, are thoroughly inflected by African-derived cultural patterns. The flexibility and gregariousness of Brazilian life, the capacity for collective enthusiasm, can be seen, in some measure, as an African inheritance. Even the expansive orality of the culture is marked by Africanism. Brazilians do not speak *Português,* puns Black liberationist Lelia Gonzales, but *Pretoguês* (Blackoguese), because African sounds and rhythms strongly imbue their speech and differentiate it from continental Portuguese.

The African influence pervades all areas of Brazilian life—literature, music, art, religion, cuisine. The most highly prized symbols of Brazilian nationhood are Black. The national dish is *feijoada,* soul food consisting of hog mawls and black-eyed peas created by Blacks in the days of slavery. Afro-Brazilian religions are omni-present and among the fastest-growing in Brazil. (Many everyday expressions—"sarava," "fazer a cabeca," "para o santo"—come from these religions.) Afro-Brazilian religion is now perfectly respectable and even fashionable among a certain White elite, and every congressman (and even the president, it is said) has his or her *macumbeiro.* And in the privileged moments of collective life, such as carnival, Brazil might be said to celebrate its own Blackness.

My purpose in this essay will be to explore the filmic presence of Afro-Brazilian culture and performance in relation to three broad themes: 1) the presence of Afro-Brazilian music and of Brazil's Africanized carnival; 2) the manifestations of Afro-Brazilian religious forms such as *candomblé, macumba,* and *umbanda;* and 3) the representation of Afro-Brazilian political struggle, which can also be seen as a kind of performance, this time on the world-historical stage. Brazilian cinema has privileged all these themes, and without them, and without Black performance, Brazilian cinema would be immeasurably impoverished. What, then, has been the contribution of Afro-Brazilian culture, specifically in the form of Afro-Brazilian music, Afro-Brazilian religion, and Afro-Brazilian political resistance, within Brazilian cinema?

Afro-Brazilian Music and the Cinema

Brazilian cinema, throughout its history, has been closely linked to Brazilian music, and Brazilian music has always been closely linked to Africa. The African presence marks both erudite music—from eighteenth-century mulatto composers like Marcos Coelho and Francisco Gomes da Rocha through Villa Lobos and his "Danças Africanas"—and, more obviously, the popular music of such Afro-Brazilian or Afro-influenced composers and singers as Gilberto Gil, Jorge Bem, Martinho da Vila, Milton Nascimento, D. Ivone Lara, Moraes Moreira, Luiz Melodia, Clementina de Jesus, Beth Carvalho, Caetano, and Djavan. Africa provided many of the instruments featured in Brazilian music—the *cuica,* the *agogô,* the *berimbau*—and, more important, it provided the cardinal aesthetic principles on which Brazilian music in general, and samba in particular, are built. These principles include a percussive performance style based on the synergistic dialectic of multiple rhythms existing in intricate and constantly changing relationships to one another; the frequency of overlapping call and response patterns, for example between a samba singer and the responding chorus; offbeat phrasing and suspended accentuation; and the assumption that music has every right to speak to its time through social allusion.

Contemporary Brazilian popular music is not only dominated by Afro-Brazilian or Afro-influenced composers and singers, but it also increasingly addresses, in its lyrics, Afro-Brazilian themes and preoccupations. In his album *Africa Brasil* (1977) Jorge Bem pays homage to the African past of Black Brazilians at the same time that he speaks of the present frustrations and aspirations of the *favelados.* In her ode to Blackness, "Abraço Negro," Ivone Lara speaks of the joys and sorrows of negritude: "Um abraco negro/um abraco negro/traz felicidade/Nego sem emprego/nego sem sossego/Negra a raiz da liberdade."

> A black embrace
> A black embrace
> Brings happiness
>
> Blacks without jobs
> Blacks without relief
>
> Black is the root of freedom.

Contemporary Brazilian music is also pervaded by affectionate references to Afro-Brazilian religions. Dorival Cayimi's "Oração a Mãe Menininha" (Prayer of Praise for Mãe Menininha) is dedicated to the beloved *candomblé* priestess from Bahia known as Menininha de Gantois, called in the song "the most beautiful star," "the mother of sweetness," and "the solace of the people."

The brilliant composer and performer Gilberto Gil, inspired both by a pan-Africanist vision and by an anthropophagic openness to world musical culture, speaks of and himself demonstrates what he calls the immense "tribalizing" power of Afro-derived music. His albums *Refazenda* (1975) and *Refavela* (1978) are clear calls for Afro-Brazilian cultural and political consciousness; indeed, Gil called the latter album a "report on the black populations of the cities." Influenced by North American soul music, the narrator of Gil's "Ilê Ayê," a carnival samba in the Bahian style, calls himself "the greatest" . . . "with kinky hair" . . . "very black pow . . ." (The lyrics of certain Gil songs, and of the more politicized sambas, make such music the equivalent of the socially conscious reggae of

Jamaica, although without the movement's Rastafarian underpinnings). At the same time, Gil constantly valorizes Afro-Brazilian religion. The song *Babá Alapalá,* subsequently incorporated into the soundtrack of the film *Tent of Miracles,* speaks of one of the deepest problems of Afro-Brazilians—the loss of ancestral references. *Babá* refers to the ancestral African *pai-de-santo* (*candomblé* priest) and Gil suggests that the link with African origins be reasserted through the *eguns,* i.e. the spirits of dead ancestors. And in *Axé Babá,* Gil asks Oxala to give peace: "Meu pai Oxalá/Dá-nos a felicidade/O Pao da vitalidade/ Do teo Axé: Do teu amor":

> My father Oxalá
> Give Us Happiness
> The Bread of Vitality
> Of Your Energy
> Of Your Love.

Apart from the participation of Afro-Brazilian musicians in composing the musical tracks of Brazilian films—Jorge Bem *(Xica da Silva),* Milton Nascimento *(The Fall),* Gilberto Gil *(Tent of Miracles, Quilombo)* the Afro-Brazilian musical presence enters the cinema in another form, through the energizing presence of carnival. It is no accident that carnival remains most alive in those countries impregnated by the African culture brought by slaves, namely in Brazil and the Caribbean. An immensely creative cultural phenomenon, carnival is the product of a profoundly mestizo culture, a protean manifestation of percussive exuberance and the "theatrical" improvisations of costumed revellers mingling in free and easy contact. While carnival admittedly has European antecedents, carnival in Brazil is a highly Africanized phenomenon. The African presence is clearly marked in the central role of polymetric music and dance, of parades and exuberant costume. It is also Afro-Brazilians who give a critical edge to the utopian impulses of carnival. Since it is they who most obviously suffer the dystopia of unemployment and marginality, it is naturally they who have the greatest investment in the liberatory mechanisms of carnival. For Brazilian anthropologist Roberto da Matta, carnival is the privileged locus of inversion, in which the socially marginalized and especially poor Blacks and mulattoes, take over the symbolic center of the city. The festival, at least in the tendencies of its symbolic system—no one is suggesting that three days of carnival actually overturn social structures produced and reproduced throughout the year—is profoundly democratic and egalitarian. In an atmosphere of gestural freedom and fantasy, revellers play out imaginary roles corresponding to their fondest desires. Poor Blacks and mulattoes appear as aristocrats in wigs and regal costumes, thus symbolically affirming their nobility within an ephemeral utopia characterized by dignity, abundance, and commonality.

Countless Brazilian films have documented the reality of carnival. The documentary shorts devoted to Rio's carnival in the first decades of this century doubtless featured the Afro-Brazilian revellers who then constituted the majority of Rio's population. Although Blacks were largely excluded from the symbolically "white" cinema of the silent period, Black culture was implicitly present in the frequent representation of carnival. The very word "carnival" figures prominently in the titles of a disproportionate number of Brazilian films, ranging from the first "views" through the series of remakes of *O Carnaval Cantado* (The Singing Carnival), first produced in 1919, through sound-era *chanchadas* such as *Alô, Alô Carnaval* (1936) to post-Cinema Novo productions such as *Quando o Carnaval* (When Carnival Comes, 1971). The *chanchadas* popular through the thirties, forties, and fifties, in fact, were designated, as if in anticipation of a Bakhtinian analysis, "filmes carnavales-cos" (carnivalesque films).

The advent of sound brought the thorough musicalization of Brazilian cinema and with it a strong Afro-Brazilian influence, even if Afro-Brazilians themselves seldom appeared on the screen. In the *chanchadas*–Rio-based comedies featuring musical numbers and carnival themes—Blacks and mulattoes tended to be featured only in minor or background roles. Despite occasional exceptions—Humberto Mauro's *Favela dos Meus Amores* (Favela of my Loves, 1935) features the musicians and dancers of the Portela Samba School—most chanchadas use Blacks merely as backdrop for the love intrigues of White stars like Carmen Miranda, Eliane Macedo or Cyl Farney. Since the *chanchadas* were partially modelled on the Hollywood musical, a genre not known generally (if one excepts "all-black musicals" like *Hallelujah* and *Cabin in the Sky*) for any exaltation of Black culture, they tended to downplay Blackness. Although samba was largely an Afro-Brazilian contribution, film after film gave the impression that samba was a White cultural product.

An exception to this rule, ironically, came from a North American director. In 1942, Orson Welles went to Brazil to film two episodes of the never-to-be finished *It's All True,* a semi-documentary aimed at fostering the Good Neighbor policy and countering Nazi propaganda in Latin America. Welles, a jazz-lover who had produced an all-Black *Macbeth* for the Federal Theatre, went to Brazil already attuned, in a certain measure, to Black performance. His guide in Rio was Vinicius de Moraes, author of the source play for *Black Orpheus,* and who later referred to himself as "o branco mais preto do Brasil" (the Blackest White man in Brazil). Welles became a samba and carnival enthusiast, to the point that he reconceptualized *It's All True,* originally intended to be a pan-American epic, so as to privilege Brazil. He organized the film around two central characters, a five-year-old boy and a samba school leader (played by Luis Sebastião Prata, commonly known as Grande Otelo) desperate over the destruction of Praça Onze, the Rio square through which the samba of the *morros* traditionally entered the city.

Welles dispatched second units to film the *desfiles* of Praça Onze and a *frevo* group from Recife, at the same time that he rented the Cinedia studios in Rio in order to supplement the more authentic footage. Originally planned to emphasize carnival itself, the Rio episode was transformed into the story of the samba when Welles resolved to explain the origins of a musical-dance form which Welles had come to see as the Brazilian equivalent of jazz. The key structural idea was to trace the itinerary of the samba from the *favelas* to its climactic explosion in carnival in the center of the city. Welles also planned to interweave sequences of *macumba* with the sequences of carnival. From the production stills and existing footage of *It's All True* it is obvious that *It's All True* would have revealed a carnival that was overwhelmingly Black. According to Brazilian film critic Paulo Emilio Salles Gomes, Welles saw the film as a hymn of solidarity to the Blacks of Rio and to their culture.[6] In fact, this sense of solidarity was not always appreciated. One member of the production crew complained in writing to RKO executives that Welles was overemphasizing the Black element and showing too much "ordinary social intercourse" between Blacks and Whites in carnival, a feature that would doubtless offend some North American viewers.[7]

One Brazilian director who has repeatedly turned to Afro-Brazilian themes is Nelson Pereira dos Santos. While most directors treated Black characters and themes with paternalistic condescension, Nelson has consistently accorded Blackness respect and pride of place. His *Rio Zona Norte* (Rio Northern Zone, 1957) focuses on the samba composer Espírito da Luz Doares, incarnated brilliantly by Grande Otelo, and on the exploitation of which he is the object. The title refers to the urban topography of Rio de Janeiro, which is divided into *zona sul* (southern zone), the wealthier (and Whiter) area of Rio embracing the beautiful beaches and lovely residential neighborhoods of Copacabana, Ipanema, Leblon and Lagoa, and the *zona norte* (northern zone), the location of the *"suburbios,"* i.e., the poorer (and Blacker) neighborhoods of lower-class working people and favelados. Rio, an agreeable city for the residents of *zona sul,* can be a hell for the inhabitants of the *zona norte,* where even the pollution seems to discriminate. (It is no accident that the Black Rio movement in music, largely inspired by North American soul and Black power influences, emerged from the northern zone). *Rio Zona Norte* shows the world of the samba as experienced by one of the many composers who used to sell their sambas for virtually nothing in the hopes of one day hearing them played on the radio or during carnival. The history of Brazilian music, like that of North American music, is replete with stories of these struggling and often anonymous musical poets. While the film is not racially schematic, we notice that Espírito's exploiters tend to be White. Maurício (Jece Valedão) pretends to be a friend and collaborator of Espírito's, but in fact he steals his sambas and sells them to the highest bidder. We see a kind of "cordial" Brazilian racism at work, in which embraces and warm words serve as camouflage for a relationship of exploitation. Espírito is manipulated because of his inexperience and due to his faith that other human beings are as uncalculating and honest as he is; he learns quickly, however, and comes to insist on his right to share in the profits and determine the musical quality of the final product.

The treatment of Espírito in *Rio Zona Norte* clearly contrasts with that given Orpheus in the Marcel Camus film. Orpheus is the "universal" mythic figure, here transported to Brazil, who knows how to make the sun rise; he is largely stripped of all social specificity. Espírito, on the other hand, is rooted in a specific time, place, culture, and even neighborhood. His exploitation too is specific; he is exploited by his "partners," by disc jockeys and by record company managers. *Rio Zona Norte* exposes the process by which samba composers lost financial and artistic control of their own musical production, whether through the crudest kind of exploitation or through a more subtle process of erudite thievery of musical ideas. (A talentless and frustrated middle-class composer speaks of writing a ballet based on Espírito's music.) *Rio Zona Norte* registers the circular process whereby samba originates in the *bactucadas* of the favelas, and then moves to nightclub and radio, gaining a layer of sophistication with every step, and finally returns to the favelas. While those who hear Espírito's music on the radio are unaware of his existence, his fellow *favelados* remember: "The *favelas* don't forget you," goes one of his sambas, "for the *morro,* the samba didn't die. . . ." The samba of the final sequence sets up a series of equations: the *morro* is the samba and the samba is Brazil: "It's my samba/and Brazil's too." Espírito's sambas speak of the people's suffering as well as their joy. They speak of favela shacks collapsing, of daily humiliations, of wells running dry; they contrast existing social imperfections with an implied utopia. And like the sambas which speak of sorrow and misery yet communicate an immense capacity for happiness, *Rio Zona Norte* itself creates a dialectical tension between social commentary and sensual pleasure, the tragic and the comic, the bitter and the sweet.

In the wake of Nelson Pereira dos Santos' pioneering efforts, the Cinema Novo directors of the sixties searched out the dimly-illuminated corners of Brazilian life—and especially *favela* and *sertao*—created a cinema in which Black Brazilians had a crucial role. The new role for

Blacks, unfortunately, continued to be a reductive one; Blacks came to be seen as the most obvious victims of an oppression that afflicted the majority of Brazilians. Racial categories were subsumed under social and economic ones. Eager to avoid the picturesque and stereo-typical vision of Brazil as the land of carnival and samba, the Cinema Novo directors tended to downplay the specifically cultural contribution of Afro-Brazilians. Carlos Diegues' short film *Escola de Samba, Alegria de Viver* (Samba School Joy of Living), subsequently incorporated into the collective film *Cinco Vezes Favela* (Fives Times *Favela,* 1962), for example, makes the didactic point that the *favelados* should abandon the escapism of samba for serious political organizing. Designed to illustrate the progressive ideas of White middle-class intellectuals, the films generally failed to confront the racial dimension of oppression or to acknowledge the rich cultural contribution of Afro-Brazilians.

This is not to say that Afro-Brazilian music and culture were entirely absent from the films of the early phases of Cinema Novo. Apart from films specifically dedicated to an Afro-Brazilian thematic, such as *O Pagador de Promessas* (The Given Word, 1962) and *Barravento* (1962)—to be discussed subsequently—many Brazilian films evoke political and social conflict through the manipulation of cultural expression, and Afro-Brazilian music is crucial in this regard. *The Given Word,* for example, sets in motion a cultural battle between the African-derived *berimbau* and church bell, synecdochically encapsulating the larger struggle of *candomblé* and Catholicism. Glauber Rocha's *Terra em Transe* (Land in Anguish, 1967) exemplifies a film featuring few Afro-Brazilian actors but where Afro-Brazilian music and culture play a primordial role. The English title misses the clear reference to the "transe" of *candomblé,* a reference which structures the entire film and even organizes a montage which periodically goes in and out of "transe." The reference is reinforced by the *candomblé* music heard during the credit sequence of the film, superimposed with aerial shots of the coast of Brazil, as if to say: this is the land of *candomblé.* The Afro-Brazilian music in the film stands in, as it were, for the people, while Rocha associates each of the representatives of the White political elite with a distinct kind of music. The rightist Porfirio Diaz is linked to the operas of Verdi and to his Brazilian imitator Carlos Gomes, while the left-intellectual protagonist Paulo Martins is associated with the music of Villa-Lobos, who, like the protagonist and like Rocha himself, is the erudite elaborator of popular themes.

The seventies witness the "recarnivalization" of Brazilian cinema, not only as a key trope orienting the filmmakers' productions, but also as a concrete presence in the films themselves. The incorporation of Afro-Brazilian music was motivated not only by a changed attitude but also by a desire to renew contact with the Brazilian audience. Carnival is a literal presence in such films as Carlos Diegues' *Quando O Carnaval Chegar* (When Carnival Comes, 1969), Walter Lima Jr.'s *Lira do Delirio* (Delirious Lyre, 1978), and Vera Fequeireido's *Samba da Criação do mundo* (Samba of the Creation of the World, 1979), based on a pageant performed by the Beíja Flor Samba School. This same period witnesses a proliferation of documentary tributes to Afro-Brazilian music and musicians: João Carlos Horta's *Pixinguinha* (1969), Carlos Tourinho's *Escola de Samba* (1978), Reinaldo Cozer's *Perola Negra* (Luiz Melodia, 1979), and Tuna Espinheira's *Samba Não se Aprende na Escola* (One Doesn't Learn Samba at School, 1979).

In other films, carnival in Brazilian films forms part of the wider circulation of popular and erudite culture. The stories of Diegues' *Xica da Silva* (1976) and Walter Lima Jr.'s *Chico Rei* (1982), for example, were first presented as samba school pageants for Rio's carnival. Indeed, Carlos Diegues has said that he conceived both *Xica da Silva* and his most recent *Quilombo* (1984) as "samba enredos" (samba plots), i.e., as films analogous to the collection of songs, dances, costumes, and lyrics that form part of that popular narrative form called a samba school presentation. The *favelados* of Fernando Cony Campos' *Ladrões de Cinema* (Cinema Thieves, 1977) steal filmmaking equipment from American tourists visiting Rio for carnival, and conceive the film they plan to make with the stolen equipment as a kind of samba school narration of a famous episode in Brazilian history—the Minas revolt against Portuguese colonialism in the eighteenth century.

Afro-Brazilian Religion and the Cinema

It is impossible to discuss the Africanization of Brazilian culture without reference to the ongoing and vital role of Afro-Brazilian religion. These religions are extremely diverse; some are of Yoruba origin, others of Kongo; some practice the cult of dead ancestors *(Egungun),* others do not; some are relatively "pure" in the sense of preserving African roots, while others, such as *umbanda,* are highly syncretic. Our purpose here is not to analyze these religions but rather to point out their central importance and their relevance to Black performance and to Brazilian cinema. The occidental vice of hierarchizing religions into the "real" religions of the sacred texts—Judaism, Christianity, and to a lesser extent Islam—and the pseudo-religions called "cults," regarded as inferior and superstitious, has blinded many western people to the beauty and profundity of what Maya Deren called, in reference to Haitian Voodoo, religions of "major stature, rare poetic vision and artistic expression."[8] Afro-derived religions are not improvised mélanges of vestigial habits and irrational superstitions; they are spiritual systems entailing a metaphysic, an ethos, and an aesthetic.

The core of most varieties of Afro-Brazilian religion is the fall of the *orixas* (spirits, saints) from ancestral Africa into the *terreirors* of Rio, Bahia and Recife. For Afro-Brazilian religion, the possession of the faithful by these spiritual beings is the ultimate sacramental expression. The orixas are worshipped in a ceremony in which they manifest themselves through the cult devotees. We have here a religion without sacred texts; the text, in a sense, is the body of the believer on which the god writes. The believers know the *orixas* exist because they reincarnate themselves in the bodies of the mediums and thus show their power. The mediums loan their bodies to the *orixas;* they are the "divine horsemen" on whom the gods ride. Rhythm is essential to this process, for it is the rhythms that summon the *orixas.* Each deity has his or her own drum rhythm, and will only come when it is played. The transe is not a delirium but an ordered system; the ecstasy of a son of Xango is not that of a daughter of Iemanja. Oxum dances voluptuously, while the *preto velho* (spirits of old slaves) limps and slouches, and the *caboclos* (indian spirits) shout and shake arrows. The medium, in this process, are little more than playthings for the gods.

The gregarious mysticism of Afro-Brazilian religious expression is strongly linked both to carnival and to Black performance. In a participatory religion in which soul claps its hands and sings, the practitioners of Afro-Brazilian religion perform; they dance and sing, the mediums above all, but also the witnesses for whole collective benefit the ritual is performed. Like carnival, *candomblé* is a kind of *festa* performed in an atmosphere of collective enthusiasm. In Bahia, there even exist "play candomblés," in which mysticism is absent, where people go uniquely in order to dance and sing. But authentic *candomblé* groups also participate in carnival in the form of *afoxés* in Bahia and *maracatus* in Recife. Apart from these concrete links to carnival, there is a deeper connection through analogy. In carnival, individuals play at being another. *Candomblé,* for its part, can be seen as a form of divinely-inspired role-playing. Both carnival and *candomblé* set in motion a complex play of identities, a creative dialectic of self and other. Just as carnival transforms a *favelado* into aristocrat, a spirit possession turns a cook or a maid into the ruler of sea or storm, into Iemanja or Iansa. In carnival, men dress up as women and women as men. In *candomblé,* a woman can be possessed by a male *orixa* or vice versa, an old woman can enact the adolescent ecstasies of young gods, and young people can incarnate the shuffling movements of older *orixas.*

It was only in the sixties that Brazilian cinema began to speak seriously of *candomblé,* and even then with a highly ambiguous attitude. The Cinema Novo directors of the early sixties, when they did not ignore Afro-Brazilian religion completely, tended to regard it as alienated and marginal, something to be tolerated or reformed by well-meaning leftists. A partial exception to this rule can be found in the films of what has been called the "Bahian Renaissance," the cinematic rediscovery of the cultural riches of Salvador, Bahia, the Africanized metropolis where *candomblé* thrives. Anselmo Duarte's *Pagador de Promessas* (The Given Word, 1962) and Glauber Rocha's *Barravento* (1962) both place *candomblé* at the center of their preoccupations. The former film centers on the vow of its protagonist, Zé-of-the-Donkey, to carry a cross into the Church of Saint Barbara in gratitude for the miraculous cure of his donkey. The priest refuses him entrance because the vow was proferred at a *candomblé* shrine. When they learn of Zé's plight, the people of Bahia, largely Black and mulatto, support Ze's right to enter the church. Anti-clerical rather than anti-Christian, Duarte contrasts the naive Christ-like faith of Zé with the petrified attitudes of the priest. The practitioners of *candomblé,* meanwhile, are shown as sympathetic but misguided. Duarte repeatedly underlines Zé's separation from *candomblé.* In the opening *terreiro* sequence, Zé is alone, kneeling at the altar, while the collectivity sings and dances. In order to emphasize that Zé is "really a Christian at heart," Duarte has him refuse repeated invitations to visit the *terreiro* in Salvador. The film evokes a kind of revolution by having the people take over an institution, yet the "revolution" consists, ultimately, only in gaining entrance to a more ecumenical and tolerant Catholic Church, and not in the endorsement of *candomblé* or in a call to political action.

Both *The Given Word* and *Barravento* open with images of *abataques* (drums) in a *terreiro,* but there the similarities end. If *The Given Word* condemns *candomblé* from the point of view of ecumenical Catholicism, *Barravento* would seem to critique Afro-Brazilian religion from the standpoint of historical materialism. The film shows the oppression inflicted on poor Black fishermen and apparently denounces the role of *candomblé* in legitimating this oppression. The initial explanatory intertitles inform us that Blacks are oppressed and that their "superstitions" keep them in a state of passive acquiescence. The protagonist Firmino, a rebel who has just come back from the city, tells his brothers and sisters to forsake religion to fight oppression.

It would be extremely simplistic, however, to see *Barravento* as a monolithic condemnation of *candomblé.*[9] Such an interpretation is based only on certain features of the film—namely the explanatory titles and on certain affirmations by the protagonist. In fact, the surface condemnation of *candomblé,* very much in the spirit of the "progressive" left of the period, is contradicted by the film's deep-structural features. First of all, *Barravento* constitutes an extremely affectionate celebration of the poetic beauty and sensuous mise-en-scène of *candomblé.* Second, the film insists on the protagonist's isolation from his fellow villagers. His clothing is urban and inappropriate, and his gestures are inordinately theatrical. Throughout the film he addresses the camera directly as if he were speaking to us, the middle-class public, and not to the fishermen of Buraquinho. The montage isolates him, alternating shots in which he appears alone with shots in which the community appears together. As for the villagers, the film celebrates their collective life, a life in which religion, music, dance and work form a harmonious whole. The fishermen, for example, perform a *puxada de rede* (pulling in of the net) in a way that combines singing, dance-like movements and a hymn of praise of Iemanja. Although they are oppressed by the owner of the net, they partially transcend their oppression by using collective strength and artistry. Rocha does not film the *puxada de rede* in a neutral way; he emphasizes close shots of bare feet in the sand, of hands on the net, of the *batuque,* all done in such a way as to picture work as a celebration of communitarian integration. *Barravento* can even be seen as an Afro-Brazilian musical which appropriates songs and dances not to entertain but rather to transform and criticize. The entire film is punctuated with music—the joyous *samba de rolda* (a circle dance foregrounding individual creativity and prowess), the combative *capoeira* with its *berimbau* accompaniment, and the literally entrancing rhythms of *candomblé.* Third, as Ismail Xavier has pointed out, the film's system of explanation is profoundly ambiguous: all the narrative events can be explained either in a materialist manner *or* as evidence of the truth and efficacity of Afro-Brazilian religion.[10]

Whereas most sixties Cinema Novo films tend to ignore or question the legitimacy of Afro-Brazilian religion, the films of the seventies are quite different. Black culture, once considered marginal, becomes the vital source of the originality of Brazilian culture. In the sixties, Cinema Novo directors working in the favelas barely noticed the presence of *macumba* shrines; their progressive vision had no place for Afro-Brazilian religion. In the seventies, in contrast, many films celebrate various forms of Afro-Brazilian religious expression. Nelson Pereira dos Santos, who made no mention of Black religion in either *Rio 40°* or *Rio Zona Norte,* in his later *Amuleto de Ogum* (Ogum's Amulet, 1975), highlights the syncretic Afro-Brazilian religion called *umbanda.* Brazil's fastest-growing religion, *umbanda* combines African elements—the *orixas,* the centrality of spirit possession—with Catholicism, Cabala, and the spiritism of Allan Kardec. Umbanda is often said to be *the* Brazilian religion because it incorporates elements from the major ethnic groups constitutive of Brazilian identity. The color breakdown in *umbanda* varies from sect to sect and *terreiro* to *terreiro,* but even when Whites are in the majority, Africa, in Roger Bastide's lovely phrase "still casts its great black shadow over Umbanda."[11]

The odyssey of the hero of *Amuleto,* Gabriel (played by the director's son Ney Sant'ana) takes him from the Northeast, where his body has been magically "closed," to the notoriously lawless Rio suburb called Caxias. He gradually absorbs the symbology and the cultural universe of umbanda. In *Amuleto,* Nelson Pereira dos Santos aims to be "popular" not only in his choice of subject matter but also in perspective, discarding all vestiges of superiority and wholeheartedly affirming the values of the milieu depicted and the spectators to whom he hopes to appeal. The film virtually obliges the spectator to become in *umbandista,* at least for the duration of the film, if only to understand the narrated events. The film simply assumes umbandista values, without explaining them to the uninitiated. A Catholic audience, the director has pointed out, need only see a priest raising the host to know a mass is being celebrated. An umbandista audience, similarly, recognizes the ceremony which "closes" the protagonist's body and recognizes his protection by Ogum, the "hard" deity of war and iron and justice. It recognizes the voice and limp of the *preto velho* and the provocative manners of the *pomba-gira* (evil in the form of a woman). At the same time, the film does not idealize *umbanda*: one umbanda priest works for popular liberation; the other is a greedy charlatan and opportunist. *Umbanda* is shown as a kind of cultural master code in which two opposing discourses, one progressive and the other retrograde, struggle for ascendancy. No longer the opium of the people, Afro-Brazilian religion has now become the scene of struggle and, potentially, the locus of resistance to cultural hegemony.[12]

The same director's *Tenda dos Milagres* (Tent of Miracles, 1976), based on the novel by Jorge Amado, also deals with Afro-Brazilian culture. Here the Black culture-hero Pedro Archanjo (a composite figure based on a number of self-taught Black intellectuals from Bahia) defends the Afro-Brazilian spiritual inheritance first against the fashionably racist Social Darwinist theories of Professor Nilo Argilo and then against the clubs and guns of the police. Deployed across a double time frame—a core story set in the past and presented as a film-within-the-film—and a frame story set in the present, *Tent of Miracles* suggests that the ideologically explicit and politically violent racism of the past has merely transmuted itself into the subtler racism of the contemporary mass-media. More interesting for our purposes, however, is the fact that Nelson Pereira dos Santos calls attention to the historical reality of the violent repression of Afro-Brazilian *cultural* manifestations. The turn-of-the-century press in Bahia was replete in denunciations of the "uncivilized" habits of Bahia's Blacks, especially around carnival time: " . . . the authorities should prohibit these *batucadas* and *candomblés* which swarm through the streets during carnival, creating an enormous din, without tone or sound . . . all this is incompatible with our state of civilization."[13] During slavery, slaves were often obliged to pretend that they were worshipping Christian saints rather than African *orixas.* But even after the abolition of slavery, Blacks were punished for practicing their religion. It was only in the seventies that Bahia became the first state to suspend the compulsory registration of Afro-Brazilian religious groups with the police. We see this process of repression in *Tent of Miracles* when the White elite forbids the Black *afoxe* groups to appear during carnival. Throughout the film the *candomblé* practitioners are threatened by a brutal and racist police. At the height of the repression, Pedro Archanjo calls on the *orixas* for help against the police; they cooperate by putting the police thug Zé into a transe that turns him around so that he attacks the police and thus saves the community.

Tent of Miracles is not, admittedly without its problems. While the film exposes the repression of Afro-Brazilian culture and thus discredits the myth of "racial democracy," it also promotes the notion of miscegenation as social panacea. The film's generational progression from Black Pedro to his mulatto son Tadei and his White bride Lou suggests a subtly prejudicial parable of whitening as "progress." Archanjo in reality embodies Jorge Anado's ideal of "mesticismo," the idea that the true Brazilian can combine European intellectualism with Afro-Brazilian mysticism and sensuality. As the protagonist himself explains: "Pedro Archanjo Ojuobá, reader of books and conversationalist . . . and *iyalorixa,* two different beings, who knows, perhaps the Black and the White? But make no mistake, professor, I'm a mixture of both, one single mulatto." In name and body, Pedro Archanjo incarnates *mesticismo.* In his unshakeable dignity, subtle humor, and above all, in the reciprocated tenderness he bears his community, he is a touching figure. By paying homage to him, and by calling attention both to the repression of Afro-Brazilian culture and to the creativity with which Afro-Brazilians have responded to that repression, Nelson Pereira dos Santos has authored a complex if problematic ode of love to Afro-Brazilian culture.

The seventies and eighties offer a proliferation of fiction and documentary films paying tribute to Afro-Brazilian religious expression. Jose Unberto Dias' *O Anjo Negro* (The Black Angel, 1972) is an homage to the power and charm of *candomblé* in Bahia. Ibere Cavalcanti's *A Forca de Xango* (The Power of Xango, 1979) interweaves amorous intrigues with the supernatural operations of the *orixas.* Marco Altberg's *Prova de Fogo* (Test of Fire, 1981), emphasizes the utilitarian and therapeutic value of *umbanda,* "the poor man's couch." This same period also witnesses a growing number of serious, respectful documentary studies of Afro-Brazilian religion: Roberto Moura's *Sai dessa, Exû* (1973), Geraldo Sarno's *Iaô,* Juana Elbein dos Santos' *Arte Sacra Negra I* (Black Sacred Art Part I, 1978), Raquel Gerber's *Ylê Xoroquê* (1981), and Carlos Blajsblat's *Egungun* (1982).

Quilombo: From Carnival to Revolt

As the example of *Tent of Miracles* shows, Black religion and Black music often acted in concert with Black political resistance and revolt. To close our discussion, therefore, we will examine some of the cinematic manifestations of Black Brazilians as embodying a revolutionary impulse of revolt against oppression. Some of the earliest historical reconstruction films interestingly were based on the episode known as the "Revolta da Chibata" (the whip revolt) in which the Black Sailor João Candido led a revolt against corporal punishment, generally applied by White officers against Black sailors, in the Brazilian navy. The incident inspired both documentaries (Revolta no Rio, 1910) and staged reconstructions like Lambertini's *A Vida do Cabo João Candido* (the Life of Commander João Candido, 1910). (The latter film, symptomatically, was the first Brazilian film to receive the ambiguous compliment of official censorship.) Our focus here, however, will be on the filmic reflections of a specific instance of Black revolt—the phenomenon of the *quilombos* or "liberated zones" founded by fugitive Blacks in the days of slavery. The official history books of Brazil, written largely by the "winners" of history, tend to privilege White rather than Black revolt. Thus Tiradentes, who was himself a slaveowner, is celebrated as a hero for having struggled against Portuguese colonialism in the eighteenth century. For many Black Brazilians, meanwhile, the *quilombos* have become a symbol of Black resistance not only against Portuguese colonialism but also against slavery. It is quite logical, therefore, that Abdias de Nascimento should name his militant movement against Brazilian racism *O Quilombismo,* or that politicized Blacks in Rio have recently founded a Black consciousness Samba school: *A Escola de Samba Quilombo.*

The most famous of the *quilombos* was Palmares, the fugitive slave republic which lasted almost a century, from 1595–1695, in the face of repeated assaults by both the Dutch and the Portuguese. Palmares at its height counted 20,000 inhabitants and an area roughly one-third the size of Portugal. Palmares was economically self-sufficient since it practiced diversified agriculture rather than the monoculture common at the time. Its "King" was a king in the African rather than the European sense, not an absolute monarch but an agreed-upon custodian of the common wealth. The Palmarinos spoke a kind of *esperanto* which fused Portuguese, African and Indian languages, a linguistic reflection of the ethnic make-up of the population. Their religion was syncretic, combining Afro-Brazilian religion with Catholicism. Given its oppositional stance, Palmares became a kind of asylum for the persecuted of colonial Brazilian society. Carlos Diegues' first feature, *Ganga Zumba* (1963), celebrates Palmares. Based on João Felicio dos Santos' novel of the same title, the film focuses on a fugitive Black slave, Antão (Antonio Pitanga), who discovers that he is a grandson of the king of Palmares. The film combines a harsh portrait of slavery involving forced labor, constant threats, whippings, rape and murder (a portrait having little to do with the idealized tableau of a benign humanized Lusitanian servitude) with a brief glimpse of the "utopia" of Palmares. A Fanonian ode to Black liberation, here taken as metaphor and inspiration for the broad contemporary struggle against neo-colonialism, *Ganga Zunba* adopts a Black perspective throughout, showing Blacks not as mere victims but as active agents who take their destiny into their own hands.

Two very recent films, not yet released in the United States, return to the subject of Afro-Brazilian political resistance. Carlos Diegues returns to the subject of his first feature in *Quilombo* (1984), this time with a substantial budget and based on more recent historical research into Palmares by Decio Freitas.[14] In the new film, Diegues presents Palmares as the "real foundation of Brazil," in which Blacks, Indians, and poor Whites created a new civilization with its own language, religion and culture—the first "great utopia of the Americas."[15] Walter Lima Jr., meanwhile, has just released *Chico Rei,* a German-Brazilian co-production based on the historical truth concerning an African king who was enslaved with his family and sold to a slave-owning gold miner in Minas Gerais. At that time, it was possible for slaves to buy their own freedom and that of their own family either by generating wealth for the mines or surreptitiously gathering mineral wealth. "Francisco," as he was called by his Portuguese name, managed in this way to liberate himself, his family, and his "tribe." Under his direction, the community purchased a mine, called Encardideira, which they worked as collective property according to traditional African communitarian forms of labor. As his power and prestige grew, he became a virtual head of state in the area and was renamed Chico Rei. Although less famous than Zumbi or Xica da Silva, Chico Rei was, in Walter Lima's view, more efficacious. His non-violent methods, a mixture of Black entrepreneurship and African communalism, led to a viable economy and the re-Africanization of one community in Brazil. The film closes, reportedly, with a festive *congada* celebrating Chico's "coronation," by which the former tribal organization is reinstituted.

In the 1980s, more than ever, many Brazilian filmmakers seem to look to Afro-Brazilian culture as a key to Brazilian nationhood, its differentiating feature from European and North American civilization, the vital source of its energy. The nineteenth-century romantics, such as Goncalves Dias and Jose de Alencar, had celebrated the "Indian," the "brave warrior," as spiritual symbol of Brazilian nationality. The Indians themselves, meanwhile, were being subjected to a process of physical and cultural genocide. The exaltation of the "brave warrior," meanwhile, implicitly operated at the expense of Blacks. The proud history of Black rebellion was ignored; the brave Indian, it was subtly insinuated, resisted slavery, while Blacks did not. Films such as *Ganga Zumba, Chico Rei,* and *Quilombo* retroactively correct this historic error by foregrounding the long, valiant and continuing history of Black resistance. Other films that we have discussed, such as *Amuleto de Ogum, Xica da Silva* and *Tent of Miracles,* highlight what Gilberto Gil calls the "vibrant," "luminous" energy inherent in Afro-Brazilian culture. All the films contribute, in some measure, to the long overdue recognition of the cultural and political role of Black Brazilians. But the danger of a new romanticism, in which the same group being hailed for its cultural importance is simultaneously the victim of the cruelest forms of oppression, is always present as long as symbolic cultural victories are not matched by the acquisition of real political and economic power.

Notes

This essay is a revised version of a talk originally given at the Center for Twentieth Century Studies, University of Wisconsin-Milwaukee in March 1983 as part of a seven-part series of public programs on "Black Performance in the New World." Organized by Patrick K. McNaughton, Department of Art History, and Carol Tennessen, Center for Twentieth Century Studies, the series was funded in part by the Wisconsin Humanities Committee, the Wisconsin Arts Board, and the National Endowment for the Arts.

1. Vinicius de Moraes described the origins of *Orfeu de Conçeiçao* as follows: "One morning, as I was reading the Orpheus legend in a French book of mythology—a story I always loved for its linking the poet and the musician—I heard a *batucada* coming from the neighboring *favela,* the Morro do Galvao. Strangely, the two ideas fused in my mind and I had the impression of a strong relation between the two. I began to reflect on the life of the Blacks in Rio and to hellenize their experience. Suddenly, I had the idea of making Orpheus a *sambista* characterized by great interior beauty very much longed after by a number of women, thus inspiring the envy of the other men from the *morro*" (translation mine). In Jose Eduardo Homem de Mello, *Musica Popular Brasileira* (Sao Paulo: UDESP, 1976), p. 59.

2. See *The New Yorker,* July 11, 1959.

3. Some of the key texts in the comparative slavery debate include: Frank Tannenbaum, *Slave and Citizen* (Knopf, 1947); Stanley Elkins, *Slavery: A Problem in American Institutional and Intellectual Life* (University of Chicago Press, 1959); Eugene Genovese, *The World the Slaveholders Made* (Vintage, 1971); Herbert Gutman, *The Black Family in Slavery and Freedom: 1750–1925* (Pantheon, 1976); Marvin Harris, *Patterns of Race in the Americas* (Walker, 1964); and Carl N. Degler, *Neither Black Nor White: Slavery and Race Relations in Brazil and the United States* (Macmillan, 1971).

4. See Abdias do Nascimento, *O Quilombismo* (Petropolis: Vozes, 1980).

5. For excellent discussions of the contemporary racial situation in Brazil, see Abdias do Nascimento, *O Quilombismo* (Petropolis: Vozes, 1980); Lelia Gonzales and Carlos Hasenbalg, *Lugar do Negro* (Rio de Janeiro: Colecao Dois Pontos, 1982); and Carlos Hasenbalg, *Race Relations in Post-Abolition Brazil: The Smooth Preservation of Racial Inequalities* (University Microfilms International, 1978).

6. Paulo Emilio Sales Gomes, "A Aventura Brasileira," *Suplemento Literario do Estado de Sao Paulo* (March 15, 1958), included in Paulo Emilio Sales Gomes, *Critica de Cinema no Suplemento Literário,* Vol. I, (Rio de Janeiro: Paz e Terra/Embrafilme, 1982).

7. I am indebted to Heloisa Buarque de Hollanda, who is currently researching the correspondence between RKO, The State Department and various Brazilian governmental agencies concerning *It's All True*. A July, 1942 letter from William Gordon, from the production team of *It's All True* to an RKO executive complains about Welles' "indiscriminate" filming of Blacks and Whites together, and asserts that he has been unable to control Welles in this matter. He also claims that Brazilians, while "completely lacking in race prejudice" may resent some of the footage because "it will tend to confirm the North American opinion that most South Americans belong to the African race."

8. See Maya Deren, *Divine Horsemen: The Voodoo Gods of Haiti* (Delta, 1970), p. 15.

9. The question of the film's attitude toward *candomblé* is complicated somewhat by the problematic conditions surrounding its production. The film was scripted, originally, and partially filmed by Luis Paulino dos Santos, a Bahian who was somewhat more sympathetic to *candomblé* than Glauber. After a series of disagreements, Luis Paulino left the production and Glauber Rocha became the director. A reading of the original script reveals that Rocha somewhat understated his debt to Luis Paulino but also that the final film is more revolutionary in tone and more avant-garde in treatment.

10. For a brilliant discussion of *Barravento* and Glauber's work generally, see Ismail Xavier's *Sertao Mar: Glauber Rocha e a Estetica da Fome* (Sao Paulo: Brasiliense, 1983).

11. Roger Bastide, *The African Religions of Brasil: Toward a Sociology of the Interpenetration of Civilizations* trans. Helen Sebba (Johns Hopkins, 1978). For more on Afro-Brazilian religion see Juana Elbein dos Santos, *Os Nago e A Morte* (Petropolis: Vozes, 1977); Robert Farris Thompson, *Flash of the Spirit: African and Afro-American Art and Philosophy* (Random House, 1983); and Sheila S. Walker, *Ceremonial Spirit Possession in Africa and Afro-America* (E. J. Brill, 1972).

12. In *The Political Unconscious* (Ithaca: Cornell University Press, 1981), Fredric Jameson argued that in the 1640s in England opposing political discourses fight it out within the shared master code of religion. Jameson's approach points the way to a "post-Marxist" attitude toward religion which would allow for a progressive role for certain religious forms, whether within left political struggle (e.g. liberation theology in Latin America) or as a locus of cultural resistance against imperialistic hegemony (e.g. *candomblé* in Brazil).

13. *Journal de Noticias* (February 12, 1901) cited in David Brookshaw, *Raca e Cor na Literature Brasileira* (Mercado Aberto, 1983). Translation mine.

14. See Decio Freitas, *Palmares: A Guerra dos Escravos* (Rio de Janeiro: Graal, 1978). For more on Black slave revolts and *quilombos,* see Lana Lage da Gama Lima, *Rebeldia Negra e Abolicionism* (Achiamé, 1981).

15. Quoted in "A Cenografia como Processo Cultural," in *Filme Cultura,* No. 43 (January–April 1984), p. 74. Translation mine.

Western Europe

Black Filmmaking in Britain's Workshop Section

Coco Fusco

One of the crucial things about media education in Britain is that you're involved in very Eurocentric theories, and if you have any sort of Black consciousness you begin to wonder where there might be room for your experience within these theories. In Roland Barthes' *Mythologies*, one of the key texts for students of semiology, the only reference to anybody Black is to the solider on the cover of *Paris Match*. The very superficial critique of colonialism found in such texts really isn't enough.

As we began to think about images and about our politics, we realized that the history of independent film and Black images was pretty dry politically speaking. And political films were also really dry stylistically, mostly straight documentary. And there is always the problem that there hasn't been much space for Black filmmakers in Britain. In terms of political film also there wasn't much room for pleasure.

Martina Attille, Sankofa[1]

In the winter of 1986, two films from the British workshops opened at downtown London's Metro Cinema.[2] One was a multilayered dramatic feature, and the other a nonnarrative, impressionistic documentary—formats usually considered to be too difficult for the theatrical market. While it was highly unusual for low budget "experimental" films to find their way to commercial venues, what made these runs even more unusual was that the films, *The Passion of Remembrance* (1986) and *Handsworth*

Afterimage, vol. 15, no. 7, 1988, pp. 11–13. Reprinted by permission of the publisher.

Songs (1986), were produced by the London based Black workshops Sankofa and Black Audio Film Collective. Those theatrical screenings were firsts for local Black film collectives, and are one of the many signs that the Black workshops are effecting radical changes in British independent cinema.[3]

Sankofa and Black Audio's intervention in British media institutions seems to have touched several raw nerves. Their insistence on shifting the terms of avant-garde film theory and practice to include an ongoing engagement with the politics of race sets them apart from long-standing traditions of documentary realism in British and Black film cultures. Black Audio's *Handsworth Songs* is a collage of reflections on the race riots that have shaken Thatcherite England, and the inadequacy of all institutional explanations of them—particularly those of the mass media. The filmmakers weave archival footage with reportage, interior monologue, and evocative music to create a gracefully orchestrated panopoly of signs and sounds that evoke Black British experiences. Sankofa's *The Passion of Remembrance* is the story of Maggie Baptiste, a young woman grappling with the problematic legacy of a Black radicalism that foreclosed discussion of sexual politics and with the differences between her vision of the world and that of her family and friends. Public and private memory reverberate through interconnected stories that take different forms: dramatic narrative, allegorical monologues, and film within film.

Critical attention to *Handsworth* and *Passion* has outstripped the response to other workshop films of the same scale. The films are at the center of polemical debates in the mainstream and Black popular press that often do little more than bespeak critical assumptions about which filmic strategies are "appropriate" for Blacks. At its best, institutional recognition takes the form of the John Grierson Award, which Black Audio received in 1987 for *Handsworth Songs*; the more common version, however, is the constant scrutiny to which the entire Black workshop sector is subjected.[4] All the Black workshops contend that they must conform much more consistently and closely to the laws that regulate them than their white counterparts.

As filmmakers and media activists, Sankofa and Black Audio question Black representation in British media from mainstream television to such bastions of liberal enlightenment as the British Film Institute (BFI), and academic film journals such as *Screen* and *Framework*. They are interrogating "radical" film theory's cursory treatment of race-related issues, and subverting the all too familiar division of independent film labor between first-world avant-garde and third-worldist activism. Sankofa and Black Audio are also concerned with mainstream images of Black identity, preconceived notions of Black entertainment, and the terminology and mythologies they inherit from the '60's-based cultural nationalism that remains allied with a realist tradition. Sankofa's reflections on the psychosexual dynamics and differences within Black British communities, and Black Audio's deconstruction of British colonial and postcolonial historiography are groundbreaking attempts to render racial identities as effects of social and political formations and processes, to represent Black identities as products of diasporic history. While these workshops are not the first or only Black filmmakers in Britain, they are among the first Black British film artists to recast the question of Black cultures' relations to modernity as an inextricably aesthetic and political issue.

Although racism is not a problem specific to Britain, the English version has its own immediate history. The existence of the Black British workshops and the nature of their production are due to the 1981 Brixton riots and the institutional responses that gave the filmmakers access to funding.[5] The newly established workshops provided the infrastructure that, combined with racially sensitive cultural policies, created conditions for them to explore and question theoretical issues. Though the chronologies of events that inform *Passion* and *Handsworth Songs* are specific to Britain, institutionalized racism, its attitudes, arguments and historical trajectories are not. In addition to institutionalized racism, we in America share the legacy of cultural nationalism, its ahistorical logic, anachronistic terms, and the scleroticizing danger of separatism. The U.S. psycho-social dilemma of belonging, which harshly affects people of color, might be offset slightly by melting pot myths and a longer history of Black American presence. But the massive influx of peoples from Latin America and the Caribbean since World War II (not to mention the abundance of mixed-race Americans) is both evidence of a similar plurality of Black cultures here and a symptom of the U.S. neocolonialist projects. The contemporary U.S. situation, then, exceeds any monolithic discourse on race, calling for strategic recognition and articulation of a plurality of racial differences. The British use of "black" as a political term for all U.K. residents of African, Afro-Caribbean, and Asian origin expresses a common social, political, and economic experience of race that cuts across original cultures, and works against politically devisive moves that would fragment them into more easily controlled ethnic minorities. As mainstream American media constitute new markets by race (the heralding of the new Hispanic moviegoer with the opening of *La Bamba* is one recent example) and as critical reflection on media culture hovers around the question of colonialism, treating it at times as if it were a phenomenon that exists "elsewhere"—we must continue a systematic, ongoing analysis of the homogenizing tendencies of both the mass media and post-structuralism, as well as the contrived segregation of post- and neocolonial subjects into folklorically infused, ahistorical, ethnic groups. Recognizing nationality's problematic relationship to the diasporic phenomenon, I will, in this article, examine the work of Black Audio and Sankofa as an instance in the development of a necessarily international critical study of race and representation.

Given the two Black workshops' stress on how multiple histories shape their presence/present, it is appropriate to begin by outlining events that led to their practice. Sankofa's and Black Audio's members are first-generation immigrants, largely from West Indian families that arrived in Britain in the 1950s and '60s. The combination of an expanding post-World War II economy in England, changing immigration laws and chronic economic hardship in newly independent colonies resulted in rapid growth of the British-based Black population into the mid-70s, when economic decline and stringent immigration policies began to close the doors. Most of the first generation of Black British subjects reached adolescence in the '70s, with little hope for decent employment, a minimal political voice, and virtually no access to media. This atmosphere of despair and foreboding was sensitively portrayed by Black British independent pioneer Horace Ove in his first feature, *Pressure* (1975) focusing on the frustrations of Black youth, and later addressed by Menelik Shabazz in his 1982 feature *Burning An Illusion*.

Britain in the last years of the Labor government before Margaret Thatcher saw the rise of neofascist groups and racially motivated attacks against Afro-Caribbean and Asian peoples, coupled with changes in policing tactics now aimed at containing the Black population. Public gatherings within the Black community, such as carnivals, were increasingly perceived and constituted as sites of criminality.[6] The Brixton riots of 1981 were not the first violent response by Blacks to their situation, but the ensuing spread of civil disturbances throughout the country generated enough fear and media coverage to prevent the explosive situation from being ignored by the government. Despite statistics indicating that the Brixton riots resulted in the arrests of more whites than Blacks, the mass media and adjunct power mechanisms had already succeeded in constituting a new Black Threat, with a new Black, male youth as its archetypal protagonist.

The independents and the Association of Cinematograph, Television and Allied Technicians (ACTT) directed many of their efforts toward the establishment of Channel 4 as a commissioning resource and television outlet for British films.[7] Its charter affirms the channel's commitment to multicultural programming. Those interest groups' lobbying, together with support from Channel 4 and BFI, also led to the Workshop Declaration of 1981, giving nonprofit media-production units with at least four salaried members the right to be franchised and eligible for production and operating monies as nonprofit companies. Workshops are expected to engage in ongoing interaction with their local communities through educational programs and training, and at the same time produce innovative media that could not be found in the commercial sector.

1981 was also a crucial year for the Greater London Council (GLC), as the beginning of its governing Labor party's six year effort at social engineering through politically progressive cultural policy.[8] This project ended with Thatcher's abolishing the council by decree in 1986. A race relations unit and Ethnic Minorities Committee were instituted largely in response to the 1981 riots and sociological studies that followed. Within the Ethnic Minorities committee was the Black Arts Division, which, under the supervision of Parminder Vir, slated monies for Black cultural activity, particularly those areas such as film and video that had previously been inaccessible due to high costs. The future members of Sankofa and Black Audio had, at this time, just completed their academic and technical training—Sankofa's members were primarily from arts- and communications-theory backgrounds, and Black Audio's members had studied sociology.[9] Funding from organizations such as the GLC and local borough councils financed their first works and made them eligible for workshop status.

By the time Sankofa and Black Audio began to work collectively, a race-relations industry had developed not only in the nonprofit cultural institutions, but in academia as well. Those theoretical debates on colonialism and postcolonialism, in which Black Audio and Sankofa actively participate, draw extensively on the work of Homi Bhabha, Stuart Hall, and Paul Gilroy.[10] Bhabha's writings combine a Lacanian perspective on the linguistic construction of subjectivity, with Fanon's investigations of racism as a complex psychic effect on colonial history.[11] These ideas provide a theoretical framework from which to investigate the unconscious dimensions of the colonial legacy, to understand racism as a dialectical encounter in which victim and oppressor internalize aspects of the other, at both the level of the individual and the social.

Passion's concern with sexual conservatism in contemporary Black communities and *Handsworth*'s poignant resurrection of the '50s immigrants' innocent faith in the "motherland" resonate with these psychological dilemmas in an expressive manner that transcends didactic illustration. They suggest alternatives to predominant forms of representation that posit the colonized as helpless victim (the liberal view) or as salvagable only through a return to an original precolonial identity (the underlying assumption of cultural nationalism). They also undermine the liberal assumption that racism is an aberration from democratic ideals of the nation-state, by bringing out their historical inextricability. In other words, the development of capitalism and the rise of the British Empire were contingent on colonial exploitation and racism—the colonial fantasy, as Bhabha puts it—is nationalism's unconscious, its dialectical negation. Black Audio's stunning reassemblage of archival images from the British colonial pantheon—*Expeditions* (1983)—is a critical reinterpretation of the fantasies that give rise to both the imperial project and its documentation.

Also influential to Sankofa's and Black Audio's aesthetics are the writings of Paul Gilroy and Stuart Hall. The two social theorists bring Foucauldian methods of institutional critique to the issue of race and the constitution of the Black subject. In their analyses of racism's many mechanisms and manifestations, they are acutely sensitive to the significance of the media and image production as means of transmitting ideas about nationality and nationalist prejudice.[12] Gilroy's interpretations of Black British culture (particularly music) as a synthesis of modern technologically influenced aesthetics and Black oral traditions theorize cultural dynamics in the Black diaspora, significantly shifting the terms of contemporary debates on postmodern eclecticism. While Hall employs Gramscian theories of hegemony to comprehend the complex power relations between institutions and the "resistance" of specific groups, he is particularly sensitive to the danger of imputing radicalism to all forms of popular expression, tempering widespread tendencies of cultural nationalism to project resistance as a leitmotif onto all popular history. The character Maggie's search for new ways of approaching past and present desires in *Passion* evokes the condition these writers address. Like them, she seeks a more nuanced political vocabulary to approach a range of subjective and collective concerns.

Before Maggie's passions, and Handsworth's songs, however, came the two workshops' earlier, more esoteric endeavors: Black Audio's *Expeditions* and Sankofa's *Territories* (1985). *Expeditions* is a two-part tape/slide show, subheaded *Signs of Empire* and *Images of Nationality*, in which archeological metaphors organize an aestheticized, ideologically charged enquiry. Drawing on images from high colonial portraiture, ethnographic photography, and contemporary reportage, Black Audio uses them as raw materials in a choreographed audiovisual performance. Over images of the past are inscribed philosophical phrases of the present. Between images of present conflict are "expeditions" that open onto a past seen through the representational genres that elide the violence of the orders with which they collude. From this new angle, maps become measurements of both distance and domination, and placid portraits take on a sinister cast. As a majestic male voice claiming that Blacks "don't know who they are or what they are," repeats over and over, it becomes a stutter-like symbol for the speaker's own incapacity to comprehend the Other's identity. Ambient sounds and manipulated voices resonate forcefully, unearthing the deep structural meaning that bind the signs together. *Expeditions* is a decidedly antirealist document; instead its makers struggle with every possible formal means of achieving a vision both poetically allusive and lucidly interpretive.

Sankofa's *Territories* also uses formal experimentation as a means of decentering thematic and structural traditions. Their first collectively produced film was made after founding member Isaac Julien's video documentary, *Who Killed Colin Roach*? (1983) about the mysterious death of a Black male youth—a case similar to that of American grafittist Michael Stewart. It is a self-conscious return to the most visible Afro-Caribbean stereotype—the carnival—examining its places and displacements within British society. Charting the intensification of policing practices over three decades, *Territories* represent carnival as a barometer of institutional attitudes. Interconnected with these political and historical developments is a critique of ethnographic representations of carnival, which reify it as a sign of "original culture," masking its evolving sociopolitical significance. The two strategies bespeak the colonialist presupposition that carnival, as an archetype of Black expression, is by nature eruptive (savage) and erotic (dangerously pleasurable and potentially explosive), and therefore calls for order imposed from without. The film's second half, a surreal collage of gay couples dancing over riots, bobbies and burning flags, is a formal rendering of that very threat of chaos, a site of excess that mocks attempts at discursive and institutional control. The film, however, not unlike the carnivalesque, is somewhat limited by its own idiom, falling back on an all too familiar avant-garde conflation of all forms of realism and narrative to add strength to its counternarrative's assertions.

The issue, however, was not central to the film's critical reception in Britain. Like *Expeditions*, *Territories* was deemed by many to be too intellectual and inaccessible. According to the filmmakers, the doubts about both works often came from white media producers who had surfaced after a decade of immersion in structuralist stylistics with a zealous new concern for "the popular." Also participating were proponents of the "positive image" thesis who argue that positive representation of black characters is the answer to racist misrepresentation. They faulted the two workshops for, in a sense, missing the point. The ironic result of this sort of social engineering is that, despite its sensitivity to media and its attempts to create new spaces, it imposes limitations that eschew any psychological complexity. As Julian Henriques puts it in his article, "Realism and the New Language,"

> The danger of this type of approach is that it denies the role of art altogether. Rather than appreciating works of art as the products of various traditions and techniques with their own distinct language, art and the media are reduced to a brand of political rhetoric.[13]

What is at stake in all these arguments, and what explains Sankofa's and Black Audio's notoriety is that their works implicitly disrupt assumptions about what kinds of films the workshops should make and about what constitutes a "proper" reflection of the underrepresented communities from which they speak. As BFI Ethnic Affairs Advisor Jim Pines put it, the overriding assumption of the debates is that Black filmmaking is a form of social work, or rather that aesthetically self-conscious film practice is too highbrow and superfluous.[14]

Clearly, there are also economic imperatives operating here. As many more established British independents gain international acclaim, arguments in support of a more commercially viable product gain momentum. And for the burgeoning collectives, the production costs of dramatic narrative are prohibitive. But the problem for the workshops remains that the combined effect of the arguments is to restrict the space they need to develop a critical voice and vision, to experiment with a variety of ready-made materials and discourses in order to "tell stories of our experiences in a way that took into account the rhythm and mood of that experience."[15]

Confronting the positive image as a problem rather than a given and defining relations to trends beyond the traditional parameters of "black communities" are issues that figure prominently in *The Passion of Remembrance.* Its dialogues are filled with questions about the images of Black identity that surround the characters and inform their behavior. The allegorical Black radical woman rebukes the allegorical radical Black man for the latent sexism in his Black Power ideology; Maggie and her family evaluate the Black couples on a prime-time TV game show; she and her brother attack one another's visions of political struggle; Maggie faces her peers' accusations that her interest in sexuality and sympathy for gay rights are not really Black concerns. Contradictions between self-image and prescribed images, between desired ones and painful ones, are repeated in the film's different generic sites, or levels. In the dramatic narrative devoted to the Baptiste family, identity conflicts are articulated as generational and cultural. The immigrant father's skills are no longer applicable in the labor market. As if to protect himself, he holds onto an outdated image of both England and the West Indies, while his son's grass-roots radicalism fossilizes into romantic nostalgia. When Maggie and her friend get ready for a night on the town, the conflicts between the men's world view and Maggie's are beautifully underscored by vivid intercutting of calypso and pop music. Indeed, what stands out most in *Passion* is the soundtrack, rich in music, poetic excerpts and charged verbal exchange. At times the filmmakers rely a bit too heavily on dialogue to carry the film's ideas, rather than exploiting the possibilities of its visual material. But even if *Passion* suffers at moments from a lack of formal cohesiveness, its intellectual strength comes from its insistence on the multiplicity of elements and images that shape Black consciousness.

Perspicacity of this sort appeared to be beyond the capacities of mainstream documentation of the 1985 Handsworth and Broadwater Farm riots.[16] In response to this conceptual lacunae came Black Audio's first film, *Handsworth Songs*, which shatters the reductivism of previous media coverage. Countering the desire of the nameless journalist for a riot "story" is the film's most often quoted line, "There are no stories in the riots, only the ghosts of other stories." In the place of monological explication are delicately interwoven visual fragments from the past and present, evoking larger histories and myths. Among the images are familiar scenes from previous riots, such as the attack of nearly a dozen policemen on one fleeing dreadlocked youth from the Brixton uprisings. With the shots of news clips, they remind us that by the time of the 1985 riots, an established and limited visual vocabulary about Blacks in Britain was in place. These references to a "riot" inconography form the synchronic dimension of Black Audio's poetic analysis of the representation of "racial" events.

The film uses archival cut-aways to reveal an uneasy relationship between camera and subject. At one point, an Asian woman turns, after having been followed by the camera, and swings her handbag at the lens; at another, the camera swoops dizzily into a school yard, holding for several seconds on children's faces, nearly distorting them. This dreamlike movement is repeated in the filmed installations of family portraits, wedding pictures and nursery school scenes, which, combined with clips from dances and other festivities, become images of the "happy past" that are a precious part of the Black immigrants collective memory. Juxtaposed against the violence and frustrations of the present, these "happy memories" brim over with pathos, but they are also set against other images from the past which betray their innocence. Newsreel images highlight the earnestness and timidity of the immigrants, while voice-overs belie the hostile attitudes expressed at their arrival. The film depicts how a Black Threat was perceived to be transforming the needs of British industry into the desires of an unwanted foreign mass. These judgmental voices are confronted by newer ones, which offer no direct explanations or responses. Refining the style they developed in *Expeditions*, the filmmakers achieve such an integration of image and sound that the voices seem as if to arise from within the scenes. We hear poems, letters, an eyewitness account of Cynthia Jarrett's death by her daughter, introspective reflections, which together create a voice-over marked by lyrical intimacy rather than omnipresence. That sense of intimacy shines throughout both *Passion of Remembrance* and *Handsworth Songs*. Rarely do such formally self-conscious projects express comparable sympathetic bonds with their characters, maintaining a delicate balance between a critique of liberal humanism and a compassion for the spiritual integrity of their subjects.

Some British critics have attempted to identify specific avant-garde influences in Sankofa and Black Audio's works, citing Sergei Eisenstein and Jean-Luc Godard as predecessors. While these assertions have doubtlessly helped to legitimate the filmmakers in the eyes of some, Sankofa and Black Audio's direct concern with current media trends and with rethinking Black aesthetics compel us to look elsewhere. The two groups, while well schooled in Eurocentric avant-garde cinema, are surrounded by and acutely aware of "popular" media forms. They can draw on the experiences of a cultural environment in which musical performance can function as a laboratory for experimenting with ready made technologically (re)produced materials.[17] They also produce films in an environment where television is the archetypical viewing experience. The fast-pace editing and nonnarrative structures found in advertising and music video—not to mention the effect of frequently flipping channels—have already sensitized television audiences to "unconventional" representation, upsetting the hegemony of the classic realist text.

The filmmakers are also concerned with how to develop an aesthetic from diasporic experiences common to Black peoples. This involves rethinking the relationship between a common language and a people, between ideas of history and nation. Paul Gilroy has pointed out that modern concepts of national identity and culture have invoked a German philosophical tradition which associates a "true" people with a place.[18] Access to historical identity as a people with a common voice is bound to the idea of a singular written language and of place. Yet centuries of capitalist and colonial development have literally displaced Black populations. Their cultures have evolved through synthesis with others as much as through preservation and resistance, forging an ongoing dialectic of linguistic and cultural transmutation. While I am wary of labelling this process a kind of proto-postmodernism, I cannot avoid noting the formal resemblances. What seems more important than ascribing terms to diasporic cultural dynamics is to be aware of the ways in which Black Audio and Sankofa have taken this dynamic into account.

> Our task was to find a structure and a form which would allow us the space to deconstruct the hegemonic voice of the British TV newsreels. That was absolutely crucial if we were to succeed in articulating those special and temporal state of belonging and displacement differently. In order to bring emotions, uncertainties and anxieties alive we had to poeticize that which was captured through the lenses of the BBC and other newsreel units—by poeticizing every image we were able to succeed in recasting the binary of myth

and history, of imagination and experimental states of occasional violence.[19]

Sankofa and Black Audio speak from Britain, with a clear focus on the conditions of racism in a country where their right to full participation in civic society is more obviously complicated by legal questions of citizenship. Given our own immigration dilemmas and chronic inequities of Black American participation in the political process, however, parallels are far from contrived. The Black British filmmakers are keenly aware of their spiritual kinship with Black American cultures, though their actual connections are primarily textual. They clearly see themselves as heirs to developments that have roots in this country, evidenced by *Handsworth*'s poignant passage devoted to Malcolm X's visit to Birmingham, and Sankofa's acknowledgment that their critique of sexual politics in Black communities draws on Black American feminist writings of the '70s and '80s. The same GLC policy-makers who funded their first works also organized Black Cinema exhibitions, introducing audiences to the cinematic endeavors of Julie Dash and Ayoka Chenzira, Haile Gerima, and Charles Burnett.

Nonetheless, there are certain distinctions between the American and British conditions for Black independents. Institutional structures such as the workshops and ACTT grant-aided division, while far from ideal, do not work against notions of shared interests the way that America's individualized, project-specific funding procedures can. And competition with the more monied, auteurist ventures of Britain's more mainstream independents is a far cry from the economic and philosophic chasms that divide marginalized independent experiments from high-budget production in the U.S. But Britain specifically, and Western Europe in general, is involved in a larger postcolonial crisis that has forced them to rethink national and cultural identity; the dilemmas touched on by Black Audio, Sankofa and others are part of that crisis. Theirs is a poetics of an era in which racial, cultural, and political transitions intersect. It is no surprise then, that their works contain references to sources as varied as Ralph Ellison and Louis Althusser, June Jordan and Jean-Luc Godard, Edward Braithwaite and C. L. R. James. On this very sensitive point I must insist that this is not a rejection of the goals of Black consciousness. This "eclecticism," aimed at theorizing the specificity of race, reflects the mixed cultural, historical, and intellectual heritage that shapes life in the Black diaspora. The sad truth is that many Blacks must live that biculturalism, while few others seek to do so. If dominant cultures' relation to Black cultures is to go beyond tokenism, exoticizing fascination or racial violence, the complexities and differences which these film artists address must be understood. Sankofa, Black Audio and many other Black media producers in Britain are mapping out new terrains in a struggle for recognition and understanding.

I would like to thank the following individuals for their invaluable assistance in providing information for this article: Julian Henriques, Parminder Vir, June Givanni, Fred D'Aguiar, Colin McCabe, Jim Pines, Stephen Philip, Dhianaraj Chetti, and of course, the filmmakers of Sankofa Film/Video Collective and Black Audio Film Collective.

Notes

1. Martina Attille, interview with the author, in "Young, British and Black: The Sankofa Film and Video Collective," *Black Film Review* 3, no. 1 (Winter 1986–87), p. 12.

2. The Metro Cinema occupies a place analogous to that of the Film Forum in New York City.

3. Menelik Shabazz was the first Black British independent filmmaker to screen his film commercially in London. *Burning An Illusion* opened in 1982.

4. I have chosen to limit my discussion of the Black workshops to Black Audio and Sankofa because of the debates around them and their filmic strategies set them apart from the rest of the Black workshop sector. Other Black workshops in England are: Cardiff, Macro, Star, Retake, and Ceddo. The last two are also London based, and I conducted interviews with their members as part of my research. I should mention here that Ceddo also produced a documentary about racially motivated riots, entitled *The People's Account* (1986). It was commissioned by Channel 4, but has not yet been aired, due to an unresolved conflict involving Channel 4 and the Independent Broadcasting Authority (IBA). The IBA found the original version of the documentary unacceptable for its accusations against the British state, even after Channel 4 lawyers had submitted requests for minor changes and had them attended to. When I was conducting research for this article last summer, the IBA was insisting on a balancing program to accompany the documentary, and on the right to cancel the airing of both if they did not approve of the balancing program.

5. Although there had been outbreaks of violence in the '70s and earlier in protest of harassment by police and right-wing groups, and in protest of the state's strategic neglect of racial injustice, the riots that took place in 1981 mark a watershed moment in the history of British race relations. The first disturbances in Britain were immediately related to the suspicious deaths of three Black youths. But what began in the Brixton area of London spread to urban ghettos in most of the industrial centers of London, lasting an entire summer. The scale of the protests, as I mention later in the article, made it impossible for the government and the media to ignore the situation. Sociological investigations into the conditions of Blacks in Britain, such as the Scarman Report, were a direct governmental response to these events. The cultural policies of the GLC and new attention to race in many British cultural institutions were other responses.

6. For an in-depth discussion of this, see Paul Gilroy, *There Ain't No Black in the Union Jack: the Cultural Politics of Race and Nation* (London: Century

Hutchinson Ltd., 1987), chap. 3; also Cecil Gutzmore, "Capital, 'black youth' and crime" in *Race & Class* XXV, no. 2 (Autumn 1983), pp. 13–30; and Lee Bridges, "Policing the Urban Wasteland" in the same issue, pp. 31–48.

7. Channel 4 started broadcasting in 1982. It is government subsidized but funded by a number of sources, including advertising and subscription payments. When it was set up it was supposed to commission and air a variety of voices, including ethnic minorities, the independent filmmaking sector, foreign programming, and nontraditional formats. The actual percentage of airtime and monies allocated to the independent sector has been exaggerated in the U.S. Most of what would be considered innovative programming is shown on two one-hour weekly slots ("Eleventh Hour" and "People to People") at off-peak hours.

8. For an in-depth discussion of this, see Franco Bianchini, "GLC R.I.P.: Cultural Policies in London, 1981–1986," in *New Formations* (Summer 1987), pp. 103–117.

9. Sociology departments in the more progressive British Polytechnics (such as Portsmouth, Middlesex, and South Bank) have quite a different course of study from their American counterparts. Theory and Research Methods are distinct branches of study, and it was within the theory rubric that Black Audio members John Akomfrah, Reese Auguiste, Lina Gopaul, and Avril Johnson encountered the critical writings that would later inform their creative work.

10. This list is not exhaustive. The work of Birmingham's Center for Contemporary Cultural Studies and London's Institute for Race Relations is also extremely important. The filmmakers are also interested in the work of many Black American essayists, particularly June Jordan.

11. For more about Bhabha's relation to Lacan and Fanon, see "Of Mimicry and Man: The Ambivalence of Colonial Discourse," in *October*, no. 28 (Spring 1984), pp. 118–124. Also see Franz Fanon, *Black Skin, White Masks*, Charles Lam Markman, trans. (New York: Grove Press, 1967).

12. See Gilroy, *There Ain't No Black in the Union Jack*, and *The Empire Strikes Back* (London: Hutchinson, 1982).

13. Julian Henriques, "Realism and the New Language," *Artrage*, no. 13 (Summer 1986), pp. 32–37.

14. Jim Pines, interview with the author, London, July 1987.

15. Attille, "Young, British and Black," p. 14.

16. In 1985, riots in Handsworth and Broadwater Farm were set off by the deaths of Cynthia Jarrett and Cheryl Groce. Police entered the Jarrett home and began to question Ms. Jarrett, who suffered from a heart condition and began to feel ill when she was questioned. The police did not respond to the oncoming heart attack. She died shortly thereafter. Ms. Groce was shot by police who were supposedly searching for someone else. The Broadwater Farm riot gained infamy from the killing of a policeman by rioters on the first night.

17. See Gilroy, *There Ain't No Black in the Union Jack*, chap. 5.

18. Ibid., chap. 6, p. 69.

19. "*Handsworth Songs*: Some Background Notes," an unpublished paper by the Black Audio Film Collective, 1987, p. 4.

The Films of Isaac Julien

Look Back and Talk Black

José Arroyo

Modes of Appropriation

> . . . all discourse is placed, positioned, situated, and all knowledge is contextual. Representation is possible only because enunciation is always produced within codes which have a history, a position within the discursive formations of a particular space and time.
>
> *Stuart Hall (1988:29)*

Black British cinema comes out of a highly charged context. Intense black uprisings against police repression and right-wing groups in Britain throughout the 70s and into the 80s are the root and some of the subject matter of Isaac Julien's films. Coco Fusco has called the 1981 uprisings, "a watershed moment in the history of race relations" (1988:21/n.5). According to Paul Gilroy the suspicious deaths of 3 young blacks in 1981 set off a chain of events that led to explosive rioting in the area of Brixton, spread throughout industrial centres of London and "provided a means to galvanize blacks from all over the country into overt and organized political mobilization" (1988:102). Kobena Mercer has argued that, "the eruption of civil disorder . . . encoded militant

Jump Cut, no. 36, 1991, pp. 98–107, 110. Reprinted by permission of the publisher and author.

demands for black representation within public institutions as a basic right (and) . . . many public institutions hurriedly redistribute(ed) funding to black projects" (1988a:6).

Isaac Julien was one of the filmmakers who benefitted from the formation of the black film workshops that were one of the results of the 1981 uprisings. As a young, black, working-class filmmaker it is questionable whether he would have had access to such an expensive form of communication/expression as film if Sankofa, the workshop he operates from, did not exist. Because black communities to a certain extent enable black filmmaking in Britain, filmmakers have often been held accountable to them, though not always by the communities themselves.[1] Their task has been seen by some to speak to and for the black communities. Many filmmakers, however, claim only to be speaking *from* a black experience in Britain rather than *for* one (Julien and Mercer, 1988:4). However, questions of filmmakers' personal expression, when they have come up, have been deemed of secondary importance.[2]

One of the implications of such discourse is that black filmmakers must communicate via a "language" which their constituencies can understand, i.e. that of dominant narrative forms. Yet, all of Julien's films eschew traditional film narrative. *Territories* (1984, 25 min.) is a short experimental documentary: recurring, discontinuous images interact with an intoning voice-over and various types of music to deconstruct, and find meaning in, Carnival and its context. *Passion of Remembrance* (co-directed with Maureen Blackwood, 1986, approx. 80 min.) is a feature which goes against dominant forms. Characters appear out of the blue. The film relies on expressionistic devices such as the indication of an audience through aural, rather than visual means. Characters look into the camera and address the audience directly. Scenes like that of the protagonist and her friends dancing do not contribute to plot development. And the main narrative is interpolated at various intervals with lengthy montage sequences. *This Is Not an AIDS Advertisement* (1988, 13 min.) is a lyric experimental film. *Looking for Langston* (1989, 40 min.), Julien's latest film, creates a non-linear narrative through the combination of archival footage, photographs, poetry, and re-enactments, to evoke the ambience, and some of the people, of the Harlem Renaissance.

Julien, along with the rest of Sankofa and the Black Audio Film Collective, has been criticized for his choice of film practices. For example, in "Two Kinds of Otherness: Black Film and the Avant-Garde," Judith Williamson obliquely pounces on this issue when she states, "Audiences do matter . . . If you're practical you do want to reach people beyond your buddies" (111). The implication is that, because of their form, films like *Handsworth Songs*, *Passion of Remembrance*, and by extension, the rest of Julien's work, are not reaching audiences.[3] Williamson then juxtaposes these films with *My Beautiful Laundrette*—"(Audiences) love it! . . . It's been a highly enjoyed film" (111). She notes that one needs a certain amount of cultural capital in order to understand and enjoy "avant-garde films," a cultural capital that is presumably not readily available to the black films'[4] "natural audience"—blacks.

Williamson's piece raises several important issues. First of all, it illustrates the degree to which minority filmmakers are marginalized through a double bind of expectation/obligation due to how little access blacks have to the means of production. As Sankofa member Martina Attille notes, "Sometimes we only get the one chance to make ourselves heard" (cited in Pines 1986:101).

Second, Williamson notes the hesitancy with which critics engage with black films. To these points we can add that these filmmakers are excluded from the mainstream, burdened with low budgets, accountable to communities, restricted to certain forms; and their films are often condemned to minimal distribution. Black filmmaking is marginalized from within by discourses of accusation/prescription. It is marginalized from without by a racist, market-driven film establishment.

Black filmmakers, however, have good reason to reject traditional narrative. To begin with, mainstream forms are most palatable when accompanied by mainstream budgets. Julien's films, with the possible exception of *Looking For Langston*, are made on budgets substantially below those of even British independent films like Stephen Frears' *Laundrette*, Derek Jarman's *Caravaggio* or Terence Davies' *Distant Voices/Still Lives*.[5] Moreover, if we agree with Stuart Hall that "all discourse is placed, positioned, situated, and all knowledge is contextual (p. 29)," then new discourses need a combination of new places, new positions, new situations, new contexts. If dominant cinema is characterized by depiction's of blacks as Mammies, Toms, and Coons, and if that is the "knowledge" that that form/context evokes, then perhaps it is not the best one to use. Likewise, following this line of reasoning, one can argue that other accepted forms of narrative have yet to represent black people as other than Other.

Salman Rushdie has written that "if you want to tell the untold stories, if you want to give voice to the voiceless, you've got to find a language."[6] This search for new "film languages" is not just characteristic of Julien or other young, black, British filmmakers. Feminists have been experimenting with innovations with language since the 70s. In her now classic essay, "Visual Pleasure and Narrative Cinema," Laura Mulvey argued that since classic Hollywood cinema condemned women to Otherness even as it created pleasure, this new feminist "language" should be one that creates displeasure (1988:58–59). This is a strategy that Julien rejects.

His work, however, does reveal various attempts at experimenting with lyric as well as documentary and fictional narrative in order to make blacks, women, gays, and sometimes black women and black gays enunciators rather than mere enoncés.[7] I hesitate to use the term avant-garde to describe the work. *The Oxford Handy Dictionary* defines the term avant-garde as meaning "innovat(ions) in art, literature, etc." (p. 48), which I think fits it well. However, in film, the term has often been associated with difficult, often formalist, "apolitical" work such as that of Fernand Léger, Germaine Dulac, Michael Snow and Joyce Wieland. Julien's work is art cinema in the sense that he often does not incorporate a linear cause-effect relation between narrative events. Its use of montage and distanciation may be considered formalist. In that it challenges dominant cinema in terms of both form and content, it is a counter cinema.

After seeing his work many times I have come to suspect that new kinds of work demand new kinds of criticism. Julien admits to having been partially influenced by avant-garde work:

> There are some avant-garde filmmakers, such as Ken McMullen and Sally Potter, whose work I am interested in. But if I were going to cite direct influences, I would look to Haile Gerima and Charles Burnett (quoted in Fusco 1988:31–32).

I am not familiar with the work of the latter two U.S. black filmmakers. This realization made me think that there is probably a whole culture of references in Julien's work that I, and other white critics, don't understand. It also made me realize that we often resort to our already acquired frames of reference rather than expand and acquire new ones to deal with new work.

With these limitations in mind I will now attempt to look directly at the films. My purpose is not to provide a "comprehensive" critique. The films are of different lengths and in different modes, and they raise many questions that will not be dealt with in this piece. What the different works have in common is that they engage with representations of race, sexuality and nation and that they are either directed or co-directed by Isaac Julien. In the films, the otherwise marginal is made central. They make a subject of various forms of Otherness. It is these strategies of representation, and their manifestation, that are the object of my gaze and the subject of this paper.

Territories

"We are struggling to begin a story," say different voices-over throughout Isaac Julien's *Territories,* "a history, a her/story of cultural forms specific to black people." Sometimes this line will be repeated almost as a chant by a woman's voice, followed a few beats later by a male voice. At other times two women will say it in unison, followed again by a male voice.

The Notting Hill Carnival, a three-day event held annually in August since 1958, is the point of departure for the film's exploration of various contested territories. Carnival has been the site of various "race riots" beginning with the year of its inception. During the 70s riots erupted in 1976, 1977 and 1978 but it is the '76 one which is most famous. That year Scotland Yard sent 1500 uniformed men to police Carnival. Such massive surveillance resulted in a riot in which black youth fought back and won. According to Paul Gilroy, the 1976 Notting Hill Carnival was "a watershed in the history of conflict between blacks and the police and in the growth of the authoritarian forms of state planning and intervention during the 1970s" (1987:93). Gilroy also writes that a concept of criminal public disorder is "central to today's racist ideology" (p. 82). The widespread media coverage of Carnival, when it did not equate blackness with exotic spectacle, equated it with crime. Both equations were widely disseminated, reinforced racist ideology and are sharply scrutinized in *Territories*.

The story *Territories* is trying to tell is not just about Carnival. Rather, it is a deconstruction of Carnival and what it represents. According to Homi K. Bhabha,

> In order to understand the productivity of colonial power it is crucial to construct its regime of "truth," not to subject its representations to a normalising judgment. Only then does it become possible to understand the productive ambivalence of the object of colonial discourse—that "otherness" which is at once an object of desire and derision, an articulation of difference contained within the fantasy of origin and identity. What such a reading reveals are the boundaries of colonial discourse and it enables a transgression of these limits from the space of otherness (1983:19).

In *Territories*, Julien precisely tries to reconstruct that regime of truth. However, he simultaneously tries to transcend it both by deconstructing it and by creating an alternative regime of truth. The film knows a silenced truth which cannot be expressed by the old language, thus, "the struggle to tell the story."

Julien's strategy for telling the story is to make the gaze black. He attempts to do this by having black people, whose faces are not visible but who speak from a diegetic space, give us their reading of the images. He also tries to defamiliarize images by manipulating them in a different way: he uses freeze frames, he reverses the images, tints them, turns them upside down.

The film begins with the word *Territories*, marked out in yellow on a black background. The outer limit of each of the letters in the word contains images, as if through a prison bar. The camera zooms into the word, explodes its limits, and gives the viewer access to what it signifies, what it contains. Then, a quasi-choral voice-over of a male and a female speaker is heard, superimposed over an image of a black man huddled against a ruin. This combination of sound and image further explores the notion of "territory" even as it sets the agenda for the film. The camera tilts down in close-up from the ruin and then tilts up to a black man, excluded even from the ruins, huddling to keep warm. The film will also end with images of ruined, bricked-up buildings. If one can read these ruins as a symbol for Britain, the place of blacks in it is then clarified by the initial voice-over, also to be repeated at various times throughout the film. The voice-over tells us,

> A new context for the political struggle for the disseminated mass of unwanted labour is provided by the streets of civil society. Its territories: the contradictory spaces which are the geographical expressions of a city. The territories of class, labour, race, sex, relations. Territory. The holding of one class' privilege in a declining system of crisis. Territories of desire, the control of one space, carnival, territory of resistance, sound systems, territories of surveillance of the mind. Territory of sexual expressions (of the body). The contradictory spaces, the geographical expressions that cohabit a city. The territory of the look. Territories of the body, conflicting with dominant demands. Behind each conflict there is a history a her/story. We are struggling to tell a story, a her/story, a history of cultural forms specific to black people.

The story is told in an experimental documentary form. The film distances itself from traditional documentary which it claims contains Carnival as, "an aesthetic spectacle." It distances itself from documentary convention by, among other devices, having two black women look at a video monitor showing these images of carnival. The spectator is made to align his/her gaze with theirs. We relegate our look as the carnival footage is reversed, freeze-framed, examined in ways which we are not used to. This image manipulation is accompanied by a woman's voice that tells us,

> Colonial fantasy requires a fixed image of the black person or the Other. But it is based on a complex kind of fixedness. The Other signifies both fear and desire and disorder due to the way in which blackness evokes both fear and fantasy on behalf of white society.

The film uses repetition and juxtaposition to shatter that fixedness. It divorces images of Carnival and of riots against the police from those images' traditional dual meaning: depicting blacks as fetishized into primitive, erotic "Other" and depicting blacks as problem.

Territories provides an incantation of sound and images. The film is layered so that images become repeated and acquire different possible meanings in different contexts. But the range of those meanings is limited through continuous reference to black voices and black looks. Close-ups of sound speakers and eyes and faces staring back at the camera are edited in at intervals throughout the film. The gaze into the audience seemingly both questions and testifies to the viewer's witnessing and interpreting.

The film conveys "a history of black forms specific to black people" through explanations of the tradition of carnival, of dancing, of music. The scenes of dub versioning provide "a deconstructive aesthetic which 'distances' and lays bare the musical anatomy of the original song through skillful re-editing which sculpts out aural space for the DJ's talk-over" (Mercer 1988b: 54–55). Such scenes can be seen as metaphoric for what the film itself is trying to do through similar editing and through the use of discontinuous gaps between image and sound, frequent fades to black and the preponderant use of jump cut montage.

The second part of the film visually introduces sexual difference, in the form of homosexuality, as one of the territories being contested in Britain. An image of two men embracing is superimposed over a burning Union Jack. At other times, a close up of one of the gay men looking at the audience from the right side of the screen will be frozen as a close up of a stern policeman is superimposed onto the left side of the screen. The dominant musical accompaniment has Joan Baez singing "The Ballad of Sacco and Vanzetti":

> Against us is the power of police . . .
> Against us is the gold . . .
> Against us is the law . . .

Territories is a film that invites viewers into critical interaction. It is open-ended and amenable to various interpretations. But it also builds boundaries through which spectatorship cannot cut. The foundation of those barriers is the periodical looking back at the audience by black subjects. It would be easy to consume the images of oppression depicted in *Territories* as our daily dose of misery, easily digested and functionally excreted. Many of us do that daily as we switch from the evening news to *Golden Girls* or another sitcom. *Territories* resists such easy dismissal by catching us in the act of looking. The look back has the effect of transforming us from passive voyeurs into conscious witnesses, however unwilling.

Passion of Remembrance

Passion of Remembrance, which Julien co-directed with Maureen Blackwood, extends many of the concerns expressed in *Territories*. Or rather it makes the agenda more inclusive. At times *Passion of Remembrance* seems crammed with every possible issue of importance to black filmmakers: racism, sexism, homophobia, police brutality, the decline of industrial society, links to civil rights struggles in the United States etc. But unlike *Territories*, *Passion of Remembrance* is a feature film, one with two directors at that; and in view of its length and its aspirations it uses different formal strategies.

Oddly enough, in spite of being a fiction film, *Passion of Remembrance* well fits Grierson's dictum for documentary: "the creative treatment of actuality." The phrase is perhaps vague enough to fit all film and photography. However, if we note that *Passion of Remembrance* is involved with representations of social processes, that it relies heavily on montage, that it is concerned with social change and that it experiments with form, the comparison might seem more apt. That is perhaps the limit of the resemblance, however. Griersonian documentary has been criticized for hiding the worker behind the machine. In *Passion of Remembrance* the individual is at the center, the form is fiction, and the point-of-view is female.

Passion of Remembrance's narrative is woven through different strands. There is "The Woman's Story," which is metaphoric. "She" represents black women activists. "He" represents black male activists. They exist in a mythical place called "Here" where "She" arrogantly reproaches and accuses "Him" for his phallocentrism and self-absorption. There is also Maggie's story. Maggie is young, black, British and working class. Her struggle is to negotiate her needs and her vision of a movement with a legacy of black activism in Britain which she finds both inspiring and, to the extent that that inheritance is also accompanied by a measure of machismo and homophobia, oppressive.

Maggie's story particularizes the debates the mythical "They" are arguing about. Each picks up the slack from the other when the form of each section seems unable to contain articulation. Maggie's story is one of the lived experiences that make up "Her" story. However, "She" is also a generation older than Maggie. "Her" history as an activist encompasses subjugation and exclusion from active decision-making by black men within the movement. Maggie's struggle is partly in honour of "Her" history, partly a struggle against that history being repeated in a new generation of black women activists. The complementary stories of Maggie, "Her" and Maggie as part of "Her" are bounded by a rich family life, an oppressive social situation, and the death and funeral of a young black man.

Passion of Remembrance attempts to reconstruct and activate a memory of a black history in Britain. This is done through a mosaic structure in which difference is marked even as the elements of the mosaic make up a whole. There are links to the civil rights movement in the United States ("the real struggle was on the other side of the Atlantic") even as the British and U.S. movements are seen as different struggles for different futures:

> she knew the site of any struggle was real. . . .
>
> [we are] staking a claim for the future [in Britain]. . . .

Black is divided from white (as shown in "His" anecdote about Sergeant Kendall) even as black and white are linked through class: "They're being treated like blacks," says Maggie about footage of police harassing coal miners. The monologues between "He" and "She" denote a gender barrier within a common struggle, while at a meeting Gary and Michael, Maggie's gay friends, bring up the question of homophobia within the Movement.

The idea of nation in *Passion of Remembrance* is just as important as it was in *Territories*. Here Britain is depicted in similar terms. After a montage of padlocks, ruins, broken windows, deserted buildings, dirt, and rubber tires, Tony (an activist involved in the Movement since the 60s) tells the unemployed Benjy (Maggie's father) that the younger generation has "grown up with this." England is shown to be a country riddled with class conflict, racism, sexism, unemployment: "This can't be my England," mocks Maggie's friend Louise, "not the England my grandfather's father fought for."

The very first shot of the film shows archival footage of black women demonstrating, their fists raised in protest. Footage of demonstrations recur throughout the film; the various montages of that footage indicate that the range of grievances citizens have against the state may be a common link. However, differences between black and white are not diminished: "Get out of our fucking country!" yell some thugs to a black family while the police witness and through inaction, condone the abuse. As "He" tells us,

> We've woken up to a situation, a country, a time, a system that is so frightening, so all embracing in the way that it lives our lives for us, or changes them or obliterates them, that we don't believe this is reality.

Passion of Remembrance is just as concerned with trying to tell its story as is *Territories*. In its attempt it uses similar strategies. The archival footage of demonstrations and police brutality are always filtered through a black gaze, mostly a black female gaze. Maggie gazes into the monitor, and we follow her gaze as the film cuts to the footage; or Maggie and Louise together guide our gaze, or Maggie's memory of her introducing a tape to a meeting sets the guidelines to our viewing. In the one instance when Tony, Maggie, and Louise watch footage together, Maggie and Louise afterwards critique Tony's interpretation of it.

Like in *Territories* we are distanced from familiar interpretations of familiar footage by having it presented in different ways; faster, slower, upside down, drained of colour, with extra colour added. *Passion of Remembrance*, however also attempts to introduce new representations, ones that fit in uneasily with traditional narrative's use of plot. For example, images such as that of Michael swimming are held for a long time. They do not further the plot. They just revel in the beauty and sensuality of black bodies in motion. Such images are unusual in the sense that in traditional dominant modes of narrative, black male sexuality is usually avoided (as in most of Sidney Poitier's films) or is presented as threatening (black exploitation films). Likewise the reason for lengthily prolonging the scene of Louise and Maggie dancing while making up and getting dressed is to depict black women reveling in each other's company, enjoying each other's sensuality.

The most interesting aspect of *Passion of Remembrance* is how it succeeds in making both "black" and "woman"—two perennial kinds of cinematic "otherness"—into a single enunciator. This is the only instance in Julien's oeuvre where this occurs. *Passion of Remembrance* is also the only film he co-directed. One can thus, without resorting to essentialism, impute that the film's privileging of a female point-of-view must in great measure be due to Maureen Blackwood.

After the initial shot of women demonstrating, the camera cuts to "Her," who addresses the audience directly. She narrates that story which is acted out mostly by Maggie. Though male characters are sometimes subjects of particular scenes in the film, their subjectivity is delegated either by "Her" (as in the case of "His" monologues) or by Maggie. In the cases where males are given a voice or made subjects without the previous benefit of female consent, women are given the last word. For example, there is a scene in which Benjy and Tony talk about unemployment and later dance together; this scene ends with Maggie telling the men off. As she closes the door on them, the film cuts to a new scene. Finally, the last shot of the film is a freeze-frame of "Her" gazing at "Him" as he disappears into the mist of the mythic "Here."

"Who will hear me now as I remember and talk of remembering," wonders Maggie at the beginning of the film. The film provides its own answer near the end—black British people. As Gary gazes at a coffin of a young black man who's been murdered, he "thinks" to him, "The media may choose to forget while we do not." The voice-over then changes its address to the audience:

> His memory lives with us . . . We're in a war that's bleeding liberation. Ghosts of people who come back again and again questioning, questioning the mockery and fictions held by centuries of traditions based upon bankruptcy and lies infested in Englishness at its worst.

Passion of Remembrance offers a black woman's point-of-view directed at a black male and female audience (a rarity though perhaps more likely from a collaborative effort co-directed by a woman and a gay man). In *Passion of Remembrance* the "us" and the "we" refer to black people. Whites are not addressed. They serve merely as the grain against which a black identity is formed, as blacks' "Other."

This Is Not an AIDS Advertisement

This Is Not an AIDS Advertisement is a short Super8 film. Like *Territories*, it was made as a reaction to a specific situation: In *Territories* it was Carnival; in *This Is Not an AIDS Advertisement* it was homophobic safe-sex ads. The dominant message in the latter is, "Feel no shame in your desire." The formal strategy to convey the message is an extension of that employed in *Territories*.

This Is Not an AIDS Advertisement, like *Territories*, relies on accretion of images, their repetition, variation, and juxtaposition to create meaning. Unlike in *Territories*, the first part employs no voice-over. The second part is accompanied by a kind of rap song made up of fragments from various sources arguing against guilt ("It's the heart afraid of breaking that never learns to dance; it's the dream afraid of waking that never learns to chance.") and for love ("Some say love it is a razor that leaves your soul to bleed; some say love it is a hunger, an endless aching need; I say love it is a flower and you its only seed.").

The salient images in the first part are these: low-angle shots of waves, their caps changing colour from shot to shot; high-angle shots of skies framed by trees; a black youth and a white youth hugging as they look at the camera, extend to it a bouquet of flowers and then bring the flowers to their chests; a face, seemingly dizzy and pained on which is superimposed a shaft of light coming into a room; police boots running up steps to stomp on the flower. In the last image, the couple laughingly kisses.

The images are amenable to various interpretations. Mine is that the filmmaker is trying to express males-loving-males as an activity that is deep, vast, ancient (there is a recurring shot of an ancient Roman relief depicting a male). And one which, in spite of official attempts to deny and stomp it out, persists. The representation of the interracial gay couple is significant in that as Kobena Mercer argues, in the war against AIDS both blacks and gays have been labeled a threat.

> Racism and homophobia activate similar psychological defense-mechanisms whereby people avoid their inner fears by projecting them externally onto some Other (1988c:152–153).

The second part begins with a male head turning, trying to face the audience as if struggling to materialize. It finally does so and stares blankly at the audience. This section is characterized by the accretion of images introduced in the first section juxtaposed against new ones. The images are cut to the beat of the soundtrack's rap. The figures in the frame invariably look back at the audience. They are aggressive objects who gain subjectivity through the matching of their gaze to that of the audience. A recurring image in the first section of a blindfolded man unblocking his eyes and gaining sight makes more forceful the power of their gaze. The film's message becomes underlined through a kind of video aesthetic of synchronously superimposing the different words that make up the phrase, "Feel no shame in your desire," onto various images. This section, like the first, ends with the laughing kiss of the interracial male couple.

"There is a Third World in every First World," writes Trinh T. Minh-ha (1986/7:3). If we take that statement metaphorically, we can see in Julien's work a cinematic mapping out of different "Third Worlds." In *Territories* oppressed groups are the State's Third World. In *Passion of Remembrance* the black gay couple and black women can be seen as black heterosexual male activists' "Third World." In *This Is Not an AIDS Advertisement* a disease is the cause of the First World's combined condemnation of race and homosexuality. Trinh, however, also notes that " 'looking back' and 'talking back' form a necessary step to the unsaying of what has been said and congealed" (1986/87:3).

I have tried to show that talking back is an integral part of Julien's work both in the discourse he creates and, integrally interlinked, in the form through which he conveys the discourses, with particular note of the device of "looking back" at the audience as a cinematic form of "talking back." I have also tried to show that *Territories* and *Passion of Remembrance* also engage in a historical "looking back" in that they both try, to different extents, to unearth and reconstruct a history of black British culture. *Looking for Langston*, Julien's latest film, takes: this "looking back" and "talking back" a step further.

Looking for Langston

A party of whites from Fifth Avenue
Came tippin into Dixie's to get a view
Came tippin into Dixie's with smiles on their faces
Knowing they can buy a dozen colored places.
Dixie grinned. Dixie bowed.
Dixie rubbed his hands and laughed out—
While a tall white woman
In an ermine cape
Looked at the blacks and
Thought of rape,

Looked at the blacks and thought of a rope,
Looked at the blacks and
Thought of a flame
And thought of something
Without a name

Langston Hughes (1942:60)

Hello Sailor Boy
In from the sea!
Hello, sailor
Come with me!

Come on drink cognac,
Rather have wine?
Come here,
I love you,
Come and be mine.

Langston Hughes (1926:74)

There is a certain kind of poetic justice in Isaac Julien's making a film about Langston Hughes and the Harlem Renaissance and dedicating it to James Baldwin. Hughes and Baldwin were black homosexual artists.[8] Like Julien, Hughes and other artists of the "Harlem Renaissance" were drawing on a history of black culture to try to create new forms that would contain and communicate new representations. Hughes tried this by creating a blues language for poetry.[9]

Speaking of some of the visual artists of the Harlem Renaissance, Mary Schmidt Campbell writes,

> Each developed a vital aspect of the Renaissance ethos—be it glorification of the Black American's African heritage, the tradition of Black folklore or interest in the details of Black life. Each broke dramatically with earlier Black art and earlier representations of

> Blacks in art. They were among the first Americans to celebrate Black history and culture and they were the first artists to define a visual vocabulary for Black Americans (p. 13).

In trying to evoke Hughes, Julien "looks back" and finds a history of black art, a history of representations of blacks, and a history of black homosexuality. In *Looking for Langston* he "talks back" these histories by collating them, interpolating a stylized critique of some current representations of black men as fetish. He sets this evocation against a mixture of '20s American blues and '80s British blues. Thus, Hughes' lines of poetry, "Why should it be my loneliness, why should it be my song, why should it be my dream deferred overlong," is followed by Blackberri, a black British singer, answering back with, "Whatever happened to a dream deferred? Things haven't changed much, I still find power in your words." Julien thus links black British culture and black American culture through time and space as a diaspora culture.

Looking for Langston tries to affirm a black gay identity. If the Harlem Renaissance was one of the golden ages of black diaspora culture, the film tells us that a lot of its brilliance came from homosexuals: Countee Cullen, Alain Locke, Bruce Nugent, Harold Jackman and others. The film links the Harlem Renaissance to contemporary Britain when a black British voice-over tells us, "We were linked by our gay desire." The same voice-over also warns that not to discuss the "moral significance of Countee Cullen, Langston Hughes, Alain Locke, choosing in the main others of their kind to love, is to emasculate and embalm their society as a whole."

In *History of Sexuality, An Introduction*, Michel Foucault writes that, in Western society, "it is through sex . . . that each individual has to pass in order to have access to his own intelligibility . . . to the whole of his body . . . to his identity" (1984:155–56). *Looking for Langston* tries to chart a course through this discourse. But in order to finish the movement from sex to identity, it must first reappropriate the image of the black male body.

The black man has often been depicted as civility's absolute "Other." As Kobena Mercer writes, "Classical racism involved a logic of dehumanization in which African peoples were defined as having bodies but no minds" (1988c:137). Black males were often not even seen to have bodies. According to Fanon, "One is no longer aware of the Negro, but only of a penis; the Negro is eclipsed. He is turned into a penis. He is a penis" (cited in Mercer 1988c:150).

In the regime described by Foucault as power-knowledge-pleasure, the black man is powerless, he is excluded from knowledge, and the pleasure his body represents to white society is one that is simultaneously imbued with danger. If we agree with Foucault that such a regime is what "sustains the discourse on human sexuality in our part of the world" (p. 11) and that such a discourse is necessary for identity, then the black man has to be empowered, given access to knowledge, and made to take pleasure in his own body. *Looking for Langston* attempts just that in order to topple the privileged regime.

In *Looking for Langston* Julien acknowledges the dominant discourse on black masculinity and simultaneously refutes it. Verbally this is done through a voice-over reading of U.S. Black gay poet Essex Hemphill's poetry in two consecutive stages. The first offers an accusation to whites—"You want his pleasure/without guilt or capture. You don't notice many things about him." The second Hemphill reading offers a refutation—"He doesn't always wear a red ski cap/eat fried chicken/fuck like a jungle."

Visually the same effect is achieved by showing white pleasure in black bodies. A montage shows us a white man in close up.[10] A cut then shows the same man caressing the photographs as they are projected onto sheets. We then see the white man give a black man money. Throughout this montage a voice set to music accusingly queries the white man's activities and motives.

> I don't suppose you ever *Hear* him clearly? / You're always too busy / seeking other things of him / His name isn't important / It would be coincidence / if he had a name, a face, a mind. / If he's not hard-on, he's hard-up.

As Julien recuperates the colonized image of the black man, he tries to infuse it with new meaning. The sexuality of the black men presented is divorced from any violence, except perhaps the one imposed by the white psyche. Moreover, the images aren't so much concerned with representing sex but with representing desire. A substantial part of the film is concerned with affirming the concept of black beauty and making it, and those who possess it, an object of desire. As the lyrics for one of the songs used in the film tells us,

> Look at me, beautiful black man
> I'm just like you
> You know that I face
> discrimination too
> Got here about ten
> I walked into this place
> But nobody here
> would look me in the face
> You're such a beautiful black man
> don't you walk with your head down bending low
> Don't you do that no more.

Beauty is depicted through a process of fetishization, and it is made desirable by making it the object of both the spectator's and the film's gaze. The latter is effected through the camera and also is designated to various characters within the diegesis.[11] The mythical "Beauty" in the film is given special lighting that delineates his cheekbones, the camera tilts and pans across his body. We are shown a close-up of his eyes that indicate his eyelashes have been perfectly curled. A poem tells us how Langston dreams of him and describes different parts of his body.

"Beauty," however is only one kind of black beauty. The end of the film shows us an image of him with his face resting on a mirror, an image of Narcissus. In the same montage, we are shown another black man, darker, stouter. He too looks into a mirror and makes a gesture as if to indicate that he's woken up, gained a consciousness of his own beauty, acknowledged his own worth.

As in his other films, Julien often prevents automatic suture.[12] The viewer is often made conscious of his/her absence through a self-conscious narration which, among other things, looks back at the audience: Langston looks straight at the audience before we enter his dream. A man looking at porn stares directly at the audience as if they were porn. Beauty undresses the white man and laughingly looks at the camera as he throws it a shirt. A shot of a man dancing and laughing while looking straight at the audience will be intercut into the film's final montage three times.

Stephen Heath asserts that the viewing subject's break with the initial relation to the image is "essential to the realization of image as signifier" (1985: 88), and the "suturing function includes the spectator as part of an imaginary production (p. 90)." Such theoretical concepts point to: how the way that Julien instigates and underlines a break with the image affects cinematic signification. If *Looking for Langston*'s form of fusion of poetry, music, archival footage and new images were not already enough, such a strategy of rupture makes the viewer think about the kind of signification s/he is helping to construct.

> Obviously, an Afro American spokesperson who wished to engage in a masterful and empowering play within the minstrel spirit house needed the uncanning ability to manipulate phonic legacies. For he or she had the task of transforming the mass and its sounds into negotiable discursive currency. In effect, the task was the production of a manual of black speaking, a book of speaking back and black.
>
> Houston A. Baker, Jr. (1987:24)

Baker is talking about another country, another culture, another medium. But the need to talk back and to talk black is just as evident in the films of Isaac Julien. I say *need*: because Julien's cinema is clearly a political cinema. Whereas others are moaning about the waning of affect (not realizing just how privileged one has to be to be able to feel that way), Julien's cinema underlines race, class, gender and sexual orientation. It's hard to take the notion of free floating signifiers seriously when black as the colour of your skin or a Hispanic name on your application form consistently means denial of work, or when two men kissing in public consistently means license to beat them up.

Western culture has for centuries provided the ideological rationale for imperialism, racism, and other forms of oppression. The present "crises of cultural authority, specifically of the authority vested in Western European culture and its institutions" (Owen 1983: 57) is no big loss for people who have historically been excluded from participating in Western meta-narratives. Marginalized peoples, especially diaspora people, brought, formed and reshaped other *récits*, grand ones made mini only by Western culture's devalorization of them as "primitive," "savage," "Other." As Paul Gilroy notes,

> *Who* is it that people like Fredric Jameson are talking about when they say, "Our grand narratives are collapsing"? Some of us, who have been denied access to some of the diachronic payoff that people like Jameson take for granted are just beginning to formulate our own big narratives, precisely as narratives of redemption and emancipation. (1988:46)

"Now that, in the postmodern age, you all feel so dispersed," writes Black British theorist Stuart Hall, "I become centered" (Quoted in Mercer 1988a: 5). Though I find Hall's comment interesting to think about, I doubt its sentiment is widely shared. However, I do think relations between the margins and the center are dynamic. I think that the way Julien demonstrates dominant depictions of race, refutes them, and offers new alternatives is a means of visualizing and helping to promote positive shifts. However, whether the strategies Julien uses to "look back and talk black"—his use of montage, non-classical continuity, direct address, his attempts at deconstruction of previous imagery and the self-conscious narration of his narratives—succeed in negotiating a discursive currency is another question, one that only time and research into audiences will answer.

Notes

This paper derives from a course on British Cinema taught by Andrew Higson at the University of East Anglia and a presentation at the 1990 SCS conference in Washington. I would like to thank Andy Medhurst, David Hall and Deborah

Regula for their feedback during the writing stage, and Chuck Kleinhans, Julia Lesage and Mark Reid for their helpful editing suggestions.

1. Martine Attille admits that "there was a sense of accountability imposed on us by our community to produce certain types of images." (Fusco, p. 26) Nevertheless, black filmmakers and perhaps any minority filmmaker, are charged with the extra responsibility of community accountability no matter who or what enabled their filmmaking. See, for example, Mahmood Jamal's "Dirty Linen," one of the many tirades against *My Beautiful Laundrette*.

2. All of these issues are clearly presented through a wide range of articles collected in Mercer, *Black Film, British Cinema*.

3. I find Williamson's piece asked many incisive questions. But I also found certain aspects disturbing. I have yet to see any white political group get the degree of flack for their choice of film practice that Sankofa and Black Audio are subject to. The form/audience equation I find particularly tiresome. Old classic realist versus avant-garde debates ignore that there is a whole generation who has grown up watching music videos who have access to, and derive various pleasure from, many forms. Practices previously the privileged domain of "High Culture" are now a flick of the switch from "Wheel of Fortune." I find a large portion of vanity sprinkled with an equally large dose of condescension in the "experts" harping about how inaccessible these films are. Moreover, I think there is an unwarranted assumption in a lot of the criticism that a more "accessible" form would automatically translate into larger audiences and thus greater political effectiveness.

I don't have access to British box-office or television ratings, but were other Workshop films more popular? And how many were less? Certainly in North America, and I don't understand why this is such a sore point with British critics, these films' form is greatly responsible for the distribution they did get. As to political affectivity, one can't talk about the films in isolation from the body of critical work that has stemmed from them. For Julien, for example, the writing and the filmmaking seem to be a part of a singular politic. Others like Stuart Hall, Jim Pines, and Salman Rushdie have contributed to very important political debates that are the result of the films. And I would argue that the work of Kobena Mercer, without which this paper could probably not have been written, has shifted some of the ground on which notions of race, representation and cinema are debated. (See the references).

4. I am not trying to impute essentiality by my usage of the term "black films." I merely mean films by black filmmakers.

5. This is a paraphrase of information given to me by Isaac Julien in March, 1989. The budget for *Langston* was approximately 100, 000 pounds.

6. Rushdie doesn't know the score, at least as regards film. The quote is taken from a piece in which he paradoxically and confusedly criticizes *Handsworth Songs* for not attempting to create a new language when in fact the problem is that he doesn't know the old one well enough to recognize the contributions *Songs* makes. See "*Songs* Doesn't Know the Score," *Black Film, British Cinema*, p. 16.

7. Lea Jacobs (pp. 157–161) defines enunciation, "as the attitude of the speaking subject in the face of his (*son*) *enoncé*, this taking part in the world

of objects. The process of enunciation, thus envisaged, is described as a relative distance that the subject puts between himself (*lui-même*) and this *enoncé*." Thus the term enunciator, as I use it in the paper, refers to the person who has the right of speech within the film.

8. I would like to follow historian John D'Emilio (pp. 110–116) in his usage of "homosexual" to denote homosexual behavior and "gay" to donote homosexual identity. Julien commenting on black homosexuals has written, "If one looks to the United States, from the perspective of a diaspora culture, one can look to Langston Hughes or James Baldwin—they're the most visible figures of our continuity in black history." See "Interview," 1988.

9. According to Arnold Rampersad, "(Hughes') finest poems . . . remained those saturated in blues language, the idiom of the black folk that Hughes had pioneered in literary verse in 1923 with his poem "The Weary Blues," then developed to its zenith as art in *Fine Clothes to the Jew* in 1927. Blues was a way of singing but above all a way of feeling, when the pain of circumstance is transcended by the will to survive—of which the most stylish token, aside from the blues song itself, is the impulse to laughter." (p. 20)

10. Speaking of Soviet Cinema, David Bordwell notes that, "the relentless presence of montage in these films aims to keep the spectator from constructing any action as simply an unmediated piece of the fabula world." (p. 239) The same could be said for the use of montage in Julien's films.

11. Metz has written, "The fetishistic prop will become a precondition for the establishment of potency and access to orgasm *(jouissance)*, sometimes an indispensable precondition (true fixation); in other developments it will only be a favourable condition, and one whose weight will vary with respect to the features of the erotogenic situation as a whole." (p. 70) In *Langston*, Beauty's fetishization is an almost metaphoric manifestation of the latter, in the sense that it is used almost as a pre-condition for the film as erotogenic situation.

12. According to Stephen Heath, the process of reading a film takes us from sheer jubilation in the image to an awareness of the frame that breaks this initial relation. We, the spectator, recognize an absence, the discontinuity of the image, its production as signifier. We then fill in the absence, sew up the shots, suture the discourse. Heath writes, "The major emphasis in all this is that the articulation of the signifying chain of images, of the chain of images as signifying, works not from image to image but from image to image through the absence that the subject constitutes . . . Thus the break in the initial relation with the image is sutured . . . across the spectator constituted as cinematic . . . subject, essential to the realization of image as signifier and to the articulation of the shots together (1981:76–113).

References

Baker Jr., Houston A. (1987). *Modernism and the Harlem Renaissance*. Chicago: The University of Chicago Press.

Bhabha, Homi K. (1983). "The Other Question." *Screen* 5(24):18–36.

Bordwell, David (1985). *Narration and the Fiction Film*. Madison: University of Wisconsin Press.

Campbell, Mary Schmidt (1987). *Harlem Renaissance Art of Black America.* New York: The Studio Museum in Harlem, Harry N. Abrams Inc.
Cham, Mbye B. and Claire Andrade-Watkins (eds.) (1988). *Blackframes: Critical Perspectives on Black Independent Cinema.* Cambridge, Mass./London: The MIT Press.
D'Emilio, John (1983). "Capitalism and Gay Identity." In Snitow, Ann, Christine Stansell and Sharon Thompson (eds.) *Powers of Desire: The Politics of Sexuality* (1983:100–116). New York: Monthly Review Press.
Fowler, F. G. and H. W. (eds.), (1978). *The Oxford Handy Dictionary.* London: Oxford University Press.
Gilroy, Paul (1987). *There Ain't No Black in The Union Jack.* London: Hutchinson.
———. (1988). "Nothing But Sweat Inside My Hand: Diaspora Aesthetics and Black Arts in Britain." In Kobena Mercer (ed.) ICA Documents No. 7, *Black Film British Cinema* (1988), London: A BFI Production Special.
Foucault, Michel (1984). *The History of Sexuality. An Introduction.* London: Penguin Books.
Fusco, Coco (1988). *Young British and Black: The Work of Sankofa and Black Audio Film Collective.* Buffalo, N.Y. Hallwalls/ Contemporary Arts Centre.
Hall, Stuart (1988). "New Ethnicities." In Kobena Mercer (ed.) ICA Documents No. 7, *Black Film, British Cinema* (1988), London: A BFI Production Special.
Heath, Stephen (1985). *Questions of Cinema.* London: MacMillan Publishers Ltd.
Hughes, Langston (1942). "Death in Harlem," *Shakespeare in Harlem.* New York: Alfred A. Knopf.
———. (1926). "Port Town" *The Weary Blues.* New York: Alfred A. Knopf.
Jamal, Mahmood (1988). "Dirty Linen." In Kobena Mercer (ed.) (1988): 21–22.
Jacobs, Lea (1988). "*Now Voyager*: Some Problems of Enunciation and Sexual Difference." In Thomas Elsaesser (ed.) *American Cinema and Its Genres.* Norwich, UEA: 157–161.
Julien, Isaac (1988). "Interview," "Race, Sexual Politics and Black Masculinity: A Dossier (by Isaac Julien and Kobena Mercer). In Rowena Chapman and Jonathan Rutherford (eds.): *Male Order: Unwrapping Masculinity.* London: Lawrence and Wishart.
Julien, Isaac and Kobena Mercer (1988a). "Introduction—De Margin and De Centre." *Screen* 4(29):2–12.
Mercer, Kobena (1988a). "Recoding Narratives of Race and Nation." In Kobena Mercer (ed.) ICA Documents No. 7, *Black Film British Cinema* (1988). London: A BFI Production Special.
———. (1988b). "Diaspora Culture and the Dialogic Imagination: the Aesthetics of Black Independent Film in Britain." In Cham, Mbye B. and Claire Andrade-Watkins (eds.) *Blackframes: Critical Perspectives on Black Independent Cinema* (1988): 50–61. Cambridge, Mass./London: the MIT Press.
———. (1988c). "Aids, Racism and Homophobia" and "Territories of the Body," "Race, Sexual Politics and Black Masculinity: A Dossier (by Isaac Julien and Kobena Mercer). In Rowena Chapman and Jonathan Rutherford (eds.): *Male Order: Unwrapping Masculinity.* London: Lawrence & Wishart.

Metz, Christian (1985). *Psychoanalysis and Cinema.* (London: MacMillan Publishers Ltd.)
Mulvey, Laura (1988). "Visual Pleasure and Narrative Cinema." In Constance Penley (ed.) *Feminism and Film Theory.* London: BFI: 57–68. Originally published in *Screen* 3(16), Autumn 75.
Owens, Craig (1983). "The Discourse of Others: Feminism and Postmodernism." In Hal Foster (ed.) *The Anti-Aesthetic: Essays in Postmodern Culture* (1983: 57–82). Seattle: The Bay Press.
Pines, Jim (1986). "The *Passion Of Remembrance*: Background Interview with Sankofa." *Framework* 32/33:92–99.
Rampersad, Arnold (1988). *The Life of Langston Hughes* Vol. I & II. Oxford: Oxford University Press.
Rosen, Phil (1986). "Introduction: Text and Subject." In Phil Rosen (ed.) *Camera, Apparatus, Ideology.* New York: Columbia University Press.
Rushdie, Salman (1988). "Songs Doesn't Know the Score." In Kobena Mercer (ed.) (1988) *Black Film British Cinema.* London: BFI: 16.
Trinh, Min-ha T (1986/7). "Introduction," "The Inappropriate/d Other," *Discourse* 8:3.
Williamson, Judith (1988). "Two Kinds of Otherness—Black Film and the Avant-Garde." *Screen* 4(29):106–113.

The Cinema of Exile

Abid Med Hondo

The person who lives in exile, not from choice but by obligation, by absolute necessity, is certainly cut off physically from his family, ethnic and cultural roots. This enforced exile is dramatic, and in the long term threatens to produce sclerosis and acculturation. Each year I feel more deeply how dangerous is the exiled person's situation. However, things and people have to be put back into their context. I want to say that, as an actor and filmmaker, I do benefit from the relative privileges attached to the person of a known, self-taught intellectual. I live in France, I live after a fashion from the cinema, as I believe it is useful to make films, and I try to show my films—not just to the European public but also to the thousands of my brothers among whom I live, or rather, who simultaneously with me live in the situation of exile.

I maintain however that with *The Nigger-Arabs (Les Bicots-Nègres)* I have established a national cinema, even though conceived and put together outside my country. For, if exile remains as the worst thing, what is essential, in the heart of that worst, is to be conscious of what has to be struggled against. And what is vital for us, here and now, is surely to struggle against capitalism under its different aspects and its multiple powers. Even if the struggle is only a brush-fire, it is a fire which will spread. And we have to make sure the fire gets to the people who are in

From John D. H. Downing, ed., *Film & Politics in the Third World* (Brooklyn, New York: Autonomedia, 1987), pp. 69–76. Reprinted by permission of the publisher and editor. (First published in French in *Dictionnaire des Nouveaux Cinémas Arabes*, Claude Michel Cluny, ed. [Paris: Sinbad, 1978]. Translation by John D. H. Downing.)

the middle of the situation, who are suffering too, who are fighting against the same phenomena of domination without anyone hearing them. They are not heard, because they do not make films—or they do so, but which of our African brothers see their films? They are well-nourished at the roots, but what they create is confiscated: the distribution of African films is zero, and if you want to make sure it happens, as in Upper Volta,[1] for example, you run up against the monopolies' networks. Here in France I am obliged to seek out where my own people are, in the slums and the shantytowns, to work with them on a film which concerns us all; in Africa, where can films be seen which are made by Africans and concern Africans? In the present situation, whether I make a film in Paris or Nouakchott,[2] it is pretty obvious that it is a film my country will not see. Over the long term, this situation becomes a latent suicide. The cinema as I practice it cannot be independent of social and political data, since it is a reflexion, a questioning of those data. If I had the chance to shoot in Mauritania, I would make a film different from the first two, since in Mauritania too everything is different. Except that it would, once more, be a multinational film: because my country has multiple cultures and three or four languages are spoken there (I do mean languages, not dialects). The fact of there being ethnic groups and different languages in a nation complicates its realities. Undoubtedly, I think about that unconsciously; all the more so because I come from parents of diverse origins, and, what is more, slaves. I feel that inside myself as a fundamental reality.

Let us keep our diversity; let us be suspicious of the concept of universalism, which is a dangerous thing. I think we do not have to copy one another, whether amongst Africans or by continent. Above all, let us avoid copying the European and American cinema. We all have our specificity. Unfortunately, certain Africans are not always conscious of their culture and realities, but they are part of them, even unconsciously, the filmmakers and the public alike. There is a problem of audience receptivity which is basic and also specific. Thus the physical time of an Arab or African film is different from a western film. Is the film too long? No, it's a matter of another mode of breathing, of another manner of telling a story. We Africans live with time, while the Westerners are always running along behind it. Here you are under pressure to tell a story in ninety minutes. There are stories which cannot be told in ninety minutes. And speeding up the narrative—notably by montage—in the "western" film, evinces the displacement of the real rhythm of social being in relation to our cinemas.

I do not believe that the same work can be received, favorably or not, in an identical way, nor can it be readable, that is to say, understood, decoded, in Senegal as in London, in Egypt as in Rome or Paris. Peoples are only known through being translated, not by having a travesty made of them: a cinema with a universal vocation would be the latter.

But words must not be allowed to deceive us either. If it is true that a film exposing realities, dealing with a people's aspirations, is by that fact a political film, it is not automatically a "revolutionary" film. This is an important demystification! When people talk about political cinema, the drama is that confusion is already being compounded. It is not pointless to repeat that a political film is not by necessity, purely, a film which deals with subjects defined as political. What is more, a political film is not necessarily a "revolutionary" film. What is revolutionary film? A film unlike those already seen? A film calling for insurrection? Which incites revolution? I have never heard of people running to look for rifles at the cinema exit, to overthrow the government or to chase out the village mayor. Revolutionary cinema without revolution: I do not understand what that means.

Let us say more simply that a committed cinema can struggle courageously and stubbornly, and also with a constant wish by the film-makers to control their own discourse. You can say everything through film. But it is appropriate to know well to whom you can speak, and to whom you want to speak. To know (or not to pretend ignorance) that all cinema has a commitment and then to say "We are not involved in politics" is only a lie and dishonesty: flight into a dreamworld, silence on everything troublesome, an evasion which gives a clear field to the forces of stagnation and subjugation. It is a political game because it works to the advantage of the existing capitalist structures—the only merit of reactionary cinema is that it can be easily recognized! On condition we do not allow ourselves to be seduced. Formally speaking, in plastic art terms, a photo, a speech, can be "revolutionary": but what do these formal "revolutions" serve? It is a question which needs asking.[3] Maybe to give the illusion of a combative cinema? And from there to create a revolutionary dynamic. . . . An illusion, to which we must add a widespread but false idea: "the public doesn't care for progressive films." A commonly accepted and maintained idea. It is a convenient pretext that you cannot suppress the public's alienation. In France, film-makers are seen willingly lining up behind this "screen" and, while still asserting leftwing ideas (elsewhere, at the dinner table), they put together a conventional and clearly conformist cinema (and thus, a reactionary cinema).

So we return to the necessity of knowing what we want to say, and to whom we have to say it. For what public has learned to read, to decode a film? An elite public. But there are other publics. Film criticism does not play its role, or rather, it plays it too well. The handful of critics we know whom I will qualify as "progressive" must then fight in place of all the others. They have no right not to be present, they must reject demagogy, paternalism, quasi-journalism. For if they desert, what remains? Criticism as practiced in the columns of the rightwing press does not interest me.

My relations with progressive critics have never been negative. I must say it is thanks above all to the western press, especially the French press, that the films which have been seen have been available, and that Africans have been informed about them. For sure, with some inadequacies on some people's part, but without undue paternalism. Criticism's influence on the conscious public, on the distributors, is an essential and often decisive support. It is very encouraging and positive that our films are taken into consideration, that they are dealt with on an equal footing, and so with the same rigor as all the others. Sembene, Tawfiq Salah, I myself and many others have been put into the festivals and some theaters thanks to some critics, whose initial battles were sometimes with their own editors.

For me, the country where for over ten years film criticism has never yielded up its responsibilities, is France. I don't forget that *Soleil O*[4] came out in a 64-seat theater because the critics fought to find a screening space when so many owners were indifferent or suspicious. Today, when an American distributor or journalist wants African films (a rare occurrence), he telephones or writes to a French journalist: it is significant, all the same.

That said, an African film criticism is indispensable. At the present time, very few critics can express themselves in our countries, and they only have a very relative power (to inform, that is) on their national, even local level. The lack of a film criticism is not Africa's special privilege. But it is a historical given that western film criticism is, today, the only one capable of reviewing our attempts; of informing the public and helping us; of studying and reflecting on African and Arab cinema. We, as African film-makers, must ourselves invent, on our own, the film language to be spoken to be able to be understood, one day, by our brothers. You are witnesses. On this account, you must not make out that our actions are in accordance with our ideas when it is not true. I mean that one does not have the right when defending progressive intentions, when one is a creative artist, a theorist, a critic, to produce or defend a consumer cinema. The public has to be awakened, or re-awakened. That demands courage. A leftwing (or so-styled) film-maker or critic, then, is only doing his duty: we don't have to award ourselves "medals."

Since *Soleil O*, I have been trying to put into practice my own special bent, and to deepen it. I am not an enemy of a simple language to convey interesting ideas. I also believe, honestly, that to relate History in its complexity, in its contradictions, in order to approach an event, sometimes you have to move beyond the first level of simplicity and obviousness, for the risk at such a point is then a dangerous Manicheanism.[5] The more deeply you go into things, the more complex the analysis. The opposition of content and form is meaningless, the theme which is chosen determines form and conditions it; the public addressed—(a milieu, a country, a period, nothing is separated, everything is bound up together)—expects the language to be understandable, which does not mean conventional. I am not an enemy of aesthetic refinement if it is integrated into a context, if it is based on something—though present-day French cinema seems to me to be adrift, bereft of any driving force, stricken with chronic mimesis. The films I have done have been produced in a given milieu, at a precise period. Were I to make a film in Mauritania tomorrow, my film language would not be the same. I certainly would not make an oversimplified film, but it would be different, with less baggage attached. Maybe I would use video. And if I discovered that cinema is still a useless activity in Mauritania, then I would do something else.

When I showed *The Nigger-Arabs (Les Bicots-Nègres)* to Mauritanians of different ethnic groups, who were immigrant workers here, I observed that they really took to the film. In their exile context, intellectual or manual workers, they reacted positively. Perhaps that was due to the fact that I did not agree to censor myself—any more than I would have done in my own country—in order to find a different level of interpretation. I did not wish to think in other people's stead. And if I am shooting in Mauritania, I would respect that as a fundamental principle, with even more vigilance: I would try to work on the film with them, in common, without putting them down by only granting them secondary status, that of a row of objects under analysis. . . .

The immigrants' strong appreciation of the film is no doubt also linked to the fact that I simply began from my own situation to pose the key questions: Why independence? To do what? Why exile yourself? Why the cinema? I then overlapped various aspects of immigration, defining (by letting the migrants define for themselves) their relations to everyday life—right up to the final utopia showing the European economy paralyzed—as has happened at regular intervals, but in short random bursts, at Renault[6] and elsewhere, for example—on the day when the immigrant workers all stop and leave for home again.

I also wanted to show that these workers aren't eating anyone else's food, and that they hardly get what is theirs by right. And to show how they live, and what their problems are, their difficulties, their contradictions, all of them things that European workers know but poorly.

Contrary to the method I used while shooting *Soleil O*, I asked a certain number of these immigrant workers if they would agree to collaborate, to participate. This was not always easy. We talked, organized gatherings. Their confidence was necessary, and it could only be true confidence if I told them to start with that I was making a film whose purpose was neither my nor their pleasure. That when a question was asked, it would be necessary to try to answer it. Once their confidence was won, I then had to navigate between a series of reefs, the first being . . . a sort of hyper-realism, which would have pushed me into spectacle and demagogy. My concern then was to avoid all revolutionary mysticism, all embroidery or prettification, where what was existing was filth and disease. I chained myself to the rigor of the image: not to let solidly established facts slip out of view simply for the benefit of a stylistic effect. But I hope at least to have written a well structured and readable film.

The second difficulty was bound up with the very nature of the method adopted. Even if I did not modify the entire structure of the film overall, I was led to change elements of the screenplay on several occasions. For example, I was shooting a scene with intellectual and manual migrant workers—on the level of the film, they were all in it together, I am no "workerist"[7] and I am suspicious of categorizations. After developing the film, I showed the rushes to the participants and we discussed them. As a result of new or complementary elements emerging from this examination, I used to shoot unforeseen takes. I believe in the effectiveness of this approach. Unfortunately, the conditions of work and the cost of film production are a limit on such experiments.

The approach in *Soleil O* had been constructed from a very elaborate script, and improvisations had remained limited and always under constraint. For *Les Bicots-Nègres*, it was appropriate from the beginning to define clearly the script's central points, the cause/effect/cause relations, knowing that in the open framework of a sequence, the actor-interpreters often drift "at will" far from the departure point, and not necessarily along the foreseen route. Shooting becomes an endless argument, complex, a passionate nightmare which is reinstated during the editing process. What do you choose, when you have the feeling that everything that is shown and spoken is essential, or important? It is a complex alchemy, which demands a lot of time, distance and reasonableness toward the people you are working with. It involves practically a second shooting parallel to the editing; which is enriching, and in my view fresher and truer, but much more risky than following a precise script.[8]

Whatever the method adopted, I believe no image, no dialogue, no linkage, should be decided once and for all . . . It is good to leave a portion to objective chance, which can enrich the purpose and the intention. I must say that this practice was only in force to a relative extent: sheer time was lacking to explore deeply the possibilities in such a method. And then, filming is a costly discourse, very costly—above all when you are shooting on a low budget, where production stretches out over a very long period: a year and a half for shooting *Soleil O*; three and a half years for *Les Bicots-Nègres!* The average shooting time for a "normal" film takes between eight and twelve weeks . . . I did not have the choice: it was doing it like that, or doing nothing.[9]

Notes

1. This nation changed its name to Burkina Faso in 1984. It continues to host the major continental festival of African films—FESPACO—every two years.

2. Nouakchott is the capital of Mauritania.

3. Hondo's remarks here are directed against the confusion in public debate on cinema in France during the 1960s and 1970s, between revolutionary styles in film-making, and the contribution to socialist revolution made by films. Often the one was assumed, by some ill-defined process, to lead to the second.

4. The title means *Sun Oh*. The film deals with the black experience, both in slavery and in labor migration.

5. This term denotes a Christian heresy flourishing in North Africa between the 4th and 5th centuries. It claimed the universe to be in thrall to two equally powerful forces, one of good, the other of evil. The term is commonly used now to indicate a very absolutist perspective on social reality, admitting of no possible shades of gray or levels of uncertainty.

6. The state-owned auto firm in France, employing many immigrant workers, and a constant center of labor unrest, given even more political significance because of its government ownership.

7. "Workerist" in Marxist political discourse denotes the view that authentic revolutionary change can only emerge from factory workers whose labor produces direct profits for the capitalist class, and thus that other classes, such as farmers, or social groupings such as professionals or students, have at best an auxiliary role to play in the revolutionary process.

8. It is well worth comparing Hondo's remarks on the experience of directing with those of Beloufa later in this volume.

9. And seven years for *West Indies,* as Mpoyi-Buatu indicates in the last chapter.

Producing African Cinema in Paris

An Interview with Andrée Daventure

Mark A. Reid

Andrée Daventure did extensive editing work in French film and television before she founded her own company, Atria. In 1953, she became an assistant film editor for a series of documentary shorts, including *La Destruction et Reconstruction de Varsovie*, a series on the Marshall Plan in 1955, a film on the United Nations, and several other educational films. By 1957, she edited features, including Louis Saslovsky's *1er Mai* (1957) with Yves Montand, Raoul André's *Clara et les Méchants* (1958) and his *Un Homme à Vendre* in 1959. Between 1959 and 1966, she worked as the assistant editor on twenty-one feature films, including C. B. Aubert's study on interracial intimacy, *Les Laches Vivent D-Espoir* (1960), Yves Allégret's adaptation of Zola's *Germinal* (1963), and René Clément's pseudo-documentary, *Paris Brule-t-Il? [Is Paris Burning?]* (1966) which she also helped to research.

In 1967, she became a chief editor on television and film productions. Several year later, she began editing films made by African filmmakers. Daventure and the French Ministry of Foreign Relations (which helps to finance Francophone African cinema, as the Ministère de las Coopération had done before 1981) have been the two most important *European* promoters of Francophone African cinema.

It was after discussions with Senegalese filmmaker Safi Faye, the first African woman filmmaker, that I asked Andrée Daventure for an interview. Daventure offered a wealth of information about the cross-cultural problems which a white French woman encounters when editing

Jump Cut, no. 36, 1991, pp. 47–51. Reprinted by permission of the author and publisher.

African cinema. She also described the production, distribution and financial relations many Francophone African filmmakers have with Atria and/or the French Ministry of Foreign Relations. These ties and problems are slightly comparable to those faced by U.S. independent video and film artists who depend on such organizations as the American Film Institute, the U.S. National Endowment for the Arts, and the U.S. Public Broadcasting Corporation.

Daventure assisted editing the following African films:

(from Mali) Alkaly Kaba's *Wallanda* (1974), Sega Coulibaly's *Le Destin* (1976) and *Kasso Den* (1979);
Souleymane Cissé's *Den Moussa* (1975), *Baara* (1978), *Finye* (1981) and *Yeelen* (which was being edited at the time of this interview);
(Cameroon) Dikongue Pipa's *Muna Moto* (1975), *Le Prix De La Liberté* (1978), *Grand Stade* (1981) and *Courte Maladie* (1984);
(Senegal) Safi Faye's *Lettre Paysanne* (1975), *Fad Jal* (1979) and *As Women See It* (a short, 1982), Ben Diogaye Beye's *Un Homme, Des Femmes* (1980);
(from Niger) Oumarou Ganda's *L'Exilé* (1980); (Burkina Faso) Sanou Kollo's *Paweogo* (1982), Gaston Kaboré's *Wend Kuuni* (1982), and Paul Zoumbara's *Jours de Lourmente* (1983);
(from Tunisia) Férid Boughedir's *Caméra d'Afrique* (1983);
(from Ivory Coast) Kitia Toure's *Comédie Exotique* (1984).

Daventure has promoted African cinema well beyond her role as editor. She has helped distribute many of these films and offers internships to some of Africa's most promising young directors. One of her interns, for example, was Burkina Faso's Idrissa Ouedraogo, who has consecutively won international acclaim for his dramatic films which visually explore urban problems. She has also been a close friend and mentor to Kitia Toure, an Ivory Coast novelist and film director.

Editing Across Cultures: Problematics and Consciousness

Mark Reid: How did you get your start in cinema?
Andrée Daventure: In 1952, I worked six months in a film lab. Then I did several internships within the traditional French film industry.
MR: But, how did you become the editor of African-directed films?
AD: One day I was thinking about Africa. I really wanted to understand what African cinema was—what it meant in Africa. I believe that people who work in film should try to understand all filmmaking styles so as to really understand cinema itself. In 1974, I looked for work with the Ministry of Cooperation [which dealt with French-African international

relations] in the African film section. I wanted to broaden my knowledge of editing in particular and of film in general.

A man asked me if I would edit a commercial for a Francophone festival. As I was editing that short, I saw some African films. On a modest little editing table, I saw Ousmane Sembene's *Borom Sarret* and found myself crying. I discovered that with all my education and exposure to film, I had not known about this important cinema. When I met with the man who had given me this editing job, I asked him to give me African films to edit. Initially my work was only fair because I recognized that films made by Africans have different film styles and pacing than the European films I had formerly edited. I knew I couldn't work in a European editing mode.

MR: Could you elaborate on this theme of cross-cultural film production and the problems which a European encounters when working on African cinema?

AD: It is very interesting to confront my own "savoir faire," which represents my personal and cultural education in France. After I have organized this perspective mentally, I listen to the African director, with whom I have a discussion each editing session.

MR: But do you understand the film's dialogue?

AD: No. That's why I pay close attention to the director, the musical soundtrack, and the characters. French editors commonly edit according to dialogue or the impact of a word, but when I edit African films, I cannot do that. So, I work according to the rhythm of the dialogue. I haven't yet been mistaken.

MR: Do you find any difference between the sense of space in the African and French films you've worked on?

AD: Sure, there's a difference. For example, I usually record eye contact. Yet in a particular African milieu, people often cannot look directly at another person.

MR: You mean that two characters are forbidden by cultural convention to exchange looks?

AD: I had not known about this taboo, but when I learned about it, a lot of my thinking and actions changed. For example, when I edited Kitia Toure's *Comédie Exotique*, I tried to convince Kitia to accept a certain scene, but he rejected it on the basis of his customs.

MR: It is important then to know the customs and to respect them by translating them into the film.

AD: But I have made errors because my culture is within me. Some things escape my sensibility when I edit African films even though I successfully edit French films without a director's guidance. I would never edit an African film without the director reviewing and discussing my job.

MR: So you and the director edit the film together?
AD: Always. It is better if the director has screened the rushes with me so we can discuss the film before the editing has been structured.

African Film Production, French Ministry of Foreign Relations, and Atriascope

MR: I would like to move from the issue of cultural aesthetics to another fundamental issue which helps to define African film—the financing and promotion of these films. What is the relation between Francophone African film production, the French Foreign Ministry, and your company, Atriascope?
AD: These are separate issues. There are the relations between Atriascope and filmmakers in general; relations between Atriascope and the French Foreign Ministry; and relations between African filmmakers and the Foreign Ministry. For example, I edited *Wend Kuuni*, which was a Burkina Faso production; this entailed a relation between Burkina Faso and Atriascope without any assistance from French governmental agencies. When I work with French governmental agencies, it is the Foreign Ministry which gives financial support while Atriascope deals with production management and film budgeting. But African filmmakers can bypass Atriascope and choose to work directly with the Foreign Ministry.
MR: I believe that Atriascope offers film production internships?
AD: We do. There's Atriascope and Atria. Atria is an association which plans and manages film budgets, disseminates publicity, markets film, and coordinates film festivals. Atriascope deals with the technical side of film production. The two activities complement each other. We try to offer editing internships, and I am glad when young African filmmakers choose to come here. We've had a Burkina Faso intern who worked on a Malian film and a Malian who worked on an Ivory Coast film. African interns are able to work on many different films at Atriascope. But this range of experience for interns is usually impossible in Africa because most African nations don't produce large enough numbers of films to allow an intern to gain post-production experience.
MR: How do African interns support themselves in Paris?
AD: Some receive scholarships and some support themselves by doing odd jobs so they can learn the filmmaking trade. I try to accept everyone regardless of whether they have a scholarship or not. This was how Kitia Toure became an intern here.
KITIA TOURE (who has just arrived): And I am continuing my internship.

AD: One day, Kitia's professor asked me to meet a student. This is how I met Kitia, who later worked on Gaston Kaboré's *Wend Kuuni*. Then, Kitia filmed and edited his own *Comédie Exotique*.
MR: Idrissa Ouedraogo told me that you are now working on his new feature *Le Choix*?[1]
AD: We'll deal with just a small part of this film's production because it is being financed by the government of Burkina Faso, not the French Ministry. A part of that film's budget is directly managed by Burkina Faso. Atria receives Idrissa's rushes, screens them at the lab, and reports any technical problems. We also send him more film stock, verify that the material has been sent out, and that French technicians were paid (because any French crew members are governed by French laws).
MR: Idrissa must pay according to a French pay scale if he hires a French production crew. In addition, he must send his footage to a French lab to be developed since there are no professional laboratories in West Africa. Does Atria do the same for Souleymane Cisse?
AD: Yes, we have edited the first part of Cisse's *Yeelen* and the French technicians have been paid. His film budget was financed by the Ministry of Foreign Relations, which received permission from the Centre National de la Cinématographie.

[Translator's note: Le Centre National de La Cinématographie—National Cinematography Center—is a governmental agency that partially funds and helps to distribute many French films. Before any French governmental agency, like the Ministry of Foreign Relations, is allowed to financially back a film, the CNC must approve the budget, then a French bank will underwrite the amount that the CNC has approved. Also see *Informations*, No. 209 (May–June, 1986).]
MR: What is the Ministère de la Coopération?
AD: It was the French Ministry that dealt with French and African relations. In 1981, when the Left came to power, these relations passed to the Ministry of Foreign Relations. Now that there is a conservative government in office, African-French relations have returned to the aegis of the Minstère de la Coopération. After five years of such reshuffling, African filmmakers have not received equal status with other international filmmakers. Now, African filmmakers must deal with still another ministry, another political administration, and deal with that group's particular cultural politics.
MR: What will become of the ties between the Ministry of Foreign Relations, Atria, and young African filmmakers who wish to do internships, now that the Right has taken power?
AD: I think that beginning today, we can no longer speak of ties to the Ministry of Foreign Relations because another ministry will deal with African cinema. I haven't formulated my feelings about the future

relations between Atria and the new French leadership. But Atria will continue to exist as a center for film production. You must understand that we started as part of the Ministère de la Coopération around 1974. I began working in a government-sponsored film production section.

Later, the government disbanded this section, and we founded Atria to continue to work with African filmmakers; at that time, we were supported by our friends in the ministry. Now, we do not have the same relations, but we have continued the work begun by the original director of the Ministère de la Coopération. We have many problems and little means but we remain optimistic. I hope that we continue and our problems with the government end, because the difficulties have become a heavy burden. I think that the African filmmakers really need us and hope we continue to exist.

Panafrican Hopes and Separate Realities: Francophone West Indian and African Filmmakers

MR: Have you ever worked with Francophone West Indian [Antillian] filmmakers?
AD: No. There isn't that much production by Antillian filmmakers.
KT: Since Antillian filmmakers are considered French, their work goes under the rubric of French film production.
MR: You mean to tell me that Julius Amede Laou is considered a French filmmaker yet his work specifically speaks from a black West Indian point of view?[2]
KT: Julius Amede Laou is considered a French filmmaker here, as is Euzhan Palcy. Her film *Rue Case Nègrès [Sugar Cane Alley]* was a French production. Since she's black, she just falls into the category of black African cinema.
MR: Why doesn't the French government recognize that black West Indian and African filmmakers have similar interests and a similar colonial history? The French establish these categories without acknowledging the historical link between these black people. One group is considered a "French" nation while the other is a "former" French colony.
AD: You can't artificially make connections. If an African meets a West Indian and they decide to work together, great.
MR: Does this happen?
AD: There's little chance. In France, a French director chooses a crew. You don't say to a filmmaker, "Would you work with such and such person?" Some filmmakers wouldn't ever change their technical crew. It is not because of skin color that a filmmaker makes such a choice. It happens because of the affinities, the meetings. There's a big difference

between a Cameroonian and an Ivorian and ditto for a Burkinian. Africa is not singular, yet sometimes you can find affinities between people from different countries. Even within the same culture you may have problems finding people who share your same sensibilities.

African Women Filmmakers

MR: You have worked with several African filmmakers but Africa has few women filmmakers. Will you comment on this?

AD: I do not ask if the person is black, white, male or female. There are not that many African filmmakers. I find great pleasure in working with women, more than with men. This comes from the rapport that two women have between them. I have enjoyed working on two films with Safi Faye but we still face that problem of cultural sensibility.

MR: Here, at Atria, you have a predominantly female staff. Does this give the rapport that you have been speaking about?

AD: Many women work at Atria because France has more female editors than male editors. It's a coincidence. Many men work as production directors, camera persons, sound technicians and typesetters.

MR: Does Atriascope solely assist feature-length productions?

AD: It's been more than a year that we've worked with shorts. For example, David Diop, Jr. finished *Poete de L'Amour* here, a short film about his father, the Senegalese poet David Diop. This young filmmaker had attended a French film school, yet the school had not taught him the technical aspects of post-production. He did an internship here at Atriascope. Meanwhile Atria developed his press packet, applied for financial assistance so David could finish his film, and managed his post-production budget.

MR: What is the title of your next film, Kitia?

KT: *Des Chiens Et Des Hommes [Dogs And Men]*. I have two projects. One is a series of shorts, which is in preparation. The other project is the feature-length film, for which I have already written the script and prepared a production budget [*Les 10 Commandments de L'Enfant*, 1989].

MR: What are your last words to an audience who is interested in African cinema?

AD: It's not a last word. Africans make their own cinema. Atria is a bureau of coordination. It is not an office which makes African cinema. Atria is not responsible for African politics. Due to the international economic situation, Africa does not have a film industry. Therefore, it is other countries' and especially France's responsibility, because of our historical relationship with Africa, to support financially the development of African filmmakers.[3]

Notes

1. Idrissa Ouedraogo's *Le Choix* won the Prix du 7e Art, the Prix de la Ville de Ouaga, the Caméra d'Or and the Meilleure Musique (Best Music) awards at the 1987 Festival PanAfricain du Cinéma.

2. For an appreciation of the black PanAfrican sensibilities of French West Indians, see my article on West Indian filmmaker and dramatist, Julius Amede Laou, and my interview with filmmaker Willie Rameau in *Black Film Review* 3.1 (1987), 8–11, 24. Reprinted in *Cinema in Caribbean Society and Culture*, ed. Mybe Cham (Trenton NJ: Africa World Press, 1991).

3. For a more critical analysis of the economic disparity between African filmmaking and that of the United States, Western Europe, the Soviet Union and Japan, see my interview with Mauritanian filmmaker Med Hondo: "An Interview With Med Hondo," *Jump Cut* 31 (March, 1986), pp. 48–49.

United States

[Black Film as Genre] Definitions

Thomas Cripps

> The agenda is set by the formula.
>
> *Erik Barnouw, in conversation*

For the purposes of this study, "black film" may be defined as those motion pictures made for theater distribution that have a black producer, director, and writer, or black performers; that speak to black audiences or, incidentally, to white audiences possessed of preternatural curiosity, attentiveness, or sensibility toward racial matters; and that emerge from self-conscious intentions, whether artistic or political, to illuminate the Afro-American experience. In the latter part of this century, this definition might be expanded to include major motion pictures and other projects made for television, as well as films that, despite foreign origins in, say, Africa, speak to Afro-American concerns.

If we were to bring this definition to a fine pinpoint, we should argue forever over who has the right to dance on the head of the pin. The Lincoln Motion Picture Company, founded in 1916, always used a white cameraman; Oscar Micheaux, especially after depression-induced bankruptcy, accepted "white" financing; *Variety* frequently evaluated so-called race movies, thereby possibly influencing their makers. Thus almost every black film, from production through distribution, was affected by whites.

Thomas Cripps, *Black Film as Genre* (Bloomington, Indiana University Press, 1979, pp. 3–12. Reprinted by permission of the publisher.

Black film taken in its narrowest sense then consists of only a tiny body of work seen by a coterie of black moviegoers, then consigned to an early death in dusty storerooms, not to be seen again until brought to light in "white" repositories like the Library of Congress.

By this standard the best single example of black film seen as pure product—produced by blacks, for blacks and with an ambition to advance the image or the cause of the race—is the fragmented evangelistic film of Eloise Gist, the traveling black preacher. She ranged over the South during the Great Depression, spreading her revivalist faith through motion pictures shot only for the specific narrow purpose defined by her own faith and spirit. Nowhere from script to screen did any white hand intrude, or any white eye observe. Neither white financing in the beginning nor white appreciation at the end affected her pristine black fundamentalism. Her films were naive, technically primitive, literal depictions of black Southern religious folklore that brought faith to life, much as an illuminated manuscript gave visual life to Christian lore in the Middle Ages.

But the entire body of such black film could be seen in a single day's session at a Steenbeck viewer. Our definition of black film must necessarily be broader so as to include the work of those self-conscious black artists who were at least as interested in the beauty of the medium as in the effectiveness of the message; the black filmmaker whose work emerged from the conventional channels of production that were lined with white money, advice, and control, even down to "final cut" approval; and finally, though rarely, film produced by white filmmakers whose work attracted the attention, if not always the unconditional praise, of black moviegoers and critics.

A broader compass also allows us to avoid the trap of claiming too much black control of certain films. In recent years many critics, inspired by the heady atmosphere of pursuit and discovery of old and presumably lost black film, have made exaggerated, unwarranted claims of certain instances of monolithic black control of the filmmaking process.

In the late 1920s, for example, the Colored Players Company of Philadelphia turned out a reputedly good film version of a famous black novel; a curiously tragic black revival of the old temperance tract, *Ten Nights in a Bar Room* (1926), starring the eminent black actor, Charles Gilpin, in his only screen appearance; and, most significant, *The Scar of Shame* (1927), an evocative, sometimes delicate drama of color caste distinctions within Afro-American circles. The recent discovery of this last film resulted in a round of showings to both black and white students, and a rash of essays saluting its presumably black originators, the creative force who looked over the shoulders of the obviously Italian technicians cited in the credits. And yet, a few moments of cursory research revealed that from top to bottom, the Colored Players were actually white, save for their front man, Sherman "Uncle Dud" Dudley, an old black vaudevillian who had dreams of a black Hollywood on the outskirts of Washington.

Perhaps because of the hazards inherent in drawing fine distinctions, black critics have avoided the task of constructing a black cinema aesthetic, at least until recently. Addison Gayle's otherwise admirable *The Black Aesthetic* (1971), for example, does not include a single sample of cinema comment. The little criticism that appears often splinters into two camps differing in focus: the one literary, the other political. The opposing sides in this argument either demand that art be a weapon against racism, or feel that art is neither bullet nor ballot. In any case, the debate has done little to help define the outlines of a black aesthetic or of black genre film.

At the center of this controversy is the "twoness" of American racial life. The term was coined by W. E. B. DuBois, the premier black intellectual of the twentieth century. It describes the anomaly of American racial arrangements, which segregate black from white, discriminate along racial lines, and yet oblige Afro-Americans to assimilate the values of white America. If films reflect the belief and behavior systems of society, we must expect that they will express these realities.

Thus, given the persistence of American racial codes, it does not seem possible, except in a unique case like Eloise Gist's film, that either a black or white filmmaker could produce a film that in some way did not suffer alteration of tone, plot, theme, pace, or character, and even benefitted from, the occasional interracial collaboration. For example, the presumably black *The Scar of Shame* was produced by a film crew that was largely white. Two years later, in 1929, King Vidor's MGM opus, *Hallelujah!*, an appreciative styling of black rural folk religion, profited from the advice of Harold Garrison, a black crew member, as well as from a panel of black Southern preachers, who gave counsel while the company shot on location along the Mississippi. The veteran black actor, Clarence Muse, performed a similar function on the set of Fox's *Hearts in Dixie* (1929) and in other Hollywood films, one of which he and Langston Hughes wrote for Sol Lesser.

But it was in the area of financing where blacks and whites really came together, even outside the realm of Hollywood. Even the most independent producers of race movies—those films made for exclusively black audiences between 1916 and 1956—relied on white sources of capital, distributors, bookers, and exhibitors. Life on the movie set was not significantly different from American life at large. Whites bossed and blacks labored, with only a little bargaining room between them.

After the film was in the can, blacks and whites still shared power over the fate of the product. From the earliest days of filmmaking at the beginning of the century, censors' scissors shaped film imagery and themes. But the Negro minister who sat on Chicago's board of censors was a rarity, so over the years, blacks tended to accommodate the wishes of white censors. In 1924 a black independent filmmaker, Oscar Micheaux, reshaped his *Body and Soul* to suit the New York state film censor. As late as 1970, Melvin Van Peebles, the most celebrated of recent black filmmakers, was outraged by the censorious and delimiting "X" rating given his *Sweet Sweetback's Baadasssss Song* by the white Motion Picture Association of America, a stigma that denied him access to a prospective black adolescent audience. Also by the 1970s, much documentary film was conditioned by its sponsorship and support by the public television network, with its charter obligation to reach all segments of the national audience. Moreover, "white" film sometimes took on a black hue as a result of favorable black responses, such as Congressman Oscar DePriest's endorsement of *Hallelujah!* in 1929 and James P. Murray's praise of Michael Roemer's *Nothing But a Man* (1964) in *Black Creation*, a magazine of Afro-American arts.

A glance at American racial history reinforces this broadened view of black cinema. In the past black creativity has been at its most clearly "black" in those endeavors *into which* blacks were most segregated from white influence—work songs, gospel songs, spirituals, and theological rhetoric. Even in these accomplishments, however, an occasional white European echo may be heard. But if James Agee is correct when he insists that black art is at its most tainted, and even corrupt, when it is exposed to white praise, we must also see that blacks in racially integrated circumstances have been equally creative. Their work, a syncretism of Africa and Europe, is given shape, substance, and meaning by both traditions. Thus blacks in the arts—architecture, symphonic music, nonrepresentational painting, and even fiction—work in European forms and conventions while preserving trace elements of Afro-American culture.

The game of basketball is an illustrative case of differing racial styles of expression within a system of formal rules and conventions. A form of competitive choreography based on rules laid out at a white YMCA college in Massachusetts, the game is leavened by contrasting styles of white and black play, the former deliberate and long-range, the latter, shaped by childhood training in constricted playgrounds, fast-paced and close-range. But both are basketball.

If black intellectuals ever hoped to break from the constraints of American racial segregation, then Hollywood liberalism with its sentimental faith in progress, goodness, and individual worth provided a strong incentive that kept many blacks pressing toward eventual racial integration. Nevertheless, this same racial liberalism held out unrealistic hopes that diverted black filmmakers away from racial independence, sometimes by enabling them to join the major studios, sometimes by imitating Hollywood genres. Particularly after 1920 as Hollywood was becoming a world cinema capital, black screen roles increased in quantity and sometimes quality.

Year after year *Uncle Tom's Cabin* (1927), *Hallelujah!*, *Hearts in Dixie*, *The Green Pastures* (1936), *Slave Ship* (1937), *So Red the Rose* (1935), *Beggars of Life* (1928), *Crash Dive* (1943), *Sahara* (1943), *Bataan* (1943), *Pinky* (1949), *Stormy Weather* (1943), *Cabin in the Sky* (1943), *Raisin in the Sun* (1961), *Lilies of the Field* (1963), *The Defiant Ones* (1958), and *Sounder* (1972) promised an ever hopeful future. Much like a happy ending, progress at least seemed possible on the screen if not in real life. Unfortunately, few pictures left Afro-Americans with a completely satisfying portrait of American life, for even so-called problem pictures and message movies of the 1940s did not offer an agenda for eventual social change. Nevertheless, as the films reinforced hope, they diverted black attention away from the goal of an independent black cinema.

It is this dual aspect of movies that forces us to view film through some critical prism that takes into account the interracial teamwork that goes into filmmaking. The team is rarely all black, and even when blacks predominate, they often come from different economic or regional backgrounds. Black film must be seen as a genre, then, for what it says and how it is said, rather than who is saying it.

Like the French semiological critics who borrow from the science of structural linguistics, we shall seek to define black genre film through social and anthropological rather than aesthetic factors. In this light, films are different from those fine arts in which the artist and his audience share a fund of common knowledge and experience. Rather, films bridge the gap between producer and mass audience, not through shared arcane tastes, but because a team of filmmakers shares a knowledge of genre formulas, more than an artistic tradition, with its audience.

Furthermore, genre films, like folktales and tribal lore, may transmit social meanings beyond the conscious intention of the filmmaker, as well as meaning brought by the audiences' own social and cultural history. Moreover, the likelihood of attracting a mass audience is further assured because such films emerged from a history that followed "the novel's way" of telling popular and easily followed stories.

In this popular sense, a shot is a sentence in a tale, as well as a value-laden poetic image. The shot-as-visual-sentence is at the heart of genre film. Shots, like sentences and unlike words, are infinite in number, and therefore unlimited in what they convey and how they are perceived by the audience. The filmmaker composes the shot, constructs the space and the figures in it (and their size), and manipulates these elements as symbols for an audience whose own cultural conditioning limits and directs perceptions.

Thus filmmakers and audiences share a few intensely powerful symbols set in an easily followed narrative form that defines the genre. In many gangster films, the action necessarily open on wet streets and darkened alleys with the hero, muffled in a trenchcoat, alone in the shadows, set apart from both cops and crooks, upholding a personal code that neither understands. In skilled hands, such repeated, codified images take on meanings larger than themselves and become powerful icons that need few words to explain them. Indeed, they become unintelligible only when the images are cluttered and the ambience is broken by extraneous materials.

When images are repeated and codified into a formula which is presented as a narrative, the resulting genre film permits instant communication between maker and audience. In the case of black film, the basic formula, by emphasizing one or another of its parts, permits expansion into subgenres that are variations on the basic traits. The viewer may see the black genre exemplified in social drama, cautionary tales, musicals, documentaries, religious tracts, and romances featuring both urbane and pastoral heroes.

Nevertheless, the subgenres share a common fund of integrated caste marks that identify the larger genre. Black genre films emerge from a segregated point of view, even when treating "white" themes, and rely on an appropriate repertoire of symbols. These might include a *folk idiom* such as black religion, an urbane *jive idiom* evolved from the lives of jazzmen, or the aloof mask of behavior that might be called *aesthetique du cool*. These idioms then form the perimeters of black social manners expressed both externally, to whites, and internally. Like so many black experiences that receive too much white attention, these modes may become less genuine as they become the subjects of show business routines, in the same way that some gospel music becomes flashy and secular removed from the church and hyped for the stage. The symbolic content of black genre film is given moral urgency by a *tone of advocacy* rather than, say, a reportorial style. Because genre film conveys shared experiences, some of which may not have touched all of the audience with equal impact, the films may employ the literary device of *anatomy* as a means of understanding the whole of black life through the depiction of its parts. Genre film acts as a *ritual* celebrating a *myth*, that is, a value-impregnated tale that is truer than mere truth. Movies made for black genre audiences may range from black versions of Horatio Alger's success stories to Southern Baptist fundamentalism. Finally, the black genre rests on *heroic figures*, either urban or pastoral, each reflecting a different focus of black experience. The urban hero, for example, corresponds roughly to the romantic white westerner, greater than his adversaries only in degree, never mythically greater in quality.

A revenge motive is often at the core of his being. He either struggles against it, redeemed and improved by the experience, or he surrenders to it, thereby avenging the pain of history, but at the price of lost innocence.

Although whites may see every black hero as a picaroon, a separation between urban and rural types is clear. The hero's response to the opportunity for revenge often marks him either as a pastoral or a picaresque hero. Much like the western loner, the pastoral hero stands apart from society, secure in his own identity and values; unlike the white westerner, however, he uses the family as his anchor. He is in Northrop Frey's "low mimetic mode" of heroism, superior to neither milieu nor men. He wins not by prevailing but by enduring. The urban picaresque hero, on the other hand, is alone, moved to vengeance, prone to violence modulated only by Shaft's cool professionalism or Sweetback's hope for eventual revolution. Not since Stepin Fetchit has there been a truly ironic black hero, inferior to both men and milieu, but nevertheless surviving by his wits, like Bre'r Rabbit, the African trickster removed to southern America.

Genre film can easily become exploitation film, through which tastes are teased but no deeper needs are met. While genre film tends to treat things as they are and avoids the trap of advocating them, exploitation film sensationalizes them. Genre film merely speaks from a segregated point of view; exploitation film prefers it. Black genre film ritualizes the myth of winning; exploitation film, at its worst, merely celebrates and dramatizes revenge as though it were a form of winning.

Black genre film celebrates *aesthetique du cool*, the outward detachment, composed choreographic strides, and self-possessed, enigmatic mask over inner urgency that have been admired in both Africa and Afro-America. In contrast, so-called "blaxploitation" film trumps *aesthetique du cool* into mere sneering and bravado. The black genre chooses hyperbole as a mode of celebrating the combination of triumph over adversity, fellow feeling, and moral superiority of the oppressed, known most recently as "soul"; "blaxploitation" film only bleats in shrill imitation. The anatomy of black life in black genre film is an instrument of communication *to* the group *by* the group, exemplified by the black CIA agents in *The Spook Who Sat by the Door* (1973), who share their nostalgia for the details of black college life, fraternal orders, and sports; "blaxploitation" film redundantly depicts only what has been done *to* blacks, not *by* them. Individual anatomical details also establish credibility, for example, when the same agents reminisce about the Penn Relays, a sporting event that only a few blacks would remember as the famed "Negro Olympics" of the 1930s and 1940s. In this sense, black genre film is like the black magazine that runs soul food recipes, or the black student newspaper that features a glossary of black argot—they are treatises for the uninitiated on the uses and beauties derived from cultivating a black identity.

Despite rivalry from exploiters, black genre film has survived for half a century, a most remarkable feat in view of the fact that cultural artifacts of an industrial society are often short-lived. To adapt Bronislaw Malinowski, black genre films have, like good tribal lore, expressed, enhanced, and codified belief; safeguarded and enforced group values; offered practical rules of conduct; and vouched for the efficacy of tribal ritual and gods. No other genre, except perhaps the American western, spoke so directly to the meaning and importance of shared values embraced by its audience.

Images of Blacks in Black Independent Films

A Brief Survey

James A. Snead

I. Black Independent Filmmakers: The First Generation

Even in the infancy of motion pictures, it was obvious that film, as a new way of perceiving reality, opened up entirely new perceptual possibilities, giving the eye an augmented sense of visual mastery over its surroundings, preserving events in motion for a seemingly unlimited number of future replays, performing a wide variety of functions: educational, propagandistic, recreational, aesthetic. Some idealists—Sergei Eisenstein and Charles Chaplin among them—even thought that film would ultimately bring about a radical improvement in human understanding and communication.

It is one of the bitter ironies of American history, then, that motion picture technology, with its singular potential for good or evil, grew to perfection during the same time period (1890–1915) that saw the systematic, determined, and almost hysterical persecution and defamation of blacks and other minority groups. Early American films depended unthinkingly on theatrical precursors, propagating racial caricatures borrowed from the popular vaudeville and minstrel shows. Black skin (often represented by black-faced whites) came to be linked with servile behavior and marginal status. The repeatability of movies—otherwise a virtue of the medium—offprinted false racial models from celluloid onto mass consciousness

From Mbye B. Cham and Claire Andrade-Watkins, eds., *Black Frames: Critical Perspectives on Black Independent Cinema* (Cambridge, MA.: MIT Press, 1988), pp. 16–25.

again and again; real viewers came to expect unreal blacks both on the screen and in the real world. Film became a hindrance rather than an aid to racial understanding, and in many cases (most notoriously *Birth of a Nation*—1915), served as a tool of the prevailing segregationist and white supremacist dogma.

Given this background, it does not seem unreasonable to connect the birth of black independent cinema with two portentous events in the year 1896 that made black independent films not only possible but inevitable. On April 23 of that year in New York, Thomas Edison demonstrated a major leap in film technology. No longer satisfied with his primitive "Kinetoscope" (which could only accommodate one viewer at a time), Edison introduced large-screen projection, an innovation that would allow movies for the first time to reach a mass audience.[1] And on May 18, less than one month later, the U.S. Supreme Court gave Constitutional assent to segregation in the *Plessy vs. Ferguson* decision, which endorsed "separate but equal" facilities for blacks and whites—a decision not to be reversed until *Brown vs. Board of Education* in 1954. Edison's promotion of movies to a communal (and no longer private) experience, the Supreme Court's division of this potential communality into black and white segments, and the growing resentment by blacks at their remorselessly negative images in mainstream features: All these factors inevitably encouraged and necessitated the first generation of black independent filmmakers.

The term "independent film" must, of course, be used with a great deal of circumspection, especially if, as is often the case, it is polemically contrasted with "Hollywood" (therefore "dependent"?) productions. If "dependence" has traditionally meant access to Hollywood's substantial financial resources, its skilled technicians and advanced film technologies, and its ready-made distribution and marketing networks, then for many black and white filmmakers, "dependence" has been something to be desired rather than scorned. In addition, even the so-called "independent" sector customarily (particularly in recent years) depends upon the keenly sought support of private investors, foundations, and public grants. One can declare one's independence, but in film—a particularly collective and capital-intensive art form—true independence is hardly attainable, even in the most modest productions.

Yet for many independent filmmakers, the word "independent" does not refer to any such one-dimensional version of economic self-reliance. Rather, the relative financial constraints under which independent filmmakers—and particularly black filmmakers—have operated for decades have often led to a certain aesthetic and creative privilege. Vincent Canby's comment in the *New York Times* on Warrington Hudlin's *Black at Yale* represents a not uncommon misunderstanding about the nature of independent film: "*Black at Yale* is a film limited only by the resources of money and time available to the director."[2] Yet it is by no means clear that more "money and time" would have made *Black at Yale* a proportionately better film. Indeed what Canby refers to as "limitations" (which Hudlin's film shares with practically the entire line of black independent films since William Foster's *The Pullman Porter* in 1910) precisely delineate the strength of the independent film. Without the incessant and confining restraints of box-office considerations, studio agenda, and censoring boards, the range of artistic choice in independent films is potentially *widened*, rather than *restricted*.

Because of, rather than *in spite of*, limited budget and screening opportunities, the adept filmmaker can exploit his or her marginal position to present the kinds of statements and images which can go against prevailing rules and codes. Such a filmmaker can choose to refute, to parody, or merely to copy Hollywood models—as in the cases of *A Black Sherlock Holmes* (ca. 1918), produced by the (mainly white) Ebony Film Company, or *By Right of Birth* (1921), produced by George P. and Nobel Johnson's Lincoln Motion Picture Company. Yet even such derivative films by black independent companies were not just slavish sepia replicas of white Hollywood; they often involved subtle, even inadvertent, critiques of white America's racial politics. Real-world *disparities* between white archetype and black copy intrude, often explicitly, to uncover unspoken political realities. Indeed, for many such films, their very technical and financial inferiority to Hollywood productions exposes the very real-life disparities between the races generally that the film's "apolitical" plot is trying to conceal.

The "independence" of the black filmmakers in the '20s was not a deliberate choice but was enforced—in every sense—by the highest legal tribunal of the land. In the first four decades of American film history, black independent films were a product of a separationist environment, which also accounted for their major market opportunity—pleasing a growing but cinematically underrepresented black audience.

The first generation of black independent filmmakers was active, albeit with uneven success, from about 1910 until the late 1930s. William Foster's Photoplay Company had produced black independent films as early as 1910, but most critics agree that the first major black independent effort was *Birth of a Race* in 1918.

Artistically one of the least successful, but one of the most powerful in terms of its political aspirations, *Birth of a Race* arose primarily in opposition to the radically ungenerous and laughably inaccurate depiction of black people in D. W. Griffith's *Birth of a Nation* (1915). To counter Griffith's propaganda, Emmett J. Scott (former secretary to Booker T. Washington) formed the Birth of a Race Company. After three years spent securing patchy and often unreliable funding both from black and white sponsors, Scott managed to turn out *Birth of a Race* (1918). Scott's film matched Griffith's epic in its pretensions, exceeded it in its length, but was woefully inferior to it in almost every other respect. The intended narrative—a comprehensive history of the Negro's past, present, and future, from Africa to America and beyond—was never finished, and the film (as mandated by some of its backers) ended up seeming like a pacifist commentary on the causes of World War One. Unfortunately, this most ambitious but technically and conceptually flawed project only seemed to show that "propagandistic rebuttals to propaganda were not yet feasible, especially from an Afro-American producer."[3]

II. Black Independent Filmmakers: 1920–1950

But there were other black artists at work as well. Perhaps because of its idiosyncratic aspirations, the failure of *Birth of a Race* did little to halt the courting of the black audience by other black filmmakers (often financed by white co-producers) and indeed may have spurred on their pursuit of that audience. The '20s helped establish black independent films as viable alternatives to the Hollywood product. Alongside the Ebony Motion Picture Company, which produced films for black audiences but allowed little creative or conceptual input by blacks, there were comparable companies with a high degree of black involvement. The Johnsons' Lincoln Motion Picture Company was one such company which, unlike the older Foster Photoplay Company, achieved a relatively high degree of success in the late teens and into the '20s with its productions *The Trooper of Troop K* (1917), *The Realization of a Negro's Ambition* (1917), and *By Right of Birth* (1921). But the fluid imprecision of the concept "independent" becomes clear when one notes that ultimately both William Foster and Nobel Johnson left their own independent production companies for the lure of Hollywood, a kind of career "crossover" that is still common among today's independents.

The Lincoln Company's productions, both under the Johnsons and their successor, Clarence Brooks, tended to emphasize black pride and consciousness, and were often explicitly political, but they were exceptions to the rule. Generally, "black" independent film companies (many of them, such as the Reol Motion Picture Corporation, or the Colored Players Film Corporation, actually organized and financed by whites) gave black audiences an image of a black middle class full of cultured, affluent, and well-mannered families more or less free from racial misery—indeed, an image conforming to the way Hollywood films typically portrayed "normal" whites as living.

The Colored Players Film Corporation is an exemplary case of a black-white joint venture which managed a high level of production integrity in its many films, including *A Prince of His Race* (1926), the temperance piece *Ten Nights in a Barroom* (1926), and perhaps most notably, *Scar of Shame* (1928). *Scar of Shame* certainly ranks among the most technically adept and thematically compelling films of the early black independent period. The film convincingly mixes black urban reference-points (slang, ghetto scenes, dress conventions) with a somewhat melodramatic Hollywood-style "social climbing" plot in which Alvin Hilliard, a middle-class composer, falls in love with Louise, a beautiful lower-class girl. Their rise and eventual demise, far from having anything to do with what Thomas Cripps calls "color-caste snobbery" (if this assertion were true, then one would expect Hilliard's mother to *embrace*, rather than reject, the lighter-toned Louise; in this case, class overrides skin color), is a fairly straightforward parable of class conflict within the black community.[4] The film's iconography (for example, the "book and lamp" motto on the title cards) is often unintentionally ironic, and its symbolism all too frequently leaden, but overall, the film commands and articulates a cinematic vocabulary that sets it apart from any but the most accomplished Hollywood productions of the '20s.

Oscar Micheaux was the dominant personality of this period, and typified better than anyone of this generation the all-around "black independent filmmaker," writing, financing, producing, directing, and distributing his own films. Starting with his first feature *Homesteader* (1919)—based on his autobiographical novel of the same title—he made thirty-three films in thirty-three years. His most provocative films include *Body and Soul* (1924)—which was Paul Robeson's first film, *The Wages of Sin* (1929—now lost), and *God's Stepchildren* (1937).

Even more than with other art forms, the prosperity and even survival of a film will require it to maintain a certain pre-existing status quo. The diversity of Micheaux's films makes generalization about them difficult, but one can safely assert that, as with the Johnson brothers, Micheaux did not shy away from politically disturbing topics, even topics that might have offended some black audiences. Yet even in his less controversial films, Micheaux encountered problems that had little to do with politics or ideology. The always insufficient financial backing for his projects led to a sloppy, "single-take" aesthetic, compounded by the often insufficient attention to heavy-handed scripts, badly directed actors, and primitive handling of lighting and camera movement. These deficiencies plagued "race films" generally, and gave Hollywood films a competitive advantage, even among black audiences. By the end of the '20s, Hollywood had already begun to incorporate elements of black culture into highly polished productions such as *Hallelujah!* (1929) and

Hearts in Dixie (1929), further siphoning off black filmmakers' potential audience. Hence a paradox arose: studio-processed and ill-informed images of blackness seemed more believable than black-sourced but technically amateurish images produced for black audiences by Micheaux and others.

Even apart from purely technical issues, the disappearance of the first generation of black independent filmmakers was hastened both by the Great Depression (which dried up sources of financial support) and by the advent of sound films (which sent the costs of production far beyond what most independents could afford). Most of their films have disappeared as well. The record of these films has been irretrievably damaged: all prints deteriorate with time, and few black independent producers could afford to make replacement copies, so time's effects have censored what even a racist environment could not, and of the hundreds of early black filmprints only a small fraction survive today.

Despite the large numbers of all-black "race films" produced in the '30s and the '40s, there were only a few remarkable examples of independent black filmmaking in the period. For the most part, with the new resources of sound film, black productions relied on routine musicals or melodramas with little political or social substance. Among the foremost black directors of the '30s was George Randol, who produced and directed (with Ralph Cooper) an independent all-black gangster film called *Dark Manhattan* (1937). The picture became a box-office hit "even though it was difficult to convince the white exhibitors that colored people would pay high prices to see their own race on the screen."[5] Million Dollar Pictures, with whom Randol was loosely associated, went on to produce a stream of quality films—mainly offprints of stock studio themes of romance, society, comedy, and crime. Randol's films, such as *Gangsters on the Loose* (1938), *Reform School* (1939), and *While Thousands Cheer* (1940), tended, like their Hollywood counterparts, to skirt controversial issues.

Another prominent black independent filmmaker of this era was Spencer Williams, who often simultaneously served as star, producer, and director. His Amegro Films produced *The Blood of Jesus* (1941) and *Go Down Death* (1944), both examples of allegorical dramas firmly based in rural black religious and poetic traditions. Yet overall, aside from these and a handful of other films (including those of Oscar Micheaux, who had survived the Depression and moved with seeming aplomb into the sound era), the horizontal and vertical monopolization of the film industry by the major studios, combined with the introduction of noted black actors (such as Mantan Moreland, Ethel Waters, and Bill Robinson) into white films, had by the late '40s all but extinguished the early vitality of black independent film.

III. Black Independent Filmmakers: 1950–1986

In this most recent phase of black independent filmmaking, technological innovations once more have conspired with political developments, but this time with happier results than in the late 1890s. The 1954 desegregation decision that overturned *Plessy vs. Ferguson* began a series of events leading from the Montgomery bus boycotts and the Little Rock crisis through the flourishing of the nationwide Civil Rights Movement and the initiation, in the mid-sixties, of the Black Power movement. Black people attained a greater political and symbolic significance in the American mind than ever before, such that within only a few years their prior images in mainstream Hollywood films were rendered inadequate. Even the pioneering Hollywood "race films" of the late '40s (*Home of the Brave, Pinky*, and *Lost Boundaries* in 1949) and the '50s (a series of films, mainly starring Sidney Poitier, between *No Way Out* in 1950 through *The Defiant Ones* in 1958) had begun, in light of the realities of the '60s, to seem sententious at best and condescending at worst. Hollywood was still unable to discern or depict the full spectrum of black American life and culture.

At the same time, technological improvements led to reductions in the price (and just as importantly, the weight) of 16mm camera and synch-sound equipment, which, together with improvements in the quality of film stock, made the 16mm format accessible to filmmakers with low budgets while making their "look" and "sound" acceptable to a broader public. The "New Wave" and *cinema verite* movements gave a certain high-cultural sanction to the use of "real-life" subjects as raw material for independent film, and in many cases blurred the distinction between "documentary" and "fictional" films altogether, creating both an appetite and a system of distribution ("art houses," public television, museums) for filmmaking that did not conform to the visual and narrative principles of Hollywood's "classical realist" tradition.

As a direct result of the Civil Rights Movement, black students began entering university film programs and film schools in large numbers in the '60s, and—in contrast to many earlier black independents—became the politically engaged beneficiaries of a stylistic and technical revolution in the world of film. Skeptical about the ability of the mainstream American film industry to put its house in order, this new generation of black independent filmmakers decided to exploit film's full aesthetic and political potential. They set about *recoding* black skin on screen and in the public realm by revising the contexts and concepts with which it had long been associated. Many of these filmmakers belonged to the black middle class and were formally trained at leading universities and film schools. But, unlike many of their precursors in the '20s and '30s, they insisted on engaging problems that addressed the diverse experience of all segments of the black community.

William Greaves was one of the pioneers of this new movement. Starting out as an actor (he appeared in *Lost Boundaries*), Greaves soon moved into film documentary work with the National Film Board of Canada, and later with the United Nations before becoming Executive Producer of WNET's "Black Journal," a television series that turned out many of the most aggressive and engaged social documentaries of the '60s. Two of his best films, *Still a Brother: Inside the Negro Middle Class* (1968) and *Ali, the Fighter* (1971) indicate Greaves' range, technical acumen, and emotional affinity with the black community. He remains an "independent," even though he has directed and produced Hollywood features as well. Documentary and *cinema verite* formats also attracted Shirley Clarke—best known for *The Connection* (1961), *The Cool World* (1963), and *Portrait of Jason* (1967)—and St. Clair Bourne (also an alumnus of "Black Journal"), whose television documentary *Let the Church Say Amen* (1973) set new standards for films of its kind.

In the '70s, the range and variety of black independent filmmaking expanded, and with it, the extent to which these films proliferated newer, authentic codings for images of black skin. There were more fictional and narrative films, and, in gaining worldwide recognition at Third World and European film festivals, black independent filmmakers now asserted their identity as a distinct and original group primed to contribute significantly to the history of filmmaking. A few independent films, such as Roy Campanella's *Pass Fail* (1978) even thematized the figure of the black independent filmmaker trying to reconcile an artistic mission (which involves breaking away from stereotyped roles) with white institutional demands to compromise or commercialize their recoding efforts. Or, in another example, the opening scenes of Haile Gerima's *Bush Mama* (1977) show black independent filmmakers in conflict with institutional authority. In this case—which was *not* staged—the L.A. Police stop and frisk a group of filmmakers merely *because* they are black men with camera equipment, graphically illustrating the threat felt by the status quo when blacks take control of their own images.

The majority of recent black independent films since the '60s have the feeling of intimate conversations between filmmaker and audience, and deal with issues *within* the black community, without special regard for a theoretical white viewer. Henry Miller's *Death of a Dunbar Girl* (1974) and *Color* (1982) by Warrington Hudlin and Denise Oliver both deal with class and color-caste discrimination more honestly, concisely, and credibly than *Scar of Shame* had in the '20s; *Suzanne, Suzanne* (1982) by James Hatch and Camille Billops emotionally mines, using Bergmanesque techniques, the story of an abused wife and her heroin-addicted daughter trying to compose a post-mortem on their recently dead husband, father, and tormentor—these are themes which the early black independent filmmakers would have treated indirectly, if at all. The viewer achieves, in the best of these films, an understanding of a complex black world from within, rather than caricature of it from without.

An important grouping of this new generation was active in Los Angeles between around 1972 and 1982. Most of these filmmakers were trained at the UCLA film school, but their films protest against the form and content of the tradition they were being taught. Their chief ambition was to rewrite the standard cinematic language of cuts, fades, frame composition, and camera movement in order to represent their own "non-standard" vision of black people and culture. Haile Gerima's *Child of Resistance* (1972) and *Bush Mama*, Larry Clark's *Passing Through* (1977), Ben Caldwell's *I and I: An African Allegory* (1977), and Charles Burnett's *Killer of Sheep* (1977) rank among the virtuoso articulations of this new stylistic sensibility. Julie Dash's *Illusions* (1982), and Alile Sharon Larkin's *A Different Image* (1982) extend the early methods and insights of the mainly male "L.A. School," integrating their unconventional stylistics and thematics with a frankly and refreshingly feminist reading of sexual and racial oppression. In their rejection of that kind of glossy technical sophistication that historically defines Hollywood's "classic realist" cinema, and in their refusal to remain on the easy surface of human relationships, L.A.'s black independent filmmakers continue in the '80s to define the possibilities and limits of this second flowering of black independent film.

IV. Black Independent Filmmakers: Beyond the '80s

Under the rubric of *entertainment*—and with the help of unprecedented investment in pyrotechnics, special visual effects, and computerized graphics and animation—the Hollywood film industry in the '80s seems determined to suppress sensitivity and three-dimensionality under a veneer of technical gloss. The future of black independent filmmaking will more and more rely on how these filmmakers handle the competitive challenge from an increasingly monopolistic and compelling industry. The mesmerizing visual enchantments of a *Beverly Hills Cop* series or a *Star Wars* trilogy have led to enormous box office success—most notably among black audiences. Yet even box-office breakthroughs by Richard Pryor, Eddie Murphy, or Whoopi Goldberg cannot disguise the fact that blacks are still being portrayed in aberrant and even bizarre roles that are mere descendants of paradigms set by Eddie Anderson, Stephin Fetchit, or Ethel Waters.

Towards the end of the '80s, a younger set of black filmmakers has achieved an unprecedented prominence, but it remains to be seen whether they will retain their prior artistic independence. The point can be illustrated by looking at three of the best recent black independents. Spike Lee's fine series of early films (produced at the NYU Film School), *The Answer* (1980), *Sarah* (1981), *Joe's Bed-Stuy Barbershop: We Cut Heads* (1983), and *She's Gotta Have It* (1986) reveal a quirky yet appealing filmic sensibility and a familiarity with the diverse social and economic condition of black Americans that few filmmakers, white or black, achieve. Mary Neema Barnette's *Sky Captain* (1984) and Reginald Hudlin's *House Party!* (1984) and *Reggie's World of Soul* (1985) share with Lee's films a fluent command of black and white cultural languages, and an insistence on counterposing them in an aesthetic dialectic. Instead of seeing blacks purely in terms of white norms and practices, these films show blacks securely positioned in their own environments, discussing and dealing with their own problems, ignoring or at best belittling the toys and games of the dominant white culture. Yet these directors, among the best of the late '80s, have received substantial attention from Hollywood, and at least two of them are actively engaged in feature film production. It will be interesting to see in the coming years whether the oppositional aesthetics and thematics of their earlier "independent" films can be adapted for mass-market consumption. Some would doubt whether white Americans can ever learn to see blacks and themselves from a black, and not a white, vantage-point. Lee's comedy *She's Gotta Have It* became, despite less-than-ideal production values, a major box office hit across the nation, attracting a considerable white audience. But whether a mass public could assimilate the messages and the methods of as somber a film as *Killer of Sheep* is another question. And one further question remains: should it? Perhaps the greatest challenge for future black filmmakers, independent or not, is to find a way to prevent an imagistic cooptation in which an insincere, ritualized tolerance of recoded images may itself become just another way of keeping blacks out of the picture.

Notes

1. James Monaco, *How to Read a Film: The Art, Technology, Language, History, and Theory of Film and Media* (New York, Oxford University Press, 1981), p. 200.

2. Quoted in The Black Filmmaker Foundation, *Black Cinema* (New York, 1982), p. 39.

3. James R. Nesteby, *Black Images in American Films, 1896–1954: The Interplay Between Civil Rights and Film Culture* (New York: University Press of America, 1982), p. 68, and passim, gives an exhaustive treatment of the entire early period of black filmmaking.

4. Thomas Cripps, *Black Film as Genre* (Bloomington: Indiana Univ. Press, 1978), p. 69.

5. James Asendio, "History of Negro Motion Pictures," *International Photographer*, January 1940, p. 16.

Making *Daughters of the Dust*

Julie Dash

I never planned a career as a filmmaker. As a child growing up in the Queensbridge Housing Projects in Long Island City, New York, I dreamed of some typical and not-so-typical career choices. None of the images I saw of African American people, especially the women, suggested that we could actually make movies. We were rarely even in them. No, I never dreamed of filmmaking when I was little. At that time, I wanted to be in the secretarial pool, typing away and having fun like the women I saw on TV and in the movies. I had no idea that the images I saw didn't depict the real life of working women.

Later, I turned my attention to something much more glamorous—Roller Derby. I was amazed by the motion, speed and power of women flying around the roller rink in competition and combat. Yeah that was going to be me. The Roller Derby queen of New York.

A child's dreams. A young girl's fantasies, shaped by the limitations imposed by my environment. My ambitions, like those of most children growing up in African American neighborhoods, in projects, in inner cities, were stifled by what I thought possible for me as a black child. My dreams were also molded by the cinema and television stories, where the likes of me didn't even exist.

I don't know if I would have survived the secretarial pool long-term. I did learn some office skills, and at different points in my career as a filmmaker I had to take temporary secretarial jobs in order to eat while I

Julie Dash, *Daughters of the Dust* (N.Y.: The New Press, 1992), pp. 1–26. Reprinted by permission of the publisher and author.

continued making films. As for Roller Derby, the television show didn't last long, and I'm not sure I would have survived that, either.

My introduction to filmmaking began at the Studio Museum in Harlem when I was seventeen. I was just tagging along with a friend who had heard about a cinematography workshop there and thought she could learn to take still photos. We joined the workshop and became members of a group of young African Americans discovering the power of making and redefining our images on the screen.

It was fun. As I became involved in the workshop, I enjoyed it and was drawn more and more to it. I'd found something that was creative and exciting and intellectually challenging, but I still didn't think this could be my work.

I made my first film when I was about nineteen. An animation film about a pimp who goes to an African village and is beaten and dragged out of the village by the people there. It was called *The Legend of Carl Lee DuVall*. I used pictures I had cut out of a *Jet* magazine, glued them to pipe cleaners, and shot them with a super8 camera. I was really beginning to love filmmaking—but still, when I went to college I initially majored in physical education. I was going to be a gym teacher.

While at City College in New York, a special program, the David Picker Film Institute, was set within the Leonard Davis Center for the Performing Arts. I went there and interviewed because it sounded like fun, and I already had some film experience from the Studio Museum. I was accepted and wound up graduating from CCNY with a degree in film production.

At that point I knew my course was set.

Los Angeles

As soon as I finished at CCNY, I moved to Los Angeles. My plan was to get into the UCLA film school. I had read about Charles Burnett, Haile Gerima, and Larry Clark making narrative films out in L.A., and that's what I was interested in. At the time, the West Coast seemed to be more involved with narrative films than the East Coast where a lot of black filmmakers, including myself, were making documentary films.

I was very excited about the prospect of UCLA. Considering my previous work and experience, as well as my degree from CCNY and good recommendations, I was sure I would get in. I didn't . . . because of a technicality.

Applicants to the film school at UCLA had to submit, among other things, three letters of recommendation. I had been promised these letters by three of my teachers (two of them white and one black). To my surprise however, one of the letters was never sent, and as a result I wasn't accepted. I was stunned.

I was even more hurt when I found out that both of the white professors had sent their letters supporting me. I thought getting into UCLA would be a triumph and an advance not just for me as a young African American, but also for other black filmmakers. Instead, I learned a bitter lesson, one that I would remember throughout my career.

A great part of filmmaking is overcoming various kinds of obstacles. What I learned then was that I would sometimes have to face sabotage, often from "my own people." I would have to feel that pain over and over again. While making *Daughters of the Dust* I encountered this to an extent that I had never suspected.

After the UCLA rejection, I had to figure out what I would do next. I had no other plan because I had been sure I would be in school. Fortunately, Larry Clark was about to begin shooting his film *Passing Through*. I joined his crew and went out into the California desert. One of the actresses I met on *Passing Through* was Cora Lee Day, whom I would cast years later as Nana Peazant in *Daughters of the Dust*.

Working on a film in the desert helped to heal me. I began to get strong again. And once again I stumbled into the next phase of my training.

One afternoon in Los Angeles, my friend and I heard about grants that were available for filmmakers at the American Film Institute (AFI). We went there looking for the grant applications. It all felt very strange to me, because the atmosphere was so relaxed, and AFI was in this beautiful house in Beverly Hills.

We were standing in the hallway looking lost when this young black man dressed in jeans came down the stairs. He said, "My sister, how can I help you?" I thought the brother must work there, you know, maybe he was the janitor (we are all infected by the stereotypes). I told him that we were there to get the grant applications. He listened, asked some questions, and then gave me some of the best advice I was ever given.

He told me to apply for a fellowship instead of a grant. He said that I should be attending AFI, not just seeking some small cash to make a short film. I took his advice, applied for the fellowship, and became one of the youngest fellows to attend AFI.

Later I discovered that the young man was Ted Lange; he would later be known as Isaac on *The Loveboat*. I also found out that he had been the co-writer on Larry's film *Passing Through*.

I always knew I wanted to make films about African American women. To tell stories that had not been told. To show images of our lives that had not been seen.

The original concept for *Daughters* was a short silent film about the migration of an African American family from the Sea Islands off the

South Carolina mainland to the mainland and then the North. I envisioned it as a kind of "Last Supper" before migration and the separation of the family.

The idea first began to wander throughout my head about 1975, while I was still at AFI. I was making notes from stories and phrases I heard around my family, and became fascinated by a series of James Van Der Zee photos of black women at the turn of the century. The images and ideas combined and grew.

In 1981, I received a Guggenheim grant to research and write a series of films on black women. In 1983, I completed my short film, *Illusions*, with Lonette McKee in the lead role as a studio executive who passed for white during World War II. This is also when I began intensive research for *Daughters of the Dust.*

Daughters of the Dust

The stories from my own family sparked the idea of *Daughters* and formed the basis for some of the characters. But when I probed my relatives for information about the family history in South Carolina, or about our migration north to New York, they were often reluctant to discuss it. When things got too personal, too close to memories they didn't want to reveal, they would close up, push me away, tell me to go ask someone else. I knew then that the images I wanted to show, the story I wanted to tell, had to touch an audience the way it touched my family. It had to take them back, take them inside their family memories, inside our collective memories.

Soon I was off, running faster and faster, trying to find more and more information that would allow me to uncover this story. I spent countless hours in the Schomburg Center for Research in Black Culture in Harlem reading and looking at images from old newspapers, magazines, and books. I went to the National Archives in Washington, D.C., as well as to the Library of Congress and the Smithsonian Institution. UCLA also has a wonderful research library that provided much needed information. And finally I went to the Penn Center on St. Helena Island, off the coast of South Carolina.

The research was fascinating. In fact, if I were not making films, I would probably be glad to spend the rest of my life digging around libraries. I learned so much about the history and experiences of African American people. One of the most fascinating discoveries I made was of the existence of over 60,000 West African words or phrases in use in the English language, a direct result of the slave trade.

As I poured through the documents, taking notes and developing the story line for *Daughters*, it became clear that a short film would not be large enough for the story. I knew I would have to make a feature. There was too much information, and it had to be shared.

The Ellis Island for Africans

The sea islands of the coast of the Carolinas and Georgia became the main drop-off point for Africans brought to North America as slaves in the days of the transatlantic slave trade. It became the Ellis Island for the Africans, the processing center for the forced immigration of millions. It also became the region with the strongest retention of African culture, although even to this day the influences of African culture are visible everywhere in America.

Many of the images seen in *Daughters of the Dust* parallel the action and behavior of African Americans today. For instance, the hand signals given by two of the men in *Daughters* is a reference to the nonverbal styles of communication of ancient African secret societies which have been passed down across thousands of years and through hundreds of generations. Today these forms are expressed in the secrets of fraternities and in the hand signals of youth gangs.

As a young girl growing up, I remember watching young men on a basketball court or at other gathering places, and before they would drink together, they always poured a little on the ground. I always thought that was a strange and funny ritual. Later, during my research for *Daughters*, I discovered the West African ritual of pouring libations, a show of respect to the ancestors, to family and to tradition. As the men on the basketball court would say, "This is for the brothers who are no longer here or couldn't be here today."

In 1984, while I was still writing, my daughter N'zinga was born. Her birth revealed to me the need to see the past as connected to the future. The story had to show hope, as well as the promise that tradition and family and life would always sustain us, even in the middle of dramatic change. N'zinga's arrival in our lives also brought the "unborn child" into the script. I hadn't seen her until I saw my own daughter.

By 1985, most of my research was completed, and I began to write the script for *Daughters of the Dust*. It would go through five complete rewrites and two polishes. In fact, I even rewrote some of it while shooting.

Although I did most of the primary research myself, I'm indebted to several people who gave me important help along the way, especially Dr. Margaret Washington Creel, Oscar Sims, and Worth Long.

Shooting *Daughters of the Dust*

By late 1986, when I was finally ready to begin shooting, I was faced with another problem: financing the film. Originally I thought it could be done for about $250,000. I had some production money from the National Endowment for the Arts, but it was not enough to begin production. I applied for and received several grants, one from the Fulton County Arts Council (GA), another from the Georgia Council on the Humanities, and another from Appalshop, Southeast Regional Fellowship (SERF). But by the standards of feature productions, it was still not enough, and I soon realized that I would just have to begin, and hope that more money would come when we had something to show.

I knew that it would be difficult to get other people to understand the vision of this unusual film. I knew it would be different from the films most people were used to, and there weren't many people willing to invest in an "untraditional" black movie.

I needed to create on screen what I had in my mind. I knew exactly what it should look like. After hours and days of discussion with Arthur Jafa (A.J.) my coproducer and director of photography, I was confident that we could capture exactly the feelings and memories I wanted to invoke in the mind of the audience. If I could show people a piece of this—literally, give them a piece of my mind—I was sure that I could raise enough money to finance the film. With the grant monies available, and after scraping together unused film stock from friends like Charles Burnett, I decided to shoot a sample of the film.

In the summer of 1987, I took a crew of ten and four cast members, Adisa Anderson, Barbara-O, Alva Rogers and VertaMae Grosvenor (who also served as a technical adviser) to St. Helena Island to shoot for five days. While there, I also conducted screen tests for Unborn Child.

After the initial shooting, we returned to Atlanta, once again broke, and faced the dilemma of trying to find money to edit the sample.

I began to do small projects for various organizations, like the National Black Women's Health Project (the people at NBWHP were extremely supportive of my work and very helpful), and put all of my earnings, beyond basic living expenses, toward editing the sample. Finally, after several months, it was ready to show.

With a completed script, a sample, and a filmography of my previous work, we renewed our search for funds. I had also revised the budget, based on our experience on St. Helena, and knew that we would actually need around $800,000 to complete the film. We sent packages to a variety of American and European sources. The responses were as interesting as their sources were varied.

Hollywood studios were generally impressed with the look of the film, but somehow they couldn't grasp the concept. They could not process the fact that a black woman filmmaker wanted to make a film about African American women at the turn of the century—particularly a film with a strong family, with characters who weren't living in the ghetto, killing each other and burning things down. And there weren't going to be any explicit sex scenes, either. They thought the film would be unmarketable. They believed that they knew better than we did about what moved black people. They figured it would be a pretty, artsy European sort of film that no one would come to see. Every major studio either passed on it or didn't respond at all.

We didn't do much better in Europe. Most of the European sources couldn't understand what we were trying to do any better than their Hollywood counterparts. One told us the film sounded too much like a typical American film. Another said it was too radical in its concept for their audience. Still others said perhaps next year, if we were still looking for funds by then.

Our most sympathetic response came from the New York-based organization, Women Make Movies Too, which held a benefit fundraiser in 1987. They raised $5,000 for the production. But these funds, as badly needed as they were, only covered the expenses of sending out all the samples. It was beginning to look like we had exhausted all our sources, when a break came that sent us flying into production.

In 1988 I was attending the PBS Rocky Mountain Retreat in Utah. There I met a woman named Lynn Holst, who happened to be director of program development for American Playhouse. Lynn was interested in the project and we spent many hours discussing it. I liked her immediately, and felt that even if nothing came out of our meetings in terms of funding, I'd made a friend. Well, I was rewarded twice. I did make a friend, *and* American Playhouse wound up providing most of the money for *Daughters of the Dust*. Finally, we would be able to make the film.

We entered a two-phase process with American Playhouse. The first was to rework and develop the screenplay even further. We added dialogue and some additional scenes. During this process, I worked closely with Lynn, and I think we learned a lot from each other. Unfortunately, a lot of what we added to the script was ultimately lost due to time and budget constraints.

The second phase was production. American Playhouse insisted we shoot a union film. (We found out later that we didn't have to.) This is often an unfortunate dilemma for the independent filmmakers, who want to respect the unions, but are rarely able to afford to complete a film in accordance with union guidelines. We wound up striking a deal with the Screen Actors Guild (SAG) under a special contract for minority, low-budget projects. Even with the special terms, we started shooting $200,000 over budget.

We entered the second phase in August of 1989. I met Steven Jones, our line producer and production manager, and Pamm Jackson our associate producer, in Beaufort, South Carolina, to scout locations. We planned to begin shooting by October but nature had something else in mind for us.

In those days of preproduction, I found myself nauseous and easily fatigued. At first I thought that it was the heat and the humidity, until I learned that I was pregnant. I had to quickly make a decision as to what I was going to do. I had two choices—to put off the production for at least another year or to have an abortion. I made my decision to go forward with the filming of *Daughters*. I flew back to Atlanta to have the abortion. This was a painful decision many women have had to face, especially women who must rely on their physical as well as mental stamina to perform professionally. Unfortunately, many women do not have the same options that I had. At least I could still make a choice. *Daughters* would become the child that I would bear that year.

The week we were to start, Hurricane Hugo slammed into the coast of North and South Carolina. We had just moved our production crew to St. Helena when we were told that the island had to be evacuated immediately, that Hugo would come crashing in on us within a matter of hours. We heard the news about four in the afternoon, and by 10:00 P.M. we had packed all of our equipment and were headed back to Atlanta in a long caravan of cars, trucks, RVs, and vans. We would have to wait out the storm. It was not our first obstacle, and I knew it would not be our last.

When Hugo finally finished feasting on the coast, we returned to shoot *Daughters of the Dust*. Fortunately, it had missed our main locations on the islands. Unfortunately, it hit Charleston and other cities, causing severe damage. Part of our good luck was that some of the relief workers who had helped in Charleston came down later to work with us on the film, some as production assistants and some as cast members. Gloria Naylor, the author of *The Women of Brewster Place* and *Mama Day*, lived in the area, and she joined the crew as a production assistant. It was great to meet her and have her on the set.

When we finally began shooting we knew we had only twenty-eight days to complete all the principal photography. Our main beachfront location (Hunting Island) was a one-mile hike to the coast from base camp. Due to environmental restrictions, we couldn't take a four-wheel-drive vehicle on the nature trail or along the coastline, so all the equipment had to be carried in each morning. We also couldn't bring in a generator, so A.J. decided to shoot with natural light—sunlight—only. Therefore we needed to squeeze in as much shooting time every day as the sun would allow. Often we would be in the middle of setting up or shooting a particular scene when the sun would suddenly cast perfect and beautiful light in another spot. We would hurriedly change directions and capture the unscheduled scene with only a moment's notice. Sometimes this would work, sometimes it failed comically; but we kept shooting.

For the most part, the crew and actors all worked in the same spirit, everyone appreciating that we were doing something different, something special. But there are always those who cannot or will not see what is being done. Two particular incidents stand out as perhaps the most damaging.

After the shooting had already begun, when we couldn't possibly stop or recast, one of the lead actresses felt that she should be paid more money. Perhaps she had heard that the budget was $800,000 and thought that we were underpaying her, or that I or the other producers were making a lot of money. In fact all of us were working practically for nothing. She and some of the others in the cast apparently didn't understand that $800,000 was an extremely low budget and that I would be heavily in debt when the film was finished. She decided to get the union to force us to pay her more money. When I found out what she was up to, I was hurt and angered. I felt that I had been ambushed by someone whose career I was, after all, helping to promote by casting her in a major role in a feature film.

In independent film, we are never able to pay top salaries. None of us are adequately compensated for the work we do, not the writers, the producers, the directors, the crew, the actors, not one. We do it to create the work. We do it to sharpen our skills. We work with the hope that if the film is good, someone will offer us a bigger budget the next time, and then we'll be able to hire and compensate adequately those who sacrificed on the low-budget projects. We work as a community of artists, collaborating to create a work of art. I was wounded, but had to stifle my emotions and get the best performances out of the actors, while trying to keep the crew focused and motivated.

On another occasion, I was confronted one morning by an actor who refused to put on his costume. We were ready to shoot a scene that included him, and for whatever reason, he decided that this was the time to assert the fact that even though I was the director, he was a man and no woman could make him do anything. This man, a Muslim, who had been telling us all about the need for unity among black people, stood there in the middle of the set, in front of the crew, and confronted me, physically. He knew that he could intimidate most people because of his size (about 6′4″) and demeanor. I knew that if I backed down from him the entire project would come crashing around me. Any authority or control I had on the set would be completely undermined. We were seconds away from actually fighting, but I made my stand. What he hadn't anticipated was my willingness to take an ass whipping rather than let him take over my film. I was ready to fight. The seconds passed by, full of tension. I could see his eyes searching my face, looking for signs of fear or weakness. The crew and cast all stood frozen, shocked, unsure. Finally, A.J. intervened and the actor took the opportunity to back away and save face. In the end, he knew that he had too much to lose if he attacked me. He put his costume on. I'd won, but secretly I was shaken for days afterward.

A twenty-eight-day shooting schedule for a feature film is incredibly punishing. We were all exhausted from working long hours day after day, in addition to our constant fight with nature. We were in areas heavily infested with mosquitoes and other biting insects, and were often caught in the middle of sudden and violent sandstorms.

But through it all, we kept shooting. We pushed on, all of us, crew, cast, everyone. We became friends, enemies, lovers, coconspirators and family. Toward the end, we ran short of money and had to wrap some of the cast early. But the spirit of the project had infected some so deeply that they stayed on and worked for nothing for days afterward. Adisa Anderson, Bahni Turpin, and Vertamae Grosvenor were some of those who stayed, helping out where they could, and their help was greatly appreciated.

When we finally wrapped the shooting of *Daughters* there was a great sense of relief, and some sadness. We all felt like we had contributed to something special, something new, something important. It would be a while, almost two years, before we would be able to measure the impact of it on an audience, but we knew it would be special.

We gave a party with food and champagne and music. We showed slides of the crew and cast at work and everybody got a kick out of seeing themselves at some pretty funny moments. I was pleased. We were a good group of hardworking filmmakers, and in spite of quite a few problems, we had all forged ahead and made something beautiful. As I enjoyed myself at the party, though, my mind began to run ahead anxiously to the next part of the process. The film now had to be edited, but we had no funds for postproduction and were already heavily in debt.

When we got back to Atlanta, I was physically and emotionally exhausted. Not only had I been consumed with the normal grind of directing a feature film, but I had been constantly fighting for money, managing personalities, and worrying about the next wave of fundraising. I also felt guilty about being away from my daughter for so long while I shot the film. Now I would be home for a while, but the pressure and stress did not end. I still had to edit the film.

I set up my living room as an editing room and watched as 170,000 feet of film was unloaded at my house—a mountain of work, an almost unsurmountable task. Keith Ward, Tommy Burns, and Angela Walker did the syncing of the film, and I began editing in January 1990.

I did what I could with whatever money became available. After the first month I brought in Joseph Burton to help with the editing, and later Amy Carey came on board to help complete it. In June, I was fortunate enough to receive a Rockefeller Fellowship for $45,000 which went immediately into the film. The National Black Programming Consortium also contributed money for postproduction. By December we finally had a fine cut. Now it was time to look for a distributor.

Amy Carey, the editor, and I took the fine cut to Los Angeles to begin producing the original music score with John Barnes. We had already scheduled and rescheduled the sound mix, so we had only two weeks to complete the music before the final sound mix. John Barnes worked throughout the Christmas holiday to compose and score seventy-two musical cues for the film.

For the soundtrack of *Daughters of the Dust*, John assembled an impressive collection of musicians and styles to evoke the film's magic and mystery. He used a myriad of instruments, including the synclavier, the Middle Eastern santour, African bata drums and African talking drums, and he successfully mixed synclavier-based percussion with authentic music from Africa, India, and the Middle East.

We wanted to depict various religions—including traditional West African worship rituals, Santeria, Islam, Catholicism, and Baptist beliefs—through musical expression.

John drew from his own spiritual beliefs, which include a respect for astrology, in composing the music. For instance, he wrote the Unborn Child's theme in the key of B, the key of Libra, representing balance and justice. "This character was coming into the world to impart justice, a healing upon her father and her mother and her family." Similarly, he wrote "Nana's Theme" in the key of A representing the Age of Aquarius, or the new age that was imminent for Nana's family.

The closing theme, called the "Elegba Theme," was written in the key of Taurus, D sharp (or E flat). John told me, "It is the key of the earth, the key ruled by love." The lyrics, "Ago Elegba . . . show the way, Elegba," he says, are about people who are moving forward after having been given love and dignity, and who are now facing the crossroads.

While we were recording the score, we began the final sound mix at Sound Trax Studios in Burbank, California. We recorded at night and mixed during the day. The whole recording session went on for ten continuous days in which we barely slept a wink.

I was certain that now that the film was completed, distribution would not be a problem. It had been hard in the early days to convey in words the idea of this film. But now that it was done, I figured there'd be no more blank looks. They wouldn't have to imagine a film about African American women at the turn of the century. Here it was, right in front of them. I was wrong. All of the distribution companies turned it down.

I was told over and over again that there was no market for the film. The distributors talked about the spectacular look of the film and the images and story being so different and thought-provoking, yet the consistent response was that there was "no market" for this type of film. Again, I was hearing mostly white men telling me, an African American woman, what my people wanted to see. In fact, they were deciding what we should be allowed to see. I knew that was wrong. I knew they were wrong.

One of the ongoing struggles of African American filmmakers is the fight against being pushed, through financial and social pressure, into telling only one kind of story. African Americans have stories as varied as any other people in American society. As varied as any other people in the world. Our lives, our history, our present reality is no more limited to "ghetto" stories, than Italian Americans are to the Mafia, or Jewish Americans are to the Holocaust. We have so many, many stories to tell. It will greatly enrich American filmmaking and American culture if we tell them.

In order to secure distribution for *Daughters* I decided to start showing it on the festival circuit. The first one we were able to present at was the prestigious Sundance Festival in Utah, in 1991.

The film was well received, as we thought it would be, and won the festival's award for best cinematography. I was very happy, not only for the success of the film, but for the recognition given to A.J. for his work as cinematographer.

I took it to festival after festival, from January through September 1991. At the Black Light Festival in Chicago, *Daughters* sold out every showing. I went to Germany, to the Munich Film Festival and the film got a tremendous response there. Everywhere we took it, whether it was the Toronto Film Festival, the London Film Festival, or the Festival of Women in Spain, the response was the same. *Daughters of the Dust* provoked the audience. Most liked it, some did not. But it provoked them, and that made me see that I had created something important, a film that caused its audience to think and react and come to grips with their own memories.

During this period I was commissioned by Alive From Off Center to direct *Praise House*, a performance film featuring the Urban Bush Women. I was glad to do the work; I was always moved and intrigued by the work of the Urban Bush Women. Also, the money helped me pay off my debt.

Finally, in September 1991, a small company operating out of New York, Kino International, agreed to act as the distributor for *Daughters of the Dust*. I was relieved, but concerned, because they only agreed to distribute the film on a staggered schedule throughout 1992. I had hoped for a simultaneous release in key markets throughout the United States. I think it is important that African American filmmakers get maximum exposure for their films during the initial release period. If not, they're often pulled before the audience has a chance to find them. This is what happened to Charles Burnett's excellent film, *To Sleep with Anger*.

The other concern that I and other African American filmmakers are faced with is the amount of money that the distributor will spend on prints and advertising. When a film is released on a staggered schedule, it often means a very small budget for promotion, thereby limiting the exposure and the potential revenues of the film.

In this case, however, I was lucky and Kino International made a big push for the opening of the film. They also had the wisdom to hire a new African American public relations firm, KJM3, to arrange publicity for the film. KJM3 worked hard, and I soon found myself swamped with requests for interviews. Suddenly, I was appearing in national magazines and newspapers all across the country.

Daughters of the Dust opened January 15, 1992, at the Film Forum in New York. It sold out every show. The day of the opening the Coalition of One Hundred Black Women of New York gave a fashion show and reception in support of the film. I was overwhelmed. People were asking me how it felt to be the first African American woman filmmaker with a feature film in theatrical release. It was a thought that had never crossed my mind. I had always considered myself one of a community of some very talented, powerful women filmmakers—women such as Neema Barnett, Ayoka Chenzira, Zeinabu Irene Davis, and Michell Parkerson. Now people were saying, "Oh you're Julie Dash."

Daughters of the Dust had finally made it to the screen. As I watched people file out of the theater on opening night, I felt all kinds of emotions. I was happy to see my work so well received; I was moved by the emotion on the faces of the people, especially older African Americans; I was proud to be contributing to the growing power of African American filmmakers, telling the stories of our people; and I was relieved that the voices of our women were finally being heard. But I didn't bask in the success of *Daughters* for too long. By the time it opened, I was already promoting the next film.

William Greaves, Documentary Filmmaking, and the African-American Experience

Charles Musser and Adam Knee

The reputations of many documentary filmmakers rest on their production of one or two ground-breaking pictures. But there are some, such as William Greaves, whose real achievements only become apparent when we look at the full accumulation of their work. In Greaves's case this was made possible by a recent retrospective of his films at the Brooklyn Museum. It included a screening of his never-released unconventional *cinéma-vérité*-ish feature *Symbiopsychotaxiplasm: Take One* (1968/1971), which is now being screened at festivals, art houses and museums. Moreover, Greaves is still turning out innovative, rigorous films—as demonstrated by his documentary *Ida B. Wells: A Passion for Justice* (1989), a historical biography of the black feminist civil rights leader Ida Wells that recently aired on PBS's *The American Experience* and that has won numerous prizes on the festival circuit.

Black independent filmmaker Bill Greaves has played a significant if not always fully appreciated role in the creation of a new post-1968 era in U.S. documentary cinema—one that is characterized by greater racial and cultural diversity among those in production. During the 1950s and early 1960s, however, Greaves endured a protracted struggle to establish himself as a documentary filmmaker of artistic and political integrity. By the mid-1960s, he increasingly achieved these goals and finally began to produce pictures on subjects of particular importance to African Americans. In 1968, while continuing to further develop his own still limited filmmaking opportunities, Greaves began to assist a new

Film Quarterly, vol. 45, no. 3, 1992, pp. 13–25. Reprinted by permission of the authors.

generation of young black documentarians through the initial stages of their professional careers—Kent Garrett, Madeline Anderson, St. Clair Bourne and others. Greaves was not only a harbinger of a new era of multi-cultural filmmaking but has emerged a pivotal figure in the history of African-American cinema.

Greaves, in addition to being an important historical force, has produced an impressive and surprisingly diverse body of work, both in approach and subject matter. This testifies, on one hand, to his inventiveness and broad range of interests and, on the other, to the numerous practical exigencies he has faced over several decades. Greaves has received much recognition for his work as executive producer and co-host of public television's "Black Journal," an Emmy-winning public-affairs series, and for his direction of such ground-breaking films as the historical documentary *From These Roots* (1974), which looks at Harlem during its cultural renaissance in the twenties and early thirties. However, the broader course of Greaves's career and the substantial contribution he has made to African-American film production—from acting in black-cast films during the nineteen-forties to serving as executive producer on Richard Pryor's 1981 hit, *Bustin' Loose*—are only now starting to receive adequate attention.

Even aside from the scores of films and television programs that Greaves has produced, directed, edited, photographed, written, and/or appeared in, his career itself deserves attention for the way it traces many aspects of African-American involvement in (and exclusion from) motion picture, television, and related industries. He was born and raised in Harlem and educated at Stuyvesant High School. While enrolled as an engineering student at City College of New York during the early forties, Greaves used his skills as a social dancer to become a performer in African dance troupes. From there he moved into acting at the American Negro Theater and was soon working in radio, television, and film. Among the films he was featured in (and sometimes sang in) at this time were the whodunit *Miracle in Harlem* (1947), one of the most technically polished of black-cast films, and the Louis de Rochemont-produced *Lost Boundaries* (1948), a highly popular film based on a true story about a black doctor who set up a practice in a New England town while "passing" for white.[1] The doctor and his family are played by white actors (in keeping with Hollywood conventions of the day), while Greaves portrays a debonair black college student who is completely comfortable with his African-American identity as he interacts with his white counterparts. It was an image seldom if ever seen in American films prior to that date. Greaves's role here clearly prefigured many of those played by Sidney Poitier in the next decade, and one is apt to wonder whether Greaves would have become one of the crossover stars of the fifties had he remained in screen acting.[2]

Greaves on *Lost Boundaries*

You have to decide when you make a movie—and it's a tough decision—how authentic, how pure, how faithful you must be to reality while at the same time making this product so that people will go to see it. This is an extremely tricky, difficult challenge for a filmmaker. And in the climate of an extremely racist society, this was a marketing problem. Now Lost Boundaries *turned out to be a massive hit. It ran for six months on Broadway, which was practically unheard of. It played at the Astor Theater and won awards and one thing or another. Mel Ferrer, the star, did a very fine piece of work. It was a very moving film. You say, Jesus, why didn't they have some light-skinned blacks in those roles? You can ask that question very aggressively today, but at the time you had to take into account the very cold temperature of the country.*[3]

Greaves himself moved easily between the white and black worlds. After acting in such ANT productions as Owen Dodson's *Garden of Time* and *Henri Christophe*, he appeared in the musical *Finian's Rainbow*, which began a two-year run on Broadway in January, 1947. As the show came to a close, Greaves joined the Actors Studio, becoming a member alongside Marlon Brando, Shelley Winters, Eli Wallach, and others. Despite this illustrious affiliation, Greaves was increasingly frustrated with the demeaning roles available to him (and blacks more generally) in theater and film. A decisive moment came in 1950, when he was slated to appear in the Broadway revival of Ben Hecht and Charles McArthur's *Twentieth Century*, starring Gloria Swanson and José Ferrer (who also directed and produced).[4] Upon reporting to the theater, Greaves discovered that he was to play a stereotypical bumbling porter and quit on the spot.

Greaves on *Twentieth Century*

All I knew was that I had built up a little reputation and my agent said "You have a part." So I reported to the theater. And then I saw this goddamn dialogue which they put in my hand and Ferrer said, "You're going to be this Uncle Tom type." I just walked out. Whenever that kind of role came up I would never play it, because it was just too demeaning. Actually that was the final straw. That was the thing that made me realize I have to get on the other side of the camera because they were messing with the image of black people with impunity.

Deciding he had to move into film production, Greaves enrolled in film-making courses at City College. With the exception of de Rochemont, who allowed him into the studio as an apprentice, no one seemed prepared to provide him the needed opportunities to achieve his goals. Like Melvin Van Peebles and many other African-American artists during the fifties, Greaves finally had to leave the country to practice his craft. In 1952, fed up with McCarthyism and the exclusionary practices of motion picture unions, Greaves moved to Canada.

Greaves on His Move to Canada

It became obvious to me that either I would stay in America and allow myself to be made a fool of, or become a very neurotic person, or be destroyed. Or leave. So I left, which was fortunate because I had a very good opportunity in Canada. The Canadians were much more liberal than Americans. Race didn't have that much meaning to them. And I was fortunate to be taken onto the production staff of the National Film Board of Canada, set up by John Grierson.

I had been reading Grierson on documentary and was very taken by his discussion of the social uses of film. He proposed ways in which film could be a social force, an educational tool, and this interested me.

Greaves worked his way up over the next six years at the National Film Board through various editing jobs to directorial work. His tenure there culminated with his directing and editing of *Emergency Ward* (1958), a production for the Canadian government that documents the events of a typical Sunday night at a Montreal hospital emergency room.

Stylistically, *Emergency Ward* falls somewhere between the "Free Cinema" of Lindsay Anderson's *Every Day Except Christmas* (1957), with its carefully prepared set-ups and tripod-dependent shooting style, and the *cinéma-vérité* style of *Lonely Boy* (1961), by Roman Kroiter and Wolf Koenig. (Not uncoincidentally, Koenig served as Greaves's cameraman on the film.) *Emergency Ward* was shot over the course of many nights and exposes us to the range of people admitted to the hospital: accident victims, people with imagined illness, people abandoned by their families, and others who are just plain lonely. Grierson's influence on Greaves is evident in this film: Greaves humanizes his subjects and reassures the viewer that the emergency ward at this institution is run as responsibly and as well as the post office in *Night Mail*. The doctors know their jobs and care; orderlies and nurses are ennobled. At the same time, this film might be seen as a forerunner of Frederick Wiseman's *Hospital* (1968), for instance in its visual sensitivity to character quirks, although it ultimately lacks Wiseman's aggressiveness and sense of style. While

Greaves found the subject matter fascinating, it offered little for him to grab hold of, given his reasons for moving behind the camera. In the all-white world of a Montreal hospital, black racial identity was not a pressing issue. Greaves learned his craft and escaped the humiliations of American racism in Canada, but it was not a place where he could readily develop the kind of distinctive voice he had displayed as an actor.

Greaves was perfectly positioned to participate in the *cinéma vérité* revolution of the early sixties—until new senior management at the Film Board decided to place him in charge of its unit making science films. Sensing a dead end, Greaves left to create and direct with a Canadian acting troupe. In 1960, he joined the International Civil Aviation Organization (I.C.A.O.), an agency of the United Nations, as a public information officer. This, in turn, led to his making a one-hour television documentary about a round-the-world flight of a major airliner (*Cleared for Takeoff*, 1963; featuring Alistair Cooke). The U.N. job eventually required Greaves to move back to New York.

Greaves was by this time eager to return to the United States, as race relations were rapidly changing: the Civil Rights movement was gaining momentum and Kennedy's New Frontier was seeking to respond to its demands. New York filmmaker Shirley Clarke had seen *Emergency Ward* and was impressed. She told George Stevens, Jr., who was head of the United States Information Agency's film division, about Greaves, and Stevens, looking for a black director, soon contracted with him to do a documentary on dissent in America. The topic quickly proved too controversial for the agency, especially when it learned that Greaves planned to include people like professed atheist Madalyn E. Murray, the "no prayers in the school" leader, in the film. U.S.I.A. subsequently decided to change the film's focus to freedom of expression. Essentially, the resulting *Wealth of a Nation* (1964) maintains that America is great in part because its citizens are allowed "to do their own thing." In this context, Martin Luther King, Jr.'s, "I have a dream" speech at the Lincoln Memorial takes on unexpected meaning, suggesting a purely personal vision rather than the expression of a larger political movement. Featuring footage of various artists and visionary architects at work, the film relies on a heavy narration to assert the potential social usefulness of individual creative expression. It ultimately becomes an essayistic paean to American myths.

Wealth of a Nation, nevertheless, established Greaves as an independent producer and with his next U.S.I.A. production he finally won the opportunity to focus on black culture from behind the camera. The African-American film-maker was originally dispatched to Dakar, Senegal, to shoot a historic gathering of black artists and intellectuals from throughout the African diaspora. The U.S.I.A. wanted a five-minute news clip. Upon arriving, however, Greaves immediately realized the value of a longer piece. After he, his cameraman, and driver shot as much footage as possible, largely without synchronous sound, Greaves utilized those editing skills acquired at the N.F.B. to put together an effective and comprehensive record of the event. This record, *The First World Festival of Negro Arts* (1966), features performances by dancers from throughout the black world and appearances by Duke Ellington, Katherine Dunham, Langston Hughes (whose poetry frames the film), and many others. Greaves's juxtapositions explore and affirm the links between African and African-American culture. It was Africans, however, rather than African Americans, who were given the opportunity to appreciate these links: while *First World Festival* proved the most popular U.S.I.A. film in Africa for the following decade, U.S.I.A. films were prohibited at the time (and until recently) from distribution in the United States.[5] Although such links could have been radicalizing for African Americans, this affirmation was more likely to serve a conservative agenda when presented to Africans—in suggesting greater identity with the United States and, by implication, with its Vietnam-era policies. If the film is considered in terms of the politics of production, however, it represents an important achievement.

Greaves on *The First World Festival of Negro Arts*

You have to realize that the reason why I went into motion pictures was to make films like The First World Festival of Negro Arts. *It was the first opportunity I had to make films that expressed a black perspective on reality. Until then I had not had access to financing which would permit that.*

The First World Festival of Negro Arts was quickly followed by another breakthrough film for Greaves, *Still a Brother: Inside the Negro Middle Class*, a 90-minute television documentary made in collaboration with William Branch for National Educational Television (NET). By the summer of 1967, the nation's inner cities were in turmoil. Television news featured rioting blacks, creating a perception among many whites that African Americans were burning down the country. As envisioned by NET, *Still a Brother* was to focus on a group of "good negroes" as a way to challenge negative stereotypes held by whites and to encourage poorer blacks to see that the system was working and creating new economic opportunities. *Still a Brother*, completed in 1967, proved more controversial than NET had expected—ultimately focusing as it does on the rise of contemporary black-pride movements. Although the film's interviews with a number of successful blacks at times suggest a preoccupation with material gains—most pointedly in the opening interview, where a man describes his version of the American dream as owning a yacht and wearing a Brooks Brothers suit—they also bespeak the extreme barriers to achieving such gains, and their great fragility once achieved. Many interviewees agree that the loss of a well-paying job often means instant loss of middle-class status to African Americans. In its emphasis on the concerns of an emerging African-American middle class, the project was an especially personal one for Greaves.

Still a Brother looks at the danger of passive wholesale acceptance of white middle-class values by blacks—a phenomenon which Greaves has referred to as a mental enslavement. The film's main contention, however, is that in the turbulent sixties, economically successful blacks were undergoing a mental revolution. Again and again, those interviewed reveal a growing understanding that the oppression of lower-income African Americans is their oppression as well. The perspective of the film, which supports black pride while stopping far short of advocating separatist politics, is one that continues to emerge in Greaves's work. It is better described as liberal than radical, but it is always questioning of liberal assumptions and sympathetic to radical goals.

Greaves on *Still a Brother*

We had difficulties once Still a Brother *was finished because NET had not expected that kind of film. They had expected an* Ebony *magazine kind of film, but we brought them this documentary that talked about mental revolution and showed increasing militancy in the black experience. People are talking about black is beautiful, the African heritage, militance, and championing Rap Brown and Stokely Carmichael. So when NET executives saw the film they sort of blinked because they didn't know whether or not they really wanted to put it into the system. They weren't clear whether or not it would be acceptable. There was a great deal of anxiety because these executives were looking at their mortgages and didn't know whether they would be tossed out of their jobs. They didn't tell me that, but it was obvious that they were really under pressure. But I must say that they rose to the occasion, which speaks well of them, and of course the film eventually received an Emmy nomination and a Blue Ribbon at the American Film Festival.*

Still a Brother was finally shown by NET on 29 April, 1968, less than three weeks after the death of Martin Luther King, Jr. The newest round of riots, sparked by King's assassination, reemphasized the urgency of the Kerner Report and its call for increased media coverage by minorities in the face of a growing separation between blacks and whites. Meanwhile, NET began to develop a national monthly magazine format "by, for and about" black Americans. Called "Black Journal," it had a predominantly black staff that included Lou Potter as "Editor," Sheila Smith as researcher, Madeline Anderson as film editor, and Charles Hobson, Kent Garret, St. Clair Bourne, and Horace Jenkins as associate or full producers. From the many who auditioned to fill the roles of co-hosts, Lou House (who later changed his name to Walli Sadiq) and Bill Greaves were selected. Alvin H. Perlmutter, who is currently known for producing various Bill Moyers specials and "Adam Smith's Money World," was at the top of the pyramid, acting as executive producer.

The series debuted in June and was broadcast during prime time by many public television stations (Wednesdays at 9 P.M. in New York City). The first program displayed remarkable promise. It begins with Martin Luther King, Jr.'s, widow, Coretta Scott King, giving a commencement speech at Harvard, and concludes with a brief, reasonably sympathetic portrait of the Black Panthers. Not only is the spectrum of black political opinion surveyed, but there is a historical segment on the black press. Even the portrait of the only black jockey in the United States is given a historical context, reminding viewers that jockeys of African descent had once been common in horse racing.

After the third program had been aired, certain contradictions within the production of "Black Journal" had crystallized. Although the series was being sold as "by, for and of the black community," the white Perlmutter was firmly in charge, and the programming was often dominated by white-produced segments. In mid-August, there was a palace revolt. Eleven of the 12 black staff members resigned in protest. NET was ready to rescind the "by, for and of" claims as deceptive. The staff members, in contrast, demanded a black executive producer—suggesting Lou Potter. NET maintained that Potter lacked adequate experience. The staff then suggested the two producers of *Still a Brother*, William Greaves and William Branch. Both were, NET claimed, unavailable.[6] Greaves, in fact, was vacationing on Cape Cod, and a quick phone call ascertained that he would take the position while retaining his role as co-host. Perlmutter became a consultant to the series, and black representation on the staff was increased.

Program No. 5 for "Black Journal" (October 1968) shows the series in full stride. In some respects, the format and aspirations of the series have changed little. Most obviously, Greaves now wears a dashiki instead of a sports coat and turtleneck, and Lou House begins and ends the program with greetings to "brothers and sisters" and a few words of Swahili. More substantively, the staff investigates controversial issues, such as the crisis surrounding the Community School Board in Oceanhill-Brownsville, in a polished and insightful manner. The producers emphasize those ways in which community control can provide better schooling that will result, for example, in dramatically improved reading scores. Black members of the school board make the case for community control, while the efforts of the United Federation of Teachers and its president, Albert Shanker, to subvert such an administrative structure are convincingly documented.

While offering a multiplicity of voices from within the African-American community, "Black Journal" presents forthright editorial comments without feeling the need to give "equal time" to extremely conservative blacks or to white spokespeople. In Program No. 5's short panel discussion, Professor Charles Hamilton, co-author with Stokely Carmichael of *Black Power*, simply states that there can be no peace in the nation until the United States gets out of Vietnam; ending the war is thus a key priority for African Americans. The program condemns the expulsion of protesting "black power" medal winners from the American Olympic team as excessive and insensitive to past racial injustices. It then documents the long-standing devaluation of African-American history and scholarship by profiling Professor William Leo Hansberry, a prominent scholar who was once denied a Ph.D. by Harvard University because no one at that institution was qualified to supervise his dissertation on African-American history.

"Black Journal" clearly deserved the Emmy it received in 1969. In a manner unique to magazine-format programming, the events of the present are situated in the context of unfolding African-American history, giving them deeper meaning and resonance.[7] Black identity is powerfully constructed. "Black Journal" consistently shows representatives of the African-American community to be reasonable, articulate, and authoritative. These spokespeople are often in the position of judging the antisocial behavior of hysterical, unreasonable whites such as Albert Shanker, Presidential candidate George Wallace, or the Oakland police chief who condemns the Black Panthers in vitriolic terms. Traditional codings of authority by race are inverted; the nature of mainstream television representations stands exposed. As the experienced Greaves told his young staff, never again were they likely to find a production situation that was so protective of their views and offered them so much freedom.

Greaves on "Black Journal"

Periodically there was a little anxiety at NET, for instance when we decided to do a show on the Black Muslims, or Paul Robeson, or Malcolm X; but quite interestingly we had a great deal of freedom on that show. That is to say I was not bugged by the management of NET for several good reasons. First, they were basically people of good will. But more importantly, perhaps, was the fact that we had developed a lot of political clout. I had purposely cultivated the black press, so they were very much behind us; I cultivated the people in the Congressional Black Caucus. And, of course, there were all these riots and demonstrations going on, so they knew if they in a sense touched us, they might get burned. I'm overdramatizing this, but the situation in the sixties was so volatile and tense that it would have made no sense whatsoever for them to have this heavy hand on a show that had been put together specifically for the purposes of expressing the concerns of the black community. That was the point made by the Kerner Commission on Civil Disorder and the Carnegie Endowment. They would have been violating the mandate of the show.

While serving as executive producer on "Black Journal," Greaves continued to operate his own production company, William Greaves Productions, which he had set up in 1964. In 1967 he applied his long-standing interest in acting and dramatic processes to a highly innovative feature film, eventually titled *Symbiopsychotaxiplasm: Take One*. Still awaiting commercial release, this picture eludes traditional generic categories, being an often humorous *cinéma-vérité*-style documentary about the filming of a screen test for a larger dramatic work—and ultimately about its own making as well. In its freewheeling camera style, its playful editing and jump cutting, its use of direct camera address and improvisation, its self-aware, tongue-in-cheek humor, and its foregrounding of the filmmaking process and the medium's materiality, *Symbiopsychotaxiplasm* shows affinities with the contemporaneous French New Wave, avant-garde American cinema, and *cinéma-vérité* documentary. In many respects, Greaves's work predates the wave of American features that were to make use of such techniques over the next few years, from Haskell Wexler's *Medium Cool* (1969), which mixes documentary and fiction about events coinciding with the 1968 Chicago Democratic convention, to Rick King's *Off the Wall* (1976), in which a counterculture youth steals a camera from the documentary crew that is filming his life and begins to make his own record of life on the lam.

One of the most distinctive aspects of *Symbiopsychotaxiplasm: Take One* is its emphasis on both film-making and acting as creative, improvisational processes. The actors do not merely play out the drama-within-the-film in New York's Central Park, they are actively involved in shaping it; the director attempts to put the rehearsing performers into a framework of tension and confrontation with each other and with himself—and then records the results. These planned dramatic conflicts sometimes spill out beyond the realm of fictional drama into actual tantrums and frustrations.

The affinity between this approach and psychodrama therapy, in which patients act out their anxieties and conflicts, is hardly coincidental. Greaves has had a long-standing interest in psychodrama, seeing it as closely allied with the techniques of method acting.[8] He focused on psychodrama sessions in two later documentaries—*In the Company of Men* (1969) and *The Deep North* (1988).

As Greaves has explained his approach, "Everything that happens in the *Take One* environment interrelates and affects the psychology of the people and indeed of the creative process itself." Greaves's shooting methods—the simultaneous use of numerous cameras to cover both the drama and the filming context—are designed to best capture this total interactive "environment." Wary of Greaves's approach, the production crew film their own meeting, a kind of mini-"revolt," over the shape (or lack of one) that the film appears to be taking. Within the film's frustrated diegesis, Greaves plays the role of a rather inept director trying to make a film tentatively entitled *Over the Cliff*. Yet through his own audacity and directorial vision, Greaves the film-maker comically upends the demeaning stereotypes of black ineptitude that haunt American cinema. The racially mixed film crew of men and women (a makeup that was then quite unusual in film-making) is itself refreshingly open and committed in its ardent questioning of creative processes, conventional aesthetic forms, and, ultimately, attitudes toward sexuality (albeit in some terms that today one may easily find off-putting). It is a film that continued Greaves's interest in issues beyond the immediate ones of African-American politics and identity—and yet the film's failure to win critical affirmation and a commercial release discouraged him from pursuing further work along these lines.

Greaves on *Symbiopsychotaxiplasm: Take One*

Symbiopsychotaxiplasm *is neither a documentary nor a traditional feature. At least I don't feel that it is. It is more of a happening. Instead of being a form of conventional art it is a piece of abstract art. Abstract in the sense that it does not obey the language of convention. It obeys the mind, the heart, the intuition, the subconscious. These are the determinants, rather than the Aristotelian approach to drama—the traditional dramatic form of Sophocles or Ibsen or whomever. You're going for—let's call it divine action, another level of insight into the human condition, using cinema.*

The fact is that we could take this event—this scene, this screen test—and throw it into a community of actors and cinema technicians and no matter how it fell, it would be a film. Before we knew it, we were dealing with some of the basic points of drama, which is conflict and development, progression, a rising conflict into some kind of crisis, climax and some resolution. It may not happen as we would like it, but some variant of that theme will occur. The problem for the film-maker is to find what the variant is and how to put it together in the editing room with the materials you have.

Ultimately Greaves recognized that he had to either become a full-time television executive or retain his independence as a film-maker and devote greater energies to independent production. In 1970 he left "Black Journal." The following year he made the feature-length "docutainment movie" *Ali, the Fighter*, about Muhammad Ali's first, unsuccessful effort to reclaim his heavyweight crown from Joe Frazier. (Ali had been stripped of his crown because of his radical politics and opposition to the Vietnam War.) Greaves deftly interweaves exchanges between fighters and their fans with scenes of press conferences, training sessions, and business discussions—then ends with the fight itself. The film's behind-the-scenes images often pertain to the economic politics—and more implicitly the racial politics—involved in the promotion of the fight. While *Ali, the Fighter* received national distribution in commercial theaters, Greaves was not so fortunate with *Nationtime: Gary*, his film of the historic first National Black Political Convention in Gary, Indiana. Attended by about 10,000 people, this 1972 gathering included representatives from the full range of African-American culture and politics, from Amiri Baraka (LeRoi Jones) to Jesse Jackson and Coretta Scott King. The film covers the efforts of participants to create a platform acceptable to numerous constituencies within the black community. Jackson steals the show with a rousing and uncompromising speech calling for black political unity in the face of white-dominated party politics. The subject matter was considered too militant by commercial broadcasters, and the film never aired.

Greaves on *Nationtime: Gary*

There was a guy who came into my office and said to me, "The Gary convention is going to be the greatest event in the history of black America, and you've got to take some cameras there. I can get you some money to make this film, and then after you make it you can sell it to television." He talked like that. "Money is no object." To make a long story short, we went down there and ended up paying our own fares because I was interested in the event anyway. We took some raw stock and filmed this event. Our company paid for that film entirely. So it practically bankrupted us (as have several other films). But we put the film together and I got Sidney Poitier and Harry Belafonte to do the narration. And I thought that with the two of them we wouldn't have any problem getting it onto the networks. But the networks and local stations wouldn't touch it. They thought the whole event was too militant and that the film was, as well. Don't misunderstand me. There were some technical problems with it that we couldn't afford to fix. But essentially it was a major, major event. How can you say you don't want to show material of the crucifixion because it is out of focus? I'm not saying it was of that magnitude but it was a very important historical moment.

At this same time, Greaves relied for much of his income on films made for the Equal Opportunity Commission (*Voice of La Raza*, 1972), the Civil Service Commission (*On Merit*, 1972), NASA (*Where Dreams Come True*, 1979), and other government agencies; but he also began to produce important historical documentaries, beginning with *From These Roots* (1974), a look at the Harlem Renaissance of the nineteen-twenties. Among other things, the film was a return to Greaves's own roots growing up in Harlem. This pioneering effort treated the major contributors to that Renaissance with much greater sympathy and insight than the then standard book on the subject—Nathan Huggins's *The Harlem Renaissance*. With little stock footage of Harlem and its intellectuals available, Greaves decided to construct a film composed exclusively of photographs. What emerged was a compelling portrait of a community made strong by relative freedom and opportunity. Within the larger context of documentary practice, the film helped to inaugurate a cycle of city neighborhood films that focused on local communities. One of the best known, William Miles's *I Remember Harlem* (1980), owed much to Greaves's earlier effort.

Greaves went on to direct a number of biographical portraits of significant figures in African-American history. Two efforts from the early nineteen-eighties were *Booker T. Washington: The Life and Legacy* (1982) and *Frederick Douglass: An American Life* (1984), both for the National Park Service. *Booker T. Washington* is notable for the way it seeks to understand the historical pragmatics of Washington's accomodationist politics, while also continuing to question them through critical responses from a committed W. E. B. DuBois and a "reporter" (played by Gil Nobel) covering his life. These two half-hour films are basically dramatizations, but they also incorporate the limited number of available archival illustrations. (The paucity of visual documentation on important figures in black history is something with which Greaves and other film-makers must frequently struggle, as did Jackie Shearer with her recent documentary *The Massachusetts 54th Colored Regiment* [1991].) At times *Booker T. Washington* and *Frederick Douglass* bear the trappings of their educational purpose quite heavily. Yet they, along with the numerous other government-sponsored films on a broad range of topics, enabled Greaves to keep working and producing films on important African-American subjects—no mean achievement during a period in which projects concerned with African-American culture were receiving little national attention or funding.

Greaves's most recent work in the area of biography is the documentary *Ida B. Wells: A Passion for Justice*, co-produced with his wife, Louise Archambault. A reporter who developed her craft in Memphis, Tennessee, Wells became an important black leader who knew how to use both the black and mainstream press to support the struggle against racism—particularly lynchings—and to fight for women's suffrage. She realized that newspapers could mobilize citizens to boycott either specific businesses or whole towns which failed to acknowledge their patronage with appropriate services and legal due process. Again Greaves encountered a paucity of visual documentation: fewer than 15 photographs of this courageous, militant woman survive, and almost all are formal portraits. Yet the film-maker succeeds in shaping these limited materials into a masterful film. Toni Morrison reads movingly from Ida Wells's autobiography, which functions not unlike a protracted interview with Wells herself. In all these biographical documentaries, Greaves explores the possibilities and responsibilities of leadership within the black community. He investigates the parameters within which these individuals operated and the social and economic forces to which they were attuned and which they mobilized. Biography is used as a way of presenting African-American history to both a general audience and, more specifically, the black community.

These historical portraits are balanced by such documentaries as *Black Power in America: Myth or Reality?* (1986), which profiles a group of successful African Americans working in professions not traditionally associated with black leadership. This hour-long program includes Franklin Thomas, head of the Ford Foundation, June Jackson Christmas, psychiatrist; Clifton Wharton, chancellor of the SUNY system; Charles Hamilton, political scientist; and Richard Hatcher, mayor of Gary, Indiana. Yet just as the critic is ready to suspect that Greaves has been overwhelmingly preoccupied with the black elite, a film like *Just Doin' It* (1976), an informal *cinéma-vérité* look at two neighborhood barbershops in Atlanta, defeats any such easy conclusion.[9]

Greaves has constantly struggled against being stereotyped in his work—as an actor and as a film-maker. His work has always displayed diversity: he has balanced his numerous documentaries with repeated forays back into fiction film-making, such as *Bustin' Loose* (as executive producer, 1981) and the never-released, hurriedly made black exploitation feature *The Marijuana Affair* (1974). Furthermore, Greaves has alternated films on contemporary subjects and issues with historical treatments. Films focusing on African-American concerns are countered by numerous films preoccupied with other issues (i.e., *Symbiopsychotaxiplasm:*

Take One); according to Greaves, roughly half his films have addressed topics other than the black experience. Industrials and government-sponsored films that operate within circumscribed parameters are offset by films in which Greaves took large artistic or financial risks.

In many respects Greaves has adapted what is most positive and progressive in Grierson's writings regarding the possibilities of and need for non-fiction films that can inform and educate the public. His approach has differed from that of many leftist or art-oriented documentarians—Barbara Kopple being one example of the former and Errol Morris an instance of the later—in that his conception of film-making avoids fetishizing the individual work and instead looks to each work as one instance in a larger struggle. It takes a pragmatic rather than a romantic approach, one that has its roots in the black film-making experience—in the race films of Oscar Micheaux, Spencer Williams, and William Alexander, which were typically made under remarkable financial constraints. Yet if Greaves's career, like that of Melvin Van Peebles, resonates with this legacy, it has done so within an entirely new social and cultural framework. This framework, characterized by the end of legally sanctioned segregation (though not of racial discrimination) and by the dominance of television, has altered the very terms of black film-making. Like those leaders that are the subject of some of his films, Greaves has had insight into the changing realities of his time, has persisted, and, often enough, has triumphed.

An examination of the career of William Greaves suggests that we need to rethink our conception and periodization of documentary film practice, which has typically been divided into two eras—the one before the *cinéma-vérité* revolution of 1960 (e.g., *Primary, Chronicle of a Summer*) and the one after. There are other turning points of equal or perhaps even greater importance, not all having to do with technology. The year 1968 can be seen as a watershed, a moment when access to the means of production and distribution began to be more open; not only "Black Journal" but "Inside Bedford Stuyvesant" and "Like It Is" also began to air in that year. These and other initiatives—such as Newsreel, Third World Newsreel, and New Day Films—began to chip away at white male hegemony in documentary film-making. Today, documentarians come from much more diverse backgrounds in terms of race, gender, and publicly acknowledged sexual orientation. Although problems of discrimination and social democracy have not been fully overcome even in this limited area, the manner in which these substantial changes have occurred needs to be better understood. Such historical reconsiderations are particularly urgent at a moment when many ideologues have launched gross polemics against multiculturalism, "political correctness," and arts funding—seemingly to taint if not obliterate our memory of these achievements.

Notes

1. Like many black-cast films, *Miracle in Harlem* had a white director (Jack Kemp)—as did *Lost Boundaries* (Alfred L. Werker).

2. In fact, Greaves was seriously considered for the part in *No Way Out* that launched Poitier to stardom. Greaves's association with renegade de Rochemont, however, may have hurt his chances of being selected for the role.

3. This and subsequent quotations come from the authors' interviews with William Greaves in April and May, 1991.

4. Both stars were hot properties, boasting Oscar nominations for that year: Gloria Swanson as Best Actress, for *Sunset Boulevard*. José Ferrer went on to win the Oscar for Best Actor in *Cyrano de Bergerac*.

5. Greaves has recently succeeded in acquiring the distribution rights to this film through his own company.

6. "11 Negro Staff Members Quit N.E.T. 'Black Journal' Program," *New York Times*, 21 August, 1968, p. 91.

7. Because the shows work effectively as unified wholes, the screening of excerpted segments was possibly the only disappointment of the Brooklyn retrospective.

8. Greaves had taught method acting for Lee Strasberg over a 12-year period, and occasionally substituted for Strassberg at the Actors Studio when he was unavailable.

9. Interestingly, Greaves's *Just Doin' It* predates Spike Lee's fictionalized portrait of a neighborhood barbershop, *Joe's Bed-Stuy Barbershop: We Cut Heads* (1982), by six years.

Black American Cinema

The New Realism

Manthia Diawara

The release of D. W. Griffith's *The Birth of a Nation* in 1915 defined for the first time the side that Hollywood was to take in the war to represent Black people in America. In *The Birth of Nation*, D. W. Griffith, later a founding member of United Artists, created and fixed an image of Blackness that was necessary for racist America's fight against Black people. *The Birth of a Nation* constitutes the grammar book for Hollywood's representation of Black manhood and womanhood, its obsession with miscegenation, and its fixing of Black people within certain spaces, such as kitchens, and into certain supporting roles, such as criminals, on the screen. White people must occupy the center, leaving Black people with only one choice—to exist in relation to Whiteness. *The Birth of a Nation* is the master text that suppressed the real contours of Black history and culture on movie screens, screens monopolized by the major motion picture companies of America.

Griffith's film also put Black people and White liberals on the defensive, inaugurating a plethora of historical and critical writings against *The Birth of a Nation*, and overdetermining a new genre, produced exclusively for Black audiences, called race films. More insidiously, however, the racial conflict depicted in *The Birth of a Nation* became Hollywood's only way of talking about Black people. In other words, whenever Black people appeared on Hollywood screens, from *The Birth of a Nation* to *Guess Who's Coming to Dinner?* to *The Color Purple*, they are represented as a

Reprinted from Manthia Diawara, ed., *Black American Cinema* (NY: Routledge, 1993), pp. 3–25.

problem, a thorn in America's heel. Hollywood's Blacks exist primarily for White spectators whose comfort and understanding the films must seek, whether they thematize exotic images dancing and singing on the screen, or images constructed to narrate a racial drama, or images of pimps and muggers. With *The Birth of a Nation* came the ban on Blacks participating in bourgeois humanism on Hollywood screens. In other words, there are no simple stories about Black people loving each other, hating each other, or enjoying their private possessions without reference to the White world, because the spaces of those stories are occupied by newer forms of race relation stories which have been overdetermined by Griffith's master text.

The relations between Black independent cinema and the Hollywood cinema just described above parallel those between Blackness and Americanness; the dichotomy between the so-called marked cultures and unmarked cultures; but also the relations between "high art" and "low art." The complexity of these relations is such that every independent filmmaker's dream is to make films for Hollywood where she/he will have access to the resources of the studios and the movie theaters. On the other hand, the independents often use an aesthetic and moral high ground to repudiate mainstream cinema, which is dismissed as populist, racist, sexist, and reactionary. Furthermore, a look at the relations between Oscar Micheaux and the Hollywood "race films," Melvin Van Peebles and the Blaxploitation films, Charles Burnett *(Killer of Sheep)*, Haile Gerima *(Bush Mama)*, and Spike Lee and the rethematization of urban life in such films as *City of Hope, Grand Canyon, Boyz N the Hood,* and *Straight Out of Brooklyn* reveals that mainstream cinema constantly feeds on independent cinema and appropriates its themes and narrative forms.

Some of the most prominent Black film historians and critics, such as Albert Johnson, Donald Bogle, and Thomas Cripps, emphasize mainly mainstream cinema when discussing Black films. With the exception of a few breatkthrough films, such as those by Micheaux, Van Peebles, and Lee, these historians are primarily concerned with the issues of integration and race relations in mainstream films, Black actors and actresses on the big screen, and the construction of stereotypes in Hollywood films. They rarely pay attention to independent cinema, which includes far more Black directors than Hollywood, and in which aesthetics, political concerns such as authorship and spectatorship, and the politics of representation with respect to Black cinema are more prevalent. Critics and historians such as Clyde Taylor, Toni Cade Bambara, Phyllis Klotman, and Gladstone Yearwood are the first to focus on Black independent cinema as a subject of study. More recently, the *Black Film Review* has assumed the preeminent role in Black film history and criticism.

Hollywood's block-booking system prevents independently produced films from reaching movie theaters and large audiences. This may be one reason why film historians and critics neglect independent cinema: some film magazines, such as *Cineaste*, adopt a policy of accepting only reviews of films that have been distributed and seen by their readers. It is also possible to argue that Black independent cinema has remained marginal until now because its language, not unlike the language of most independent films, is metafilmic, often nationalistic, and not "pleasurable" to consumers accustomed to mainstream Hollywood products. Black independent cinema, like most independent film practices, approaches film as a research tool. The filmmakers investigate the possibilities of representing alternative Black images on the screen; bringing to the foreground issues central to Black communities in America; criticizing sexism and homophobia in the Black community; and deploying Afrafemcentric discourses that empower Black women. The narratives of such films are not always linear; the characters represent a tapestry of voices from W. E. B. DuBois, Frantz Fanon, Toni Morrison, Malcolm X, Martin Luther King, Jr., Karl Marx, Angela Davis, Alice Walker, and Zora Neale Hurston. Even what passes as documentary in Black independent films, like *The Bombing of Osage Avenue* (Louis Massiah), is an artistic reconstruction of archival footage and "real" events.

What is, therefore, the Black independent cinema, and what constitutes its influence on mainstream cinema? The French appropriately refer to independent cinema as *cinema d'art et essai*. In France, the government sponsors such a cinema by imposing a distribution tax on commercial films. The *cinema d'art et essai* is less concerned about recouping its cost of production and making a profit; its main emphasis is toward artistic development, documenting an area of research, and delineating a certain philosophy of the world. In the late 1950s, a group of French youth, who were dissatisfied with commercial films and wanted to make their own films, mobilized private and personal funds along with government funds to produce low-budget films. The result is well known today as the French New Wave, considered by some as one of the pivotal moments in film history.

As an alternative to commercial cinema, which emphasized the well-made story, acting, and the personality of the actor, the New Wave put in the foreground the director, whom it raised to the same artistic level as the author of a painting, a novel, or a poem; the New Wave also demystified the notion of the well-made story by experimenting with different ways of telling the same story, and by deconstructing the notion of actor and acting. Jean-Luc Godard's *Breathless* (1959), for example, is famous for its reinsertion of the "jump-cut" as a valid narrative device. The jump-cut, which was avoided in Hollywood films in order not to disrupt the spectator with "unnecessary" repetitions, has today become a powerful narrative device used by directors such as Spike Lee, who redefines it and uses it to describe the repetition and the sameness in racial and sexual stereotyping. In *Do the Right Thing* (1988) Lee uses the same angle to repeat several shots of Blacks, Italians, Jews, and Koreans repeating racial stereotypes, unlike Godard, who uses the same image twice from the same angle. Lee practices the same device in *She's Gotta Have It* (1985) to construct sexual stereotypes among young Black males.

This example of the New Wave reveals that independent filmmakers come to their vocation for at least two reasons: one political, and the other artistic. Politically, they are dissatisfied with commercial cinema's lack of courage to address certain issues. They feel that they have to make their own films if they want to see those issues on the screen. Artistically, they want to explore new ways of telling stories; they want to experiment with the camera, the most powerful invention of modern times, and engage the infinite possibilities of storytelling. There are other examples of alternative or independent cinemas that occupy important places in the history of film. The Italian Neorealism, the Brazilian Cinema Novo, and the Argentinian Third Cinema have all created alternative narrative techniques that were at first unknown to commercial cinemas, but are claimed today as part of traditional narrative practices.

Similarly, the cloning of Hollywood's mind to Black history and culture, which do not revolve around White people, is the reason why most Black filmmakers since Oscar Micheaux have turned first to the independent sector. Since Oscar Micheaux, Black independents have pioneered creating alternative images of Blacks on the screen, constructing new narrative forms derived from Black literature and folklore, and denouncing racism, sexism and homophobia in American culture.

This is not, however, to romanticize the independent practice. Micheaux made his films by selling personal property and borrowing money from friends. Still today, independent filmmaking causes many people to become poor. It take more than six years for some filmmakers to gather the money for one film. Charles Burnett's *To Sleep With Anger,* and Julie Dash's *Daughters of the Dust* came only after arduous years of fundraising. Haile Gerima has been trying to raise funds for *Nunu* for several years now. We have not seen second features by talented directors such as Billy Woodberry *(Bless Their Little Hearts),* Larry Clark *(Passing Through),* Alile Sharon Larkin *(A Different Image),* and Warrington Hudlin *(Street Corner Stories).* Spike Lee sums up the harsh reality of independent production as follows:

> When I went to film school, I knew I did not want to have my films shown only during Black History Month in February or at libraries.

> I wanted them to have a wide distribution. And I did not want to spend four or five years trying to piecemeal together the money for my films. I did my first film, *She's Gotta Have It,* independently for $175,000. We had a grant from the New York State Council on the Arts and were raising money the whole time we were shooting. We shot the film in twelve days. The next stage was to get it out of the lab. Then, the most critical part was when I had to hole up in my little apartment to get it cut. I took about two months to do that. I had no money coming in, so I had to hold off the debtors because I knew if I had enough time to at least get it in good enough shape to show, we could have some investor screenings, and that's what happened. We got it blown up to 35mm for a film festival. What you have to do is to try to get a distributor. You enter as many film festivals as you can.[1]

Black independent cinema is any Black-produced film outside the constraints of the major studios. The filmmakers' independence from Hollywood enables them to put on the screen Black lives and concerns that derive from the complexity of Black communities. Independent films provide alternative ways of knowing Black people that differ from the fixed stereotypes of Blacks in Hollywood. The ideal spectators of the films are those interested in Black peoples' perspectives on American culture. White people and Whiteness are marginalized in the films, while central positions are relegated to Black people, Black communities, and diasporic experiences. For example, the aesthetics of uplifting the race in a film like *The Scar of Shame* (1928, The Colored Players) concern particularly Black spectators, whom the filmmakers' stated mission is to entertain and educate. The film posits Black upper-class culture as that which should be emulated by lower-class Blacks in order to humanize themselves. Unlike Hollywood films of that time, which identified with the ideal White male, the camera in *The Scar of Shame* identifies with the position of the Black bourgeoisie. The film is precious today as a document of Black bourgeois ways of being in the 1920s and 1930s. Crucially, it constitutes, with Oscar Micheaux's films, a genre of Black independent cinema which puts Black people and their culture at the center as subjects of narrative development; in these films, Black people are neither marginalized as a problem, nor singled out as villainous stereotypes such as Hollywood constructs in its films.

Contemporary independent films continue the same effort of inquiring into Black subjectivities, the heterogeneity of Black lives, the Black family, class and gender relations, and diasporic aesthetics. Recently, independent Black women filmmakers such as Kathleen Collins *(Losing Ground),* Alile Sharon Larkin *(A Different Image),* Ayoka Chenzira *(Zajota: the Boogie Spirit),* Julie Dash *(Daughters of the Dust),* and

Zeinabu Davis *(A Powerful Thang)* have explored such themes as Black womanhood and spirituality, diaspora art and music, and Afrocentric aesthetics. Black manhood, the urban landscape, unemployment and the Black family are thematized in films like *Sweet Sweetback's Baadasssss Song* (Van Peebles), *Killer of Sheep* (Burnett), *Bless Their Little Hearts* (Woodberry), *Serving Two Masters* (Tim Lewis), *Street Corner Stories* (Warrington Hudlin), *Chameleon Street* (Wendell Harris), and *Ashes and Embers* (Haile Gerima). The themes of sexuality and homophobia are depicted in *Tongues Untied* (Marlon Riggs), *Storme: Lady of the Jewel Box* (Michelle Parkerson), *She's Gotta Have It* (Spike Lee), *Ganja and Hess* (Bill Gunn), *Splash* (Thomas Harris), and *She Don't Fade* (Cheryl Dunye). The major Black documentary artists, such as William Greaves, Louis Massiah, Camille Billops, and St. Clair Bourne, have also enriched the documentary genre by focusing their cameras on Black people in order to reconstruct history, celebrating Black writers and activists, and giving voice to people who are overlooked by television news and mainstream documentaries.

Two Paradigms of Black Cinema Aesthetics

Jane Gaines defines Oscar Micheaux's editing style as follows: "Perhaps to elude any attempt to essentialize it, we could treat this style as more of an ingenious solution to the impossible demands of the conventions of classical Hollywood style, shortcuts produced by the exigencies of economics, certainly, but also modifications produced by an independent who had nothing at stake in strict adherence to Hollywood grammar." Gaines goes on to posit that Micheaux's "freewheeling cinematic grammar" constitutes both a misreading and an improvement upon Hollywood logic. Clearly, Micheaux's "imperfect" cinema (to borrow a term from Julio Garcia Espinoza), which misreads and improves upon Hollywood logic, is a powerful metaphor for the way in which African Americans survived and continue to survive within a hostile economic and racist system, and used the elements of that survival as raw material to humanize and improve upon American modernism. Micheaux's "loose editing," like the improvisation of jazz, surprises and delights the spectator with forbidden images of America that Hollywood's America conceals from its space. In so far as the classical Hollywood narrative proceeds by concealment of space, Micheaux's "imperfect" narrative constitutes an excess which reveals the cheat cuts, the other America artificially disguised by the Hollywood logic. It is in this sense that Gaines writes of improvement of film language by Micheaux. In *Black American Cinema* Ron Green compares Micheaux's film style to Black English, and to jazz. His cinema is one of the first to endow African Americans with cinematic voice and subjectivity through his uncovering of new spaces at the threshold of dominant cinema.

The first step in interpreting a Black film aesthetic must therefore be directed towards an analysis of the composition of the new shots discovered by Micheaux, and their potential effects on spectators. In *Black American Cinema* Micheaux's films are discussed in an in-depth manner for the first time by Jane Gaines and Ron Green. Micheaux's legacy as an independent filmmaker not only includes his entrepreneurial style in raising money and making films outside the studios. He also turned his cameras towards Black people and the Black experience in a manner that did not interest Hollywood directors of race films. Crucially, Micheaux's camera positioned Black spectators on the same side as the Black middle-class ideology, acquiring for his films an aesthetic that was primarily specific to the ways of life of that class.

Similarly, in the 1970s, Melvin Van Peebles and Bill Gunn positioned spectators with respect to different imaginaries derived from the Black experience in America. In *Sweet Sweetback's Baadasssss Song,* Van Peebles thematizes Black nationalism by casting the Black community as an internal colony, and Sweetback, a pimp, as the hero of decolonization. In *Black American Cinema*, Toni Cade Bambara refers to *Sweet Sweetback* as "a case of Stagolee meets Fanon or Watermelon Man plays Bigger Thomas?" *Sweet Sweetback* is about policing and surveillance of Black communities, and the existentialist struggle of the film's main character, a Black man. As Bambara notices, Bigger Thomas is not the only literary reference in the film; it also draws on the theme of the running Black man in *Invisible Man*, which is collapsed into a transformed Hollywood stereotype of the Black stud. As such, *Sweet Sweetback* is famous as the paradigmatic text for the 1970s Blaxploitation films. The theme of the Black man running from the law or from Black-on-Black crime, which links Van Peebles to such Black American writers as Richard Wright, Ralph Ellison, and Chester Himes, is also echoed in 1990s films like *Juice, Straight Out of Brooklyn*, and *Boyz N The Hood*, not to mention *New Jack City*, a film directed by Van Peebles's son Mario Van Peebles.

Sweet Sweetback's aesthetic draws on the logic of Black nationalism as the basis of value judgment, and defines itself by positioning the spectator to identify with the Black male hero of the film. Bambara rightly criticizes the centrality of Black manhood at the expense of women in *Sweet Sweetback*, but recognizes nationalist narratives as enabling strategies for survival, empowerment, and self-determination. As Sweetback is helped to escape from the police by one Black person after another, the nationalist discourse of the film transforms the ghetto, where Black people are objects, into the community, where they affirm their subjecthood. To put it in Bambara's words, "Occupying the same geographical terrain are the *ghetto*, where we are penned up in concentration-camp horror, and the *community*, where we enact daily rituals of group validation in a liberated zone."

In *Ganja and Hess*, Bill Gunn aestheticizes the Black imaginary by placing the spectator on the same side as the Black church. The spectator draws pleasure from the film through the confrontation between the ideology of the Black church and vampirism, addiction to drugs and sex, and materialism. *Ganja and Hess* is perhaps the most beautifully shot Black film, and the most daring with respect to pushing different passions to their limits. The Black artist, Meda (played by Bill Gunn himself), is a nihilist who advocates total silence because, as a Black person, his art is always already overdetermined by race in America. The love scenes in the film are commingled with vampiristic gestures that are attractive and repulsive at the same time. At the Cannes Film Festival in 1973, Gunn's daring camera angles during one of the love scenes brought spectators to joy, applauding and screaming "Bravo! Bravo!" in the middle of the film. *Ganja and Hess* also pushes the classical narrative to the threshold by framing a frontal nude image of a Black man coming out of a swimming pool and running toward a window where a woman, Ganja (Marlene Clarke), smilingly awaits him.

What is radical about both *Ganja and Hess* and *Sweet Sweetback* is their formal positioning of Black characters and Black cultures at the center of the screen, creating a sense of defamiliarization of the classical film language. The two films also inaugurate for Black cinema two narrative tracks with regard to time and space. While *Ganja and Hess* is cyclical, going back and forth between pre-Christian time and the time after Christ, *Sweet Sweetback* is a linear recording of the progress of Black liberation struggle.

With regard to Black aesthetics, it is possible to put in the same category as *Ganja and Hess* such films as *A Powerful Thang* (Davis), *Daughters of the Dust* (Dash), *Losing Ground* (Collins), *Killer of Sheep* and *To Sleep with Anger* (Burnett), *Tongues Untied* (Riggs), and *She's Gotta Have It* (Lee). These films are concerned with the specificity of identity, the empowerment of Black people through mise-en-scène, and the rewriting of American history. Their narratives contain rhythmic and repetitious shots, going back and forth between the past and the present. Their themes involve Black folklore, religion, and the oral traditions which link Black Americans to the African diaspora. The narrative style is symbolic.

Sweet Sweetback on the other hand, defines its aesthetics through recourse to the realistic style in film. The story line develops within the logic of continuity editing, and the characters look ordinary. The film presents itself as a mirror on a Black community under siege. The real effect is reinforced throughout the film by events which are motivated by racial and gendered causes. The sound track and the costumes link the film to a specific epoch in the Civil Rights Movement. Unlike the first category of films, which uses the symbolic style and concerns itself with the past, *Sweet Sweetback* makes the movement toward the future-present by confronting its characters with the obstacles ahead of them. Other films in this category include *Cooley High* (Michael Schultz), *House Party* (Reginald Hudlin), *Chameleon Street* (Harris), *Passing Through* (Clark), *Do The Right Thing* (Lee), *Straight Out of Brooklyn* (Rich), *Juice* (Ernest Dickerson), and *Boyz N The Hood* (Singleton). These lists are neither exhaustive nor fixed. The realist category has more in common with the classical Hollywood narrative, with its quest for the formation of the family and individual freedom, and its teleological trajectory (beginning, middle, and end). The symbolic narratives have more in common with Black expressive forms like jazz, and with novels by such writers as Toni Cade Bambara, Alice Walker, and Toni Morrison, which stop time to render audible and visible Black voices and characters that have been suppressed by centuries of Eurocentrism.

The comparison of the narrative styles deployed by *Sweet Sweetback* and *Ganja and Hess*[2] is useful in order to link the action-oriented *Sweet Sweetback* to modernism, and the reflexive style of *Ganja and Hess* to postmodernism. *Sweet Sweetback* defines its Afro-modernism through a performative critique of the exclusion of Blacks from reaping the fruits of American modernity and liberal democracy. *Ganja and Hess* is a postmodern text which weaves together a time of pre-Christian Africa, a time of Christ's Second Coming in the Black church, and a time of liberated Black women. Crucially, therefore, the repetition of history as played out on the grid of the Black diaspora is important to the definition of Gunn's film language. Through the repetition of these Black times in the film, Bill Gunn defines a Black aesthetic that puts in the same space African spirituality European vampire stories, the Black church, addiction to drugs, and liberated feminist desires.

The New Black Films

It is easy to see the symbolic, reflexive, and expressive styles in films such as *Killer of Sheep* and *Daughters of the Dust*, and the active, materially grounded, and linear styles in *Boyz N the Hood.* But before looking more closely at these films, it is important to put into some perspective the ways in which Black films posit their specificity by challenging the construction of time and space in Hollywood films. It is only in this sense that arguments can begin about whether they displace, debunk, or reinforce the formulaic verisimilitude of Hollywood.

The way in which a filmmaker selects a location and organizes that location in front of the camera is generally referred to in film studies as mise-en-scène. Spatial narration in classical cinema makes sense through a hierarchical disposition of objects on the screen. Thus space is related to power and powerlessness, in so far as those who occupy the center of the screen are usually more powerful than those situated in the background or completely absent from the screen. I have described here Black people's relation to spatially situated images in Hollywood cinema. When Black people are absent from the screen, they read it as a symbol of their absence from the America constructed by Hollywood. When they are present on the screen, they are less powerful and less virtuous than the White man who usually occupies the center. Hollywood films have regularly tried to resolve this American dilemma, either through token or symbolic representation of Blacks where they are absent—for instance, the mad Black scientist in *Terminator 2*; or through a substitution of less virtuous Blacks by positive images of Blacks—for instance, *Grand Canyon* or *The Cosby Show*. But it seems to me that neither symbolic representation nor positive images sufficiently address the specificity of Black ways of life, and how they might enter in relation to other Americans on the Hollywood screen. Symbolic representation and positive images serve the function of plotting Black people in White space and White power, keeping the real contours of the Black community outside Hollywood.

The construction of time is similarly problematic in the classical narrative. White men drive time from the East to the West, conquering wilderness and removing obstacles out of time's way. Thus the "once upon a time" which begins every story in Hollywood also posits an initial obstacle in front of a White person who has to remove it in order for the story to continue, and for the conquest ideology of Whiteness to prevail. The concept of beginning, middle, and end, in itself, is universal to storytelling. The difference here is that Hollywood is only interested in White people's stories (White times), and Black people enter these times mostly as obstacles to their progress, or as supporting casts for the main White characters. "Once upon a time" is a traditional storytelling device which the storyteller uses to evoke the origin of a people, their ways of life, and the role of the individual in the society. The notion of *rite de passage* is a useful concept for describing the individual's separation from or incorporation into a social time. The classical narrative in cinema adheres to this basic ideological formula in order to tell White people's stories in Hollywood. It seems that White times in Hollywood have no effect on Black people and their communities: whether they play the role of a negative or positive stereotype, Black people neither grow nor change in the Hollywood stories. Because there is a dearth of Black people's stories in Hollywood that do not revolve around White times, television series such as *Roots,* and films such as *Do the Right Thing*, which situate spectators from the perspective of a Black "once upon a time," are taken out of proportion, celebrated by Blacks as authentic histories, and debunked by Whites as controversial.

To return again to the comparison between *Sweet Sweetback* and *Ganja and Hess*, it is easy to see how important time and space are to defining the cinematic styles they each extol. The preponderance of space in films such as *Ganja and Hess* reveals the hierarchies of power among the characters, but it also reveals the preoccupation of this style of Black cinema with the creation of space on the screen for Black voices, Black history and Black culture. As I will show later with a discussion of space in *Daughters of the Dust*, Black films use spatial narration as a way of revealing and linking Black spaces that have been separated and suppressed by White times, and as a means of validating Black culture. In other words, spatial narration is a filmmaking of cultural restoration, a way for Black filmmakers to reconstruct Black history, and to posit specific ways to being Black Americans in the United States.

The emphasis on time, on the other hand, reveals the Black American as he/she engenders him/herself amid the material conditions of everyday life in the American society. In films like *Sweet Sweetback* and *Boyz N the Hood*, where a linear narrative dominates, the characters are depicted in continuous activities, unlike the space-based narratives, where the past constantly interrupts the present, and repetitions and cyclicality define narration. Crucially, whereas the space-oriented narratives can be said to center Black characters on the screen, and therefore empower them, the Black-times narratives link the progress of time to Black characters and make times exist for the purpose of defining their needs and their desires. Whereas the space-based narratives are expressive and celebratory of Black culture, the time-based narratives are existentialist performances of Black people against policing, racism, and genocide. I would like now to turn to *Daughters of the Dust* and *Boyz N the Hood* to illustrate the point.

Space and Identity: Black Expressive Style in *Daughters of the Dust*

I am the first and the last
I am the honored one and the scorned one
I am the whore and the holy one
I am the wife and the virgin
I am the barren one and many are my daughters. . . .
I am the silence that you cannot understand. . . .
I am the utterance of my name.

Daughters of the Dust

I have argued that the Hollywood classical narrative often articulates time and space through recourse to a discriminating gaze toward American Blacks. When the story is driven by time and action, it is usually White times. I'll say more about this in my discussion below of *Boyz N the Hood*. Similarly, when spatial considerations dominate the production of the story, the purpose is usually to empower White men. Common sense reveals that characters that are more often on the screen, or occupy the center of the frame, command more narrative authority than those that are in the background, on the sides, or completely absent from the frame. By presence, here, I have in mind first of all the literal presence of White characters in most of the shots that constitute the typical Hollywood film, which helps to define these characters as heroes of the story. There is also the symbolic presence through which narrative authority for the organization of space is attributed to certain characters in the story. These devices of spatial narration are effective in linking characters with spaces, and in revealing space occupancy as a form of empowerment. For example, through the character played by Robert Duval in *Apocalypse Now*, Francis Ford Coppola parodies the power associated with White male actors such as John Wayne as they are framed at the center of the screen.

There is a preponderance of spatial narration in Julie Dash's *Daughters of the Dust*. Black women and men occupy every frame of the film, linking Black identity to a place called Ibo Landing in the Sea Islands of South Carolina, and, more importantly, empowering Black women and their ways of life. On a surface and literal level, the wide appeal of the film for Black women depends on the positioning of the women characters as bigger than life in the middle of the screen, which mirrors the beautiful landscape of Ibo Landing. Black women see themselves on the screen, richly adorned, with different hues of Blackness and Black hair styles, and flaunting their culture. In *Daughters of the Dust*, the screen belongs to Black women. At a deeper level, where space and time are combined into a narrative, Julie Dash emphasizes spatial narration as a conduit to Black self-expressivity, a storytelling device which interrogates identity, memory, and Black ways of life. *Daughters of the Dust* stops time at 1902, when the story was set, and uses the canvas of Ibo Landing in the Sea Islands to glance backward to slavery, the Middle Passage, African religions, Christianity, Islam, the print media, photography, moving pictures, and African-American folkways, as elements with which Black people must come to terms in order to glance forward as citizens of the United States. In other words, the film asks us to know ourselves first, know where we came from, before knowing where we are going. To put it in yet another way, Ibo Landing is a symbolic space in which African Americans can articulate their relation to Africa, the Middle Passage, and the survival of Black people and their ways of life in America. Crucially, the themes of survival, the memories of African religions and ways of life which enter into conflict with Christianity and European ways of life, and the film's proposal of syncretism as a way out, are narrativized from Black women's points of view. I want to take more time here to show how Julie Dash uses women's voices to make these themes compatible with the space of Ibo Landing.

The conflict in the film concerns the migration of the Peazant family from Ibo Landing of the Sea Islands to the North. At first the conflict is set in binary terms. For those who support the migration North, the space of Ibo Landing is primitive, full of people who worship the sun, the moon, and the river. The North therefore promises literacy, Christianity and progress. For Grandma Nana and the Unborn Child who link their identity to the space of Ibo Landing, the North represents the destruction of the family, disconnection from the ancestors, and the loss of identity for the children. For Grandma Nana, Ibo Landing is where the ancestors watch over the living, protect them, and guide them. It is in this sense that Nana does not want the family reunion to be a farewell party between those who are leaving and those staying. She prepares herself to give them something that they "can take North with [them] along with [their] big dreams."

As filmic space, Ibo Landing is the link between Africa and America. Or, to put it another way, Ibo Landing is Africa in America. According to the film, it is where the last slaves landed. *Daughters of the Dust* also argues that it is where African Americans remained isolated from the mainland of Georgia and South Carolina, and "created and maintained a distinct imaginative and original African-American culture." The Peazant family must therefore learn the terms of their belonging to Ibo Landing, which will be an example of African-American belonging to America, and must use the space of Ibo Landing to validate their identities as Americans of a distinctive culture. It is interesting to notice here that, unlike the Hollywood narratives which claim space only as a process of self-empowerment, *Daughters of the Dust* acknowledges through the letter that Iona receives from her Indian lover that the space belonged to the Indians first.

Weaving the voices of Grandma Nana, the Unborn Child, and Eula (the mother of the Unborn Child) through the spaces of Ibo Landing, Julie Dash creates a narrative that connects Africa to America, the past to the present. Using African ancestor figures as her narrative grid, she places Grandma Nana at the center of her story, and constructs oppositional characters around her. On the one hand we have Haagar, Viola, the Bible lady, and Eli, who is Eula's husband; on the other hand we have Yellow Mary, Eula, and Iona, who is Haagar's daughter. We have characters who are alike and who constitute reincarnations of ancestor figures with similar dispositions; and characters who are contraries of one another, and therefore require the intervention of the ancestors to bring peace and harmony.

Grandma Nana is the oldest person on the island. She spends most of her time visiting the graveyard where the ancestors are buried, and by the water which is a dwelling place of the spirits of the ancestors. I do not have enough space here to discuss the significance of water in *Daughters of the Dust*. But it is crucial to point out the recurring Middle Passage theme of Africans walking on water to go back to Africa. As an intertextual religious space, the use of water by Grandma Nana to communicate with the gods echoes *Yeelen* by Souleymane Cissé, where the mother bathes with milk in the middle of the river and asks the Goddess to protect her son. *Daughters* also reminds us of *Testament* by Black Audio Film/Video Collective, in which the characters walk into the middle of the river or visit graveyards in order to unlock the secret of the past. It would also be interesting to investigate the use of water in vases and on altars as a representation of Voodoo in *Daughters* and in *Dreaming Rivers* by Sankofa Film/Video Collective.

Daughters depicts the survival of African religious practices in Ibo Landing through Grandma Nana in other ways as well. She can hear the calls of the spirits, and, therefore, works with the Unborn Child to keep the family together. She teaches Eli about the core of African ancestor worship: "It's up to the living to keep in touch with the dead, Eli. Man's power don't end with death. We just move on to another place; a place where we go and watch over our living family. Respect your elders, respect your family, respect your ancestors."

A recourse to religion is central to the understanding of *Daughters of the Dust*. For Grandma Nana, ancestor worship provides the strongest stability for the Black family in America and Africa. Unlike Christianity and Islam, which are teleological and reserve the final reward for the end in Heaven, the ancestors in Grandma Nana's belief system just move to another world and watch over their living descendants. The children are the reincarnation of the ancestors, and this makes them precious to the adults whose fathers and grandfathers have joined the land of the ancestors. The Unborn Child in the film is one such reincarnation. She is doubled not only in the figure of Grandma Nana herself, but also in the young girl with tribal scars who appears with her mother in one of the flashbacks. She travels through time, and she is present at different settings in the film: we see her among the first generation of Africans working with indigo dye, and we see her in 1902 setting among children playing in the sand. Like the ancestors, her role is one of a mediator in the family. It is in this sense that Grandma Nana states that for Africans, the ancestors and the children are the most sacred elements of society.

Julie Dash also uses the religious theme of reincarnation, and links the Unborn Child to African-American survival during slavery, genocide, and the rape of Black women. In the film, the theme of the Peazant family's disintegration entailed by the migration to the North is replayed in the subtheme of Eli's self-exile from his wife, Eula, because she's carrying a child that Eli does not consider his. Eli's first reaction to Eula's pregnancy is to become an iconoclast toward the ancestor belief system that Grandma Nana wants to maintain. He puts into question the religion and culture he has received from childhood to adulthood. In other words: How can this happen to him, who has played by the rules? How come the gods are not avenging his misfortune? Subsequently, he picks up his ax and proceeds to smash all the fetishes that he had previously revered.

Grandma Nana finds an answer to Eli's blasphemous questions in her belief system. She links Eula's pregnancy to the condition of Black women in slavery who were raped, denied motherhood rights, and treated like animals. At the same time, the power and complexity of Black people come from their ability to maintain the sacredness of the womb by restoring to the group the children of interracial rape. Grandma Nana uses ancestor worship, and the place of children in it, to appropriate the baby Eula is carrying. By doing so, she bends the filiative and patriarchal rules Eli maintains in order to disavow the Unborn Child. For Grandma Nana, Eli, too, must learn the process of cleansing rape from the child's name, and making it his own child. Grandma Nana argues that the womb is as sacred as the ancestors, and that the Unborn Child is sent by the ancestors, precisely at this critical juncture in Ibo Landing's history, to ensure survival: "You need this one, Eli, to make the family stronger like it used to be." It is interesting to note the spatial organization as Grandma Nana talks to Eli. As the oldest person in the Peazant family, her role is that of a teacher. As she speaks to Eli, the space revealed on the screen is that of children playing games on the beach. The narrative implication here is that the children are the audience of her teaching. At one point during the children's game, the film changes to a slow motion. As the children fall on top of one another, we hear screaming and groaning, which reminds us of the Middle Passage during which hundreds of Africans were piled on top of each other in the cabins of slave ships. The implication of Grandma Nana's teaching is that, just as captured Africans were thrown together during that painful time of the Middle Passage, Blacks today must see themselves in the same boat, and fight together to "make the family stronger."

Eli's questions about the paternity and, therefore, the race of the Unborn Child also touch on the issues of light skin and dark skin, pure blood and mixed blood, superior and inferior; in short, we are dealing with racism among Blacks. It is in this sense that Yellow Mary is ostracized by Haagar and Viola, who use her light complexion as a sign of betrayal and try to banish her from Ibo Landing. For Grandma Nana, Yellow Mary and the Unborn Child contribute to the survival and maintenance of Black people in America, because their presence makes Blackness diverse and complex. Black survival in America confounds and embarrasses both Whiteness and essentialist notions of pure Africans. Julie Dash puts onto stage one of the most beautiful and powerful scenes in the film to illustrate this point. Haagar and others have been chastising Yellow Mary for not being Black enough, when Eula stands up and delivers a speech worthy of an ancestor figure. The mise-en-scène of this sequence reveals Black women in all their powers, as Eula reminds Haagar that no one is Blacker or purer than anyone else, and warns her and Viola about the wrath of the gods, if they were to continue their gesture of expelling Yellow Mary out of the race. Spatial representation again becomes paramount, because Eula's speech is directed to the on-screen audience of the Peazant family, as well as the off-screen spectators.

I have discussed so far the ways in which *Daughters of the Dust* uses African belief systems as the center which enables Black women and men to articulate their identities on the space of Ibo Landing. Grandma Nana, particularly, posits the ancestor worship system as a text which holds together the world of the Ibo Landing and provides answers to practical daily problems. A crucial question remains: whether the belief in ancestors can coexist with other belief systems, such as Christianity and Islam, on and off the island? At first, religious systems seem to be opposed in *Daughters of the Dust*. Bilal, who is Muslim, is opposed to the Baptists, who think that their God is better. Viola and Haagar use Christianity to elevate themselves above Grandma Nana. They see ancestor worship as an idolatry which is confined to Ibo Landing. They look to the North as a sign of enlightenment and Christian salvation.

Clearly, Julie Dash represents all these belief systems on the space of Ibo Landing not to show the fixity of different religions, and their essentialist nature, but to propose all of them as part of what makes Black people in America complex. Toward the end of the film Grandma Nana brings together the different belief systems, when she ties together the Bible and a sacred object from her own religion, and asks every one to kiss the hybridized Bible before departing from the island. This syncretic move is her way of mixing up the religions in Ibo Landing, and activating their combined power to protect those who are moving North. Earlier in the film she commands Eli to "celebrate our ways" when he goes North.

The syncretic move is therefore also a survival tactic for the African ways of life up North.

Arguably, another reason for deploying ancestor worship (and casting Grandma Nana at the center in the film) is to reveal its usable power in holding the Black family together. Placing women at the center of the frame is also Julie Dash's way of creating space for Black people in modernity, and is her redefinition of Black images in their relation to such modern tools as still photography, newspapers, and moving pictures. Julie Dash's spatial narrative style inextricably combines the identities of her characters with the landscape of Ibo Landing. Her mise-en-scène of Grandma Nana, Haagar, Yellow Mary, and Eula in the center of the frame makes the space theirs, and their possession of the space makes them bigger than life. They become so associated with the space of the Ibo Landing, through close-ups of various sorts, that it becomes difficult to imagine Ibo Landing now without the faces of these Black women. Analogically speaking, it is like imagining America in Western films without the faces of John Wayne, Kirk Douglas, and Gary Cooper.

The spatial narrative style of *Daughters of the Dust* enables Julie Dash to claim America as the land of Black people, to plot Africanism in American ways of life, and to make intelligible African voices that were rendered inarticulate. To return to the thematization of religion in the film, Julie Dash has made manifest an Africanism that was repressed for centuries, but that refused to die. As Grandma Nana states, "those African ancestors sneak up on you when you least suspect them." With her revival of ancestor worship as a narrative grid, as a point of reference for different themes in the film, Julie Dash has ignited the fire of love and caring among Black people. The path between the ancestors and the womb constitutes a Black structure of feeling, a caring handed down from generation to generation, which commands us to care for our children. In an article entitled "Nihilism and Black America," Cornel West proposes "a politics of conversion" as a way out of the carelessness of Black-on-Black crime, and as a protection against "market-driven corporate enterprises, and white supremacism." For West,

> The genius of our black foremothers and forefathers was to create powerful buffers to ward off the nihilistic threat, to equip black folk with cultural armor to beat back the demons of hopelessness, meaninglessness, and lovelessness. . . . These traditions consist primarily of black religious and civic institutions that sustained familial and communal networks of support.[3]

Perhaps Julie Dash's theory of ancestor worship should be among those institutions that constitute Black structures of feeling; as Grandma Nana puts it, let the ancestors guide us and protect us.

Black Times, Black Stories: *Boyz N the Hood*

> Either they don't know, or don't show, or don't care about what's going on in the 'Hood.
>
> *Boyz N the Hood*

To return now to *Boyz N the Hood*, I would like to illustrate its emphasis on time and movement as a way of defining an alternative Black film language different from the spatial and expressive language of *Daughters of the Dust*. Like *Daughters of the Dust, Boyz N the Hood* begins with a well defined date. But unlike *Daughters of the Dust*, which is set in 1902 and looks into the past as a way of unfolding it story, *Boyz N the Hood* starts in 1984, and continues for more than seven years into the future. *Daughters of the Dust* is about Black peoples' reconstitution of the memories of the past; it is a film about identity, and the celebration of Black ways of life. *Boyz N the Hood*, on the other hand, is a rite of passage film, a film about the Black man's journey in America. The story line is linear in *Boyz N the Hood*, whereas *Daughters of the Dust* unfolds in a circular manner.

In films like *Boyz N the Hood, Juice, Straight Out of Brooklyn,* and *Deep Cover*, the narrative time coincides with the development of the lives of the characters of the films. Many of these films begin with the childhood of the main characters, who then enter into adulthood, and face many obstacles in their lives. These films produce an effect of realism by creating an overlap between the rite of passage into manhood and the narrative time of the story. The notion of rite of passage, which defines the individual's relation to time in terms of separation from or incorporation into society, helps us to understand the use of narrative time in a film like *Boyz N the Hood*. The beginning, middle, and end of *Boyz N the Hood* constitute episodes that mark the young protagonist's incorporation into the many levels of society. In fact, the structure of the film is common to African-American folktales, as well as to the classical cinema. It is as follows: A boy has to go on a journey in order to avert an imminent danger. He travels to the home of a relative or friend (uncle, aunt, father, mother, wise man, and so on) who teaches him, or helps him to overcome the obstacle. At the end, he removes the danger, and his nation (or community, or family) gets stronger with him. This skeletal structure is common to texts as diverse as *The Epic of Sunjata* (D. T. Niane), the *Aeneid* (Virgil), and *The Narrative of the Life of Frederick Douglass* (Douglass), as well as to the Hollywood Western genre, the martial art films, and the Rocky films with Sylvester Stallone. The literal journey in time and space overlaps with the symbolic journey of the rite of passage. Typically, this type of storytelling addresses moments of crisis, and the need to build a better society.

The moment of crisis is symbolized in *Boyz N the Hood* by the opening statistical information, which states that "One out of every twenty-one Black American males will be murdered in their lifetime. Most will die at the hands of another Black male." Thus, *Boyz N the Hood* is a cautionary tale about the passage into manhood, and about the development of a politics of caring for the lives of Black males. More specifically, it is about Tre Styles (Cuba Gouldings), the main character, and his relation to the obstacles that he encounters on his way to manhood. Crucially, the major distractive forces in the film are the police, gang life, and the lack of supervision for the youth. To shield Tre from these obstacles, his mother sends him to live with his father, whose teaching will guide him through the many rites of passage toward manhood.[4]

The film is divided into three episodes, and each episode ends with rituals of separation and transition. In the first episode the ritual ends with Tre leaving his mother (first symbol of weaning) and friends behind. The story of this episode implies that most of the friends he leaves behind will not make it. On the way to his father's house, Tre's mother says, "I don't want you to end up dead, or in jail, or drunk standing in front of one of these liquor stores." The second episode ends with Doughboy's (Ice Cube) arrest by the police, who take him to the juvenile detention camp. The third episode ends with the death of Ricky Baker, Doughboy, and many other Black males. At the end of each episode, Tre moves to a higher understanding of life.

Let us know focus on one of the episodes in order to show its internal conflicts, and the specific elements that enter into play to prevent the passage of young Black males into manhood and caring for the community. I will choose the first episode because it introduces the spectator to most of the obstacles which are complicated and repeated in the other episodes. The film opens with a shot in which the camera zooms in on a stop sign until it fills the screen. We see a plane flying over the roofs, and the next shot reveals Tre and three other young kids walking to school. The subtitles say: "South Central LA, 1984." The children walk by a one-way street sign. This sign, too, is depicted in close-up as the camera travels above to establish the crossroad. Then the four kids take a direction facing a wrong-way sign. They travel on that road and see a crime scene that is circled by a plastic ribbon with the words: "Police Line Do Not Cross." Inside the police line there are three posters of President Ronald Reagan with a sign saying: "Reagan/Bush, Four More Years." The kids cross the police line, as one of them moves closer to the Reagan posters. At that moment a rhythmic and violent editing reveals each of the posters in a close-up with the sound of a gunshot. There are bullet holes in the poster. In the next scene, the kids are in a classroom where the students' artworks on the wall reflect the imagery of policing: drawings of a Los

Angeles Police Department helicopter looking down on people, a police car, a coffin, and a poster of wanted men. Tre disrupts a lesson on the Pilgrims, and when the teacher asks him to teach the class, he points to the map of Africa and states that: "Africa is the place where the body of the first man was found." This is a reference to the multiculturalism debate not only across the curriculum, but also in rap music, and in the press. Tre's lesson ends with a fight between himself and another boy. The following shot begins with Tre walking home. He passes a group of young Black males shooting dice. They break into a fight. As Tre crosses the street to go home, he is almost run over by a blue car which presumably is driven by gang members. His mother is on the telephone talking to the teacher about the fight and Tre's suspension. The editing of the soundtrack is interesting in this scene. As Tre walks past the men shooting dice, their noise is placed in the background, and we hear in the foreground the conversation between Tre's mother and the teacher. This editing device unites different spaces through their sharing of the same sound. For example, later in the film, the community is shown as one when people in different places listen to the same rap song. (Similarly, in *Do the Right Thing*, Spike Lee uses the DJ and his music to unite the community.) The last scene in this episode involves Tre and his mother driving to his father's house. They pass by liquor stores and junkies standing by the doors. The mother reassures Tre that she loves him, and will do anything to keep him from ending up in jail, or standing in the streets in front of liquor stores.

Signs (Stop, One Way, Wrong Way, LAPD, Liquor Store, POLICE LINE DO NOT CROSS, and so on) play an important role in limiting the movement of people in South Central Los Angeles. Showing the airplane flying over the roofs not only indicates where we are in LA, but also suggests the freedom associated with flying away from such an enclosed space. Black American literature often draws on the theme of flying to construct desire for liberated spaces: Bigger Thomas of *Native Son* (Richard Wright) sees flying as a way out of the ghetto of South Side Chicago; Milkman of *Song of Solomon* (Morrison) reenacts the myth of flying Americans in order to free himself from an unwanted situation.

The signs become control tools for the police, in the way that they limit individual freedom of movement in the "hood." They also define the hood as a ghetto by using surveillance from above and outside to take agency away from people in the community. In fact, *Boyz N the Hood* is about the dispute over agency and control of the community that pits the protagonist and his allies against gang members and the police. The drawings of helicopters, police cars, and wanted men show how the police surveillance has penetrated the imaginary even of schoolchildren in the hood. Later on in the film, helicopter noise, police sirens, and police brutality are revealed to be as menacing and distracting to people in the hood as drugs and gang violence.

The dispute over the control of the hood is also a dispute over images. The police need to convince themselves and the media that every Black person is a potential gang member, armed and dangerous, in order to continue the policing of the hood in a terroristic and militaristic manner. For the Black policemen in the film, the life of a Black person is not worth much: "one less nigger out in the street and we won't have to worry about him." It is by making the gang members and other people in the hood accept this stereotype of themselves that the community is transformed into a ghetto, a place where Black life is not worth much. It seems to me that *Boyz N the Hood* blames the rise of crime and the people's feeling of being trapped in the hood on a conspiracy among the gang members, the police, the liquor stores, and Reagan. Indeed, the film raises questions of human rights violation when gang warfare and police brutality collude to prevent people from moving around freely, sleeping, or studying.

On the other hand, Tre's struggle to gain agency also coincides with his passage to manhood, and the development of a politics of caring for the community. *Boyz N the Hood*, in this respect, is one of the most didactic Black films. The other contenders are *Deep Cover*, and perhaps some rap videos which espouse a politics of identification with lawbreakers against the police.[5] The didacticism of *Boyz N the Hood* emanates from the film's attempt to teach Tre not to accept the police's and the media's stereotype of him and other young Black males as worthless; and to teach him to care for his community and reclaim it from both the gangs and the police. Didactic film language abounds in the film. We see it when the camera lingers on the liquor stores and homeless people, as Tre and his mother drive to his father's house. The mother, in one of the first instances of teaching Tre in the film, states that she loves him and that is why she is taking him out of this environment. Earlier in the same episode, we also saw the Reagan posters interpreted in a didactic manner, so as to blame him for the decay of the urban community. The posters are situated in the same environment as the murder scene.

However, Tre's father, more than the didactic camera and editing styles, is the central figure of judgment in the film. He calls the Black policeman "brother" in order to teach him, in the presence of Tre, how to care about other Black people; he delivers lessons on sex education, Black-on-Black crimes, the dumping of drugs in the Black community, gentrification, and the importance of Black-owned businesses in the Black community. He earns the nickname of preacher, and Tre's friends describe him as a sort of "Malcolm/Farrakhan" figure. Crucially, his teachings help Tre to develop a politics of caring, to stay in school, and more importantly, to stay alive. It is revealing in this sense that a didactic and slow-paced film like *Boyz N the Hood* can be entertaining and pleasurable at the same time.

The New Black Realism

Realism as a cinematic style is often claimed to describe films like *Boys N the Hood, Juice,* and *Straight Out of Brooklyn.* When I taught *Boyz N the Hood*, my students talked about it in terms of realism: "What happened in the film happens everyday in America." "It is like it really is in South Central LA." "It describes policing in a realistic manner." "The characters on the screen look like the young people in the movie theater." "It captures gang life like it is." "It shows Black males as an endangered species." "I liked its depiction of liquor stores in the Black community." "I identified with Ice Cube's character because I know guys like that back home."

Clearly, there is something in the narrative of films like *Boyz N the Hood* and *Straight Out of Brooklyn* that links them, to put it in Aristotelian terms, to existent reality in Black communities. In my class, some students argued that these films use hip hop culture, which is the new Black youth culture and the most important youth culture in America today. Thus, the characters look *real* because they dress in the style of hip hop, talk the lingo of hip hop, practice its world view toward the police and women, and are played by rap stars such as Ice Cube. Furthermore, the films thematize an advocacy for Black males, whom they describe as endangered species, in the same way that rap groups such as Public Enemy sing in defense of Black males.

It seems to me, therefore, that the films are about Black males' initiation into manhood, the obstacles encountered that often result in death and separation, and the successful transition of some into manhood and responsibility toward the community. In *Juice*, for example, of the four young boys who perform the ritual of growing up, two die, one is seriously injured by a gun shot, and only one seems to have been successfully incorporated into society. Removing obstacles out of Black males' way is also the central theme of *Chameleon Street, Straight Out of Brooklyn, Deep Cover* and *Boyz N the Hood.*

In *Deep Cover,* the ritual of manhood involves the main character's exposure of a genocide plotted by drug dealers in Latin America and the highest officials in the US government against the Black community. The real "deep cover" in *Deep Cover* is the recipe for caring for the community against genocidal forces like White supremacists, drugs, and Black-on-Black crime. The removal of obstacles out of the main character's way leads to the discovery of the politics of caring to the Black community. In this film, as in many new Black realism films, to be a man is to be responsible for the Black community, and to protect it against the aforementioned dangers. John (Larry Fishburne), a cop working undercover as a drug dealer, enters in an intriguing relationship with a Black detective (Clarence Williams), who plays the born-again policeman. The religious policeman keeps reminding John of his responsibility to the community, and John laughs at him. Toward the end of the film, when the character played by Clarence Williams gets shot, John is united with him by the force of caring, and realizes that he must fight both the drug dealers and the police to protect his own.

A key difference between the new Black realism films and the Blaxploitation series of the 1970s lies in character development through rites of passage in the new films. Unlike the static characters of the Blaxploitation series, the characters of the new realism films change with the enfolding of the story line. As characters move obstacles out of their way, they grow into men, and develop a politics of caring for the community. The new realism films imitate the existent reality of urban life in America. Just as in real life the youth are pulled between hip hop life style, gang life, and education, we see in the films neighborhoods that are pulled between gang members, rappers, and education-prone kids. For the black youth, the passage into manhood is also a dangerous enterprise which leads to death both in reality and in film.

Notes

For all references to *Black American Cinema*, see Manthia Diawara, *Black American Cinema* (NY: Routledge, 1993).

1. Janice Mosier Richolson, "He's Gotta Have It: An Interview with Spike Lee," in *Cineaste*, Vol. 28, No. 4, (1992), p. 14.

2. For more on the aesthetics of *Sweet Sweetback* and *Ganja and Hess*, see the important book, *Black Cinema Aesthetics: Issues in Independent Black Filmmaking*, edited by Gladstone L. Yearwood, Athens: Ohio University Center for Afro-American Studies, 1982; Tommy L. Lott "A No-Theory Theory of Contemporary Black Cinema," in *Black American Literature Forum* 25/2 (1991); and Manthia Diawara and Phyllis Klotman, "*Ganja and Hess:* Vampires, Sex, and Addictions," in *Black American Literature Forum* 25/2 (1991).

3. Cornel West, "Nihilism in Black America," in *Dissent* (Spring 1991), 223.

4. Clearly, there is a put-down of Black women in the rhetoric used to send Tre to his father's house. For an excellent critique of female-bashing in the film see Jacquie Jones, "The Ghetto Aesthetic," in *Wide Angle*, Volume 13. Nos. 3 & 4 (1991), 32–43.

5. See Regina Austin, " 'The Black Community,' Its Lawbreakers and a Politics of Identification," in *Southern California Law Review* (May 1992), for a thorough discussion of Black peoples' identification with the community and its lawbreakers.

The State and Future of the Black Film Movement: An Exchange

The Paradox Of Black Independent Cinema

Clyde Taylor

"Produced, Written and Directed by Omnibus Artifactor. Music by Omnibus Artifactor. Starring Omnibus Artifactor." Screen titles like these, as the lights go down, are a symptom of a dimness in the development of Black cinema.

The major crisis in the independent Black film movement is the drought in production. The flutter of publicity around Spike Lee's success has blinded many to the decline in recent years of features completed by Black independents. But the need survives to bring critical assessment to the independent scene, now some two decades old in its recent incarnation, apart from our concern with this crisis in production. The relative lull in production might in fact be taken as an opportune moment for self-examination.

It is past time now to ask, why has the movement not shaken its auteurism? The auteur concept of cinema, launched by Parisian critics in the '50s, argues that films get their distinctive qualities from the dominant hand of a single author, or *auteur*, usually the director. Hence the influential critical shift in focus from stars to directors. But its real intent and effect was to draw movies more deeply into the orbit of the western aesthetic, where individualistic, authorial creativity prevails over performance or collaborative art.

Of the many reasons why Black independents have become mired in authorism, the one most cited is economic need. "I couldn't get the budget for a script writer," they've said repeatedly. To which I often want to reply,

Black Film Review, vol. 4, no. 4, 1988, pp. 2–3, 17–19. Reprinted by permission of the publisher.

"But you got one for a camera person." The economic explanation is used to excuse a kind of Third World Renaissance man syndrome. In repressed environments, those who have acquired any literacies are forced to act as pan-literates, functioning as physicians, poets, and political leaders all at once, much like Aimé Césaire. Marvelous though this may be, we must ask what the cost in perpetuating a psychology of cultural Robin Hoods wandering around in Sherwood Forests of prolonged underdevelopment? Or whether complementary literacies for Black filmmaking are really so hard to come by?

The biggest loss to the author complex in Black cinema may be in narrative development. This despite the high level of writing included in the movement at its best. With few exceptions, the films that have made the most interesting impact come off as good writing: Charles Burnett's *Killer of Sheep*, Julie Dash's *Illusions*, etc. Parts of Haile Gerima's *Bush Mama* and *Child of Resistance* call up comparison to Baraka's plays. But few films from the movement, including those above, could not have gained from the kind of collective critical examination Black playwrights have given each other for years at readings before the Frank Silvera Workshop.

The issue of extending the literary input into Black films was raised at a UNESCO conference preceding FESPACO '85 in Burkina Faso. There, some of Africa's most accomplished writers reminded their film-directing brothers of their failure to seek their collaboration, or to draw on the existing wealth of African novels, plays, and short stories. The same debate applies with equal force to the Black American scene. At a meeting of Black writers and critics, I recently heard Pulitzer Prize-winning author Toni Morrison remark, "I can write anything. But I have never been asked to write a screenplay."

It is not a matter of giving up creative control, but of opening the film-making process to complementary creative-critical voices. One would think this imperative would be compelling to filmmakers who have expressed the determination to reach large Black audiences. One would think this need would be immediately felt by filmmakers whose rhetoric embraces collectivism over Hollywood individualism.

If the object of Black independent filmmaking were truly to reach the broadest Black audience with narrative explorations probing their realities with dialogic depth and historical resonance, then dialogue about the film might well begin before the film is made. Would it be preposterous to air narrative possibilities before audience panels for their feedback on the story, to be modified or rejected as the director saw fit? Hollywood, for different motives, is far more diligent in pre-searching public response through test screenings previews and the like. By contrast, the fictional films of Black independents are not usually storied enough to be discussed beforehand; and at the other end, their audiences are likely to be asking what the story was.

Do we see evolution toward wider narrational collaboration over the last decades? If anything, we probably see a retreat from the fraternal exchange that took place within the L.A. Rebellion when its filmmakers worked together at UCLA. One exception is St. Clair Bourne. I recall his perception of the value of collaboration with writers in an early 1980s interview. He was charged up by the prospect of working with a writer on the set, to get another view of the material, even for a cinema verite documentary where the object is to record direct action unrehearsed. Since *Let the Church Say Amen!* he has used writers on most of his documentaries, with decidedly positive effects in *In Motion: Amiri Baraka* and *Langston Hughes: The Dream-Keeper*. (The writer in both instances was Lou Potter, unless memory fails.)

But the point is not to go beyond auteurism just to make the film "better," that is, more smoothly palatable and coherent. The point is to enlarge the magnitude of productive ideas exchanged through the filmmaking medium of communication. It is a point that has something to do with a filmmaker's capacity and willingness to grow.

And to this point should be directed the observation that Black independents as a group show superb indifference to seeing the films of their contemporaries. I have left some of them holding court at the dinner table to see a film of their colleagues that they hoped to catch, some other time. Dozens of them have still not seen Charles Burnett's *Killer of Sheep* (1977), probably the most decorated film of the movement. I can count on one hand the times, outside of festivals, I've seen an independent at the screening of another.

And more than once they have lamented, in Paris or Amsterdam, say, that they had to be brought this far to meet each other and see each other's work. That irony could be laid to the door of U.S. cultural imperialism. Maybe. But does Ma Bell turn down calls between Black filmmakers? Or, by what feats of internal determination do they expect Black viewers to resist inertia to go see their films in random venues when they cannot themselves overcome inertia when it gets down to films of other Black independents? Might not the revenue from Black independent films rise, even approach profit, if all filmmakers came out and paid, to see each other's works?

The temptation toward author-centered filmmaking is fed by the lack of chances for independents to practice their craft. They may feel, during the interminable delays, a pressure toward realizing a dense self-expression through a few works rather than aiming for a more dispersed personal contribution to a wider cinematic context. If Black independent cinema had some institutional foundations, the strategy of spreading one's talents and remaining open to those of others might appear more attractive. But while several filmmakers have shown initiative in programming film series and festivals, organizing discussion panels and workshops, publishing newsletters, the interest in joining efforts toward organizational collaborations, collective fundraising, lobbying, the much needed work of cultural politics, has never been slacker than at the fatiguing end of the Reagan era.

The seeds of authorism may have been sown among Black independents in film school. Formal cinema training is one of the defining characteristics of the current film movement. And most film schools train their students to be auteurs, demanding from them projects where they work as solo honchos. One-brain filmmaking, it must be admitted, also has a vigorous history among filmmakers who sharpened their chops outside of universities. Gordon Parks, Melvin Van Peebles (who went to film school at the San Francisco Art Institute, but acts like he didn't), and Robert Townsend come to mind as examples. And early independent Oscar Micheaux, the most self-made Black director of them all, was equally attached to his creative isolation.

But a certain character has been given the central flow of recent Black independent films from its origins within the American university. The conspicuous body of Black independents has been called, not without reason, intellectual filmmakers. It helps to see this description as one half of an antagonistic dialogue over intellectual versus populist Black cinema, that has gone on, quite uninspired, for many years.

Looking at the products of the last two decades, what do we find? That the intellectual, university-trained cinema has had some success in winning restrained respect in art and culture quarters. But does not that limited success mask, and cushion the lack of success in its stated aim—to reach the Black community in ways that advance self-definition through cinematic self-examination? Have the efforts of these directors to break the hold of official (museum, university, foundation) culture been whole-hearted?

The other half of this false dilemma, false because its terms are given by the fixed definitions of the dominant media order, is drawn from a glance at the populist argument. Quite simply and expectedly, the populist persuasion, led by directors like Melvin Van Peebles, Fred Williamson, Jamal Fanaka, and Robert Townsend, while it may have titillated more Black people, can claim little gain in projecting any social dialogue of weight, or, less difficult to assess, in elevating group self-esteem, establishing positive images endowed with dignity, etc. (with the exception of *Sweetback*, which remains a powerful evocation of police-state brutality against the Black community, about as timely today as when it was made). Beneath the glib pronouncements of social commitment, revealingly like the tabloid babble of Black entertainers, it is too easy to spot the features of the archetypal Black entrepreneurial hustler.

So there is reason to approach this false dichotomy from the intellectualist side, considering the mega-incredibility of the populists. Even as we do we should guard against the danger of accepting its premises, of seeing both avenues as fixed tendencies rather than dynamic, changeable ones which, as social circumstances and history alter, might produce different results. It is not likely that those who have established their identities through one of these avenues will suddenly switch to another. (Maybe this is too bad; a bit of switching might be a good idea.) The view from the street is that the intellectual filmmakers all harbor some inner desire to head for Sunset Boulevard. But this hidden aspiration, to the extent that it is there, and I haven't seen much evidence of it, competes with several others, including the gathering of fame and glory in the international film community.

The problem is that without more production, we cannot begin to guess what direction the intellectual directors would move toward if they could extend and multiply their film concepts. Would they, given more opportunities and resources, try to broaden the range of their filmic reference, and Black audience appeal, while holding focus on the identities of Black people desperately in need of a better understanding of their contemporary history, as are all publics? Would they diversify their service to undominated Black representation, by drawing on the repositories of storytelling virtuosity within Black culture, or perhaps by acting as producer for other Black independents (as St. Clair Bourne has done and Robert Gardner has expressed intent to do)?

Suppose the game-plan of the intellectual independents were realized: to produce a succession of films each more popular than the last, earning them bigger budgets and opportunities? What then? Bigger canvases for self-motivated personal expression within a lip-serving context of social responsibility? Do we have evidence that anything would happen other than, say, a bigger film about Black hair from a director mystically liberated by dreadlocks, or grander humanizing portraits of Black lesbians from directors seeking room for their own sexual expression, or laments of the anxieties of exile from geographically displaced directors (this last applying to Black British films), or portraits of misunderstood artists from directors who seek themselves as such? Projects like these might be well-received in a genuinely democratic, pluralistic film culture. But they seem marginal, to me, to the historical, collective need of the Black community to express itself.

One of the few directors who has had enough of a career to throw light on the possible trajectories of the intellectualist-populist debate is Spike Lee. Retrospectively, what's of interest here is the controlled, deliberate ambiguity in his films between intellectualist and populist representations. His first film *The Answer* (1980) was his thesis film at New York University, and does not set itself apart from other intellectualist beginnings of the period. Its narrative premise dealt with a Black man hired by a major studio to direct a remake of *Birth of a Nation* (which didn't go down well with the NYU faculty, probably because it ruffled the department's reputation for teaching the film without reference to its racism). The man's "creative differences" with the studio result in the Klan burning a cross on his lawn. A typically spoof-humorous story concept of the sort that university trained filmmakers hopefully grow out of. If it bears a conceptual family resemblance to *Hollywood Shuffle*, that may be because media repression is a likely theme among beginning Black independents, more, one begins to suspect, out of a sense of injury to their prospects than concern over the political inequity involved. However that may be generally, *The Answer*'s unbuoyant attempt at satirical comedy carries signs of future populist possibilities unsuccessfully fused with an intellectualist concern with the politics of representation. The film's interesting but undeliverable implication is that Black imagery remains as negatively controlled as it was in D. W. Griffith's epic of racism.

A slighting different ambiguity frames each of his subsequent films. *Joe's Bed Stuy Barbershop: We Cut Heads* is a tonal puzzle, a comedy about communal pathology and corruption that is amusing and interesting but decidedly unfunny, unless read with healthy doses of New York cynicism. Yet in *Bed Stuy* you can see the move to appreciate with affection the young bloods coping in their funky, troubled environment, warts and all, that gets fuller play in his two following hits. The intellectualist-populist ambiguity of *She's Gotta Have It* lies in its unresolved tension between sexploitation and fashionable gender issues. Here, Lee seems to be saying something about those issues, obliquely, but what?

Up to this point, which had brought him further than most Black independents in breaking out of the museum film circuit, at least as far as the art houses, Lee had offered two initiatives worth contemplating as escape routes out of the limited intellectual film bag. One was the need to seriously move on, from film to film not only in matter and motif, but also in coding for different audiences. *Bed Stuy*, well-liked at film festivals, really didn't fit well into the museum repertory. Not arty enough for the art houses, too short and tonally ambivalent for the popular houses, it was still too focused on entertainment and popular culture, that is, not library-solemn enough for the museum-university film series-Public Broadcasting System circuit where most Black independent films end up. From *Bed Stuy* on, Lee was bringing sensibilities and moves into his films that came more genuinely from Black street culture than from the cinema traditions of film school.

The other contribution he made to the search for self-identity of the intellectualist orientation is respecting the value of complexity and obliqueness in telling a story. Compared to his output, some other Black independent films seem to suffer from an uninflected sincerity that aims them unintentionally toward the children's hour, or seem otherwise burdened by the weight of their social thesis. The flip side of Lee's risks in this direction is the uncertainty of his filmic statements.

Unlike most of its white reviewers, for me this was not a problem in *School Daze*. Remarkably, he moved beyond the art house parameter, itself another halfway house for the rehabilitation of Black consciousness, and made a studio-backed film with more potent social-political bite than the vast majority of Black independent films, which is what that posture is supposed to be about. (I need to confess here that my mind was closed to this historical possibility, and therefore am indebted to Spike Lee for opening my sense of the possible.) He made it more explicitly directed to Black audience consumption than any big theater movie since *Sweetback*. And reading it as part of that audience, and part of it that went to a Black college, I found its statements neither too ambitious, numerous or anything but clear.

But *School Daze* has its ambiguity, even though it doesn't strike centrally at its social-political text. By lacing his narrative so splashingly with marvelous doses of Black youth funk culture, with its genital-posterior sexism blatantly flaunted (a kind of mix you might call hip-pop), Lee has once again, as with *She's Gotta Have It*, sweetened reflection with entertainment values impinging on exploitation and the indulgence of personal erotic fantasy.

Spike may stumble, Spike may fall. Spike may go for the okey-doke. The iron hand of Hollywood capitalism and "market forces," in short, the media industry, have a history of making sure innovators will. The kind of media hype Lee whips up demands to be pacified, finally, with abject, apolitical inanity. There is something in this society that has a need for Joe Louises to end up as door-openers at Las Vegas casinos. Spike's own personal moment of truth will come when his mass media voice is silenced. Will he sulk into a seat on the New York Stock Exchange, or will he have the heart to pick himself up, brush off the "I told you so's" and make smaller, independent films again?

How Lee navigates these hassles may offer more light on the intellect-populist tangle. But he has already demonstrated the artificiality of its problematic. As have *Stand and Deliver, The Garbage Boys* and *Sweetback*, not to mention *Battle of Algiers*. By coming from their own ranks, Lee poses to the intellectualist tradition the possibility of the obsolescence of its self-confining purity. He has raised the possibility that it has shrunk, so chilled from the Sambo syndrome as to surrender to Hollywood without a fight the right of Black people to enjoy conflict, drama, humor, poetry, while being challenged or edified.

The university-trained Black independents were into filmmaking, several of them told me in the '70s, because Black people don't read. But they read a bit more, it turns out, than they go to see independent films. One reason is that the films, many of them, even in concept, lack those qualities that encourage one to turn from book to screen. You can read the five-line synopsis of many of the fiction films coming from newer Black independents and know that ain't it, the marks of amateur literary invention and self-indulgence are so plain to see.

In a medium that was built to handle it, the Black-I filmmakers have seldom looked for the vitality, style, and pageantry of Black life, the culture's appreciation of spectacle, performance, and virtuosity. The passion of *Passing Through* (unchained in its music but almost matched in its cinematography), the stylist theatricality of *Illusions*, the freewheel antics of *A Place in Time*, the rhythmic chant-raps of *Clarence and Angel*, all inexpensively captured, have too few counterparts in other Black independent efforts. Livelier and more politically potent results would emerge from just sticking a camera in Black people's faces and asking them how they feel, about anything.

The issue is not simply the cult of authorship or of intellectualism, which, broken down, is more accurately seen as isolated preoccupations with purist cultural nationalism on the one hand and (western) aestheticism on the other. The issue is whether these understandable attachments can be assimilated and transcended toward a more representative, relevant Black media representation. The imperative has always been the freedom of Black people from cultural and political repression, not the success, as cultural trophies, of a self-selected generation of cinema artists. Right now the question seems to be whether that generation can put aside its pet, personal ideologies, pushed in their films like a celebrity's favorite charity, to make serious breaks toward the pressing interests of their communicants. Or will the restless demands of history prompt another self-selection of Black image-makers from the ranks of creators less mystified by cinema pedagogies and less intimidated by the aesthetic criteria of funding panels?

No one familiar with them at work in their process can fail to be braced by the courage of the Black independent filmmaker, even when the basis of that courage is an ego-driven lust for the rewards of successful creative self-expression. (That goes with the territory.) It is hard to think of a tougher creative situation than the one they face. But they fail to understand some of the (oral) criticism directed to Black cinema when they reply that such shows an ignorance of the difficulties of the process. Short-sighted criticisms are sometimes voiced which imply that the limitations of Black cinema are the fault of individual film directors and films. Short-sighted reception of critical commentary fails to see that its real significance is directed to concern for the expressive destiny of Black people at a particular juncture of time and not to the relative brilliance or lack of it of a particular effort.

The time may have arrived, or come past due, to rethink the basis and rationales of Black filmmaking. The case may be that the discursive assumptions that legitimated Black independence for the most recent stretch are in danger of lapsing into impotent truisms and justifications that have been largely bypassed by a new social spectrum.

This rethinking might begin with a realization of the present situation of Black independence as trapped within discursive boundaries patrolled by such un-popular institutions as the university, the national government foundations, the museums, the western-dominated international film festivals, the public libraries, and PBS. Black independence is further threatened, I believe, by the trap of the "intellectual" or artistic posture that has been coded into it, which has become a means of shielding filmmakers from ideas, from new historical formations, perspectives and challenges.

Having bogarded these assertions, I am called on to back them up with some sample scrap of the kind of rethink I have in mind.

Since the present generation of Black independents picked up and then froze a few dusty McLuhanisms to serve as theoretical armament, the prospects for renovating cultural and communicative thought have drastically changed. One way of seeing that change is as the deinstitutionalization or decolonization of knowledge. Which means simply that the evidence has piled up overwhelmingly that "knowledge," received ideas, social consensus, and cultural images have been shaped by the powerful institutions that control them. Which nobody in the Black community ever had any excuse for not knowing. But this general understanding of the frame-up of knowledge has reached the stage where each person in an information occupation must now be either busy rewriting knowledge or automatedly reproducing, photocopying really, the official story that oppresses them. It has reached the stage where the stances for re-presenting cultural information that became influential in the 1960s are thrift shop collectibles today.

It is possible and meaningful now to think of "history" as the continual constitution and reconstitution of the present, as well as the past. And that this process of establishing meaning for both the present and the past, a major function of any culture, is one that repressed peoples have been forcibly excluded from, and to which they usually return as late and bashful participants. But the exceptional confrontations from repressed populations through the 20th century have dramatically increased the opportunities for such participation. When the information workers from such populations are laggard in making the most of these opportunities, they reshackle their sisters and brothers. It is to the wider participation of such communities, to the continual rewriting of the present as well as the past, that any liberative politics of representation must contribute. And it is through such contributions that Black cinema has most to offer to its times.

When we ask ourselves what the emerging body of Black independent films brings to this process of writing and rewriting the informational space that Black people share with the rest of the world, might we not come away from such an inquiry convinced that some rethinking is in order?

Whatever the reason, whatever the discursive borders and constraints responsible, many Black films, even in conception, politely accept their confinement outside the extraordinary reconstruction of knowledge taking place, the one most stimulating conversation going on in the world. This implies that it is not enough for a film to say that Blacks had this neglected cultural hero, or that one, too. Such recuperations establish icons of museum interest, while leaving it to the insane prevailing information structure to orient these icons into context as knowledge. What the new, activist role of the repressed information work demands is not only the production of fresh information but a demonstration of how it invalidates not only old information but the inherited structure of information itself, within the film, or play, or whatever.

Much can be learned in this respect from two Black British films, *Territories* and *Handsworth Songs*. Both adopt the strategy just described, of questioning the rooted order of information and history-making at the same time that they present new readings of present-past history. These films interestingly break out of current notions of authorized documentary style where the making the film without a narrator has become a fetish. It is time to see the inapplicability of this fetish to Black creators who are not merely positing images for interpretation through the prisms of the same tired regime of social integration, but who recognize the need to take over the role of interpreter, or at least questioner themselves.

It may be—and here I am not falling into the unexamined Anglophilia surrounding the Black Brit phenomenon—that the university training of this crop of filmmakers has not shielded them from some ideas from which U.S. Black filmmaking could profit. (Black Brits have their own problems with auteurism.) But the best use of these two films is to alert us to what is already obvious—the need to rethink the goals and strategies of independent Black film. A decent place to start is to ask what kind of mobilizing and agitation is appropriate to free Black independents, those who want to be free, from the stranglehold of the western and U.S. cultural empire.

Something is called for, and it is not just the responsibility of the filmmakers to close out the damaging paradox of the present Black independent scene. Namely, that situated in the most advanced and powerful medium, and exploding a decade ago with transgressive energy, it has allowed itself to become the tamest, least expressive, least dangerous form of contemporary Black expression.

Which Way the Black Film Movement?

David Nicholson

Four years ago, I began *Black Film Review* with the intention of providing a forum in which to evaluate Hollywood's images of Blacks in commercial film and television. That first issue was a single page, prepared on my home computer, pasted up, photocopied, and mailed to about 50 friends.

That the current magazine is of such improved, professional quality marks a number of significant steps forward for *Black Film Review*. Many people have helped make that a reality: writers, editors, artists, photographers, filmmakers, volunteers who gave up their Saturdays and Sundays to re-type articles.

A few months ago, I decided to step down as editor and publisher. For four years, *Black Film Review* has been a significant part of my life; indeed, at times it has been my life. I want now the free time to do other things: high on the list is completing my novel. But I am also aware that too often Black organizations fail because those who found them burn out or find themselves unwilling to let go. With this issue, we publish the second number of our fifth volume, publishing from an office outside my home, and with a paid staff. Institutionalizing the magazine and turning it over to others seems the best way to ensure it lives to publish its 10th volume.

In the four years since its founding, the focus of *Black Film Review* has changed. While *BFR* continues to report on aspects of Hollywood film, it now covers independent film from throughout the African Diaspora. The United States, the Caribbean, Great Britain, France and, of course,

Black Film Review, vol. 5, no. 2, 1989, pp. 4–5, 16–17. Reprinted by permission of the publisher.

Africa. *BFR* has become the sole consistent source of information about these films and the men and women who make them.

Black Film Review will continue to support the Black independent film movement by bringing readers news and information about the films and filmmakers. After four years, however, it seems an opportune time to assess the state of the movement. Some may argue that the dearth of production, so that too often we find ourselves talking about films that are 10 to 15 years old, is the movement's biggest problem.

Yet despite all the hardships Black filmmakers face in raising funds and, once their projects are completed, getting their films out to an audience, it seems to me much more significant that the movement lacks direction, lacks the cohesiveness at its center that marked (at least in retrospect) the Black Arts movement of the 1960s. Some time ago, I was talking to a man who has been long familiar with the movement and who, in the course of his career, has assisted Black filmmakers by showing their films in numerous venues. "The problem," he said to me, "is that most of the films just aren't very good."

Having now seen many more Black independent films than I had when I first started *Black Film Review,* I am inclined to agree. That is the dirty little secret of the movement: Too many of the films aren't very good.

In the Fall 1988 issue of *Black Film Review,* Assistant Editor Clyde Taylor pointed out that Black filmmakers seem to suffer from an auteur complex. They want to write, produce, direct, edit, and act in their films, offering as an excuse the limitations of their budgets, yet few are energetic and talented enough to do all these tasks and make good films.

I believe this is true, but I find the auteur complex too limited an explanation for the poor quality of so many films. There remain, in addition, a number of other factors that combine to make the movement the paradoxical creature it is today: a creature both promising and problematic.

Black Americans have unparalleled access to the technology, to the means of production, if you will, of cinema for the first time in our history in this country. In part this is due to the number of Blacks who attend film school. There may be fewer Blacks than whites in film school, but there are more Blacks in film school than ever before.

This is both good and bad. We are learning the techniques of cinema, but film school training tends to produce film school films, especially when the student is an undergraduate, or comes to film school immediately after college. It is a phenomenon akin to one I observed at the University of Iowa Writers Workshop. Younger students, fresh out of undergraduate school, tended to be imitative writers, while older students who had worked, been married, had children, wrote out of their experience and, in the best cases, out of their fresh unique vision of the world.

An appalling number of the shorter independent films I have seen are bland tone poems, not without some visual appeal, startling juxtapositions, or daring camerawork, but utterly without content or point.

I am not suggesting, of course, that aspiring filmmakers not attend film school, for I have also seen bad films made by filmmakers with years of professional work as cameramen and camerawomen for commercial houses or television stations. Bad films are bad films because they are poorly conceived and thus poorly structured and poorly written. They are often films made in adherence to a particular political or cultural stance. The result is that they are films that ignore the real experience of Black people in this country for some easier, more convenient view. When these filmmakers make fictional, narrative films, they fail to give us folk we know and can identify with (even if we don't like them), people forced to choose between imperfect alternatives (and who sometimes make the wrong choice). Instead, the filmmakers give us upright ideas, mouthpieces for their misguided enthusiasms.

Before I go on, I want to make it clear here that I am talking, for the most part, about narrative films. And I want to acknowledge, again, that the filmmaker who wants to make a feature film has a hard way to go. Government funding agencies such as the National Endowment for the Arts, the National Endowment for the Humanities, the Corporation for Public Broadcasting, and private agencies and foundations all suffer from the "Dead Negro Syndrome." Thus it is easier, say, to get a grant to make a film about Frederick Douglass (a radical in his time, but now safely dead and buried), than it would be to get a grant to make a film about Louis Farrakhan or a narrative film that tried to explore some aspect of contemporary Black life through fictional characters.

The irony is that, based on the films I have seen, the documentary would likely be far superior to the narrative film.

In their quest for "art," or for some mistaken ego fulfillment, Black filmmakers seem to have forgotten (or perhaps think themselves above learning) elementary rules of plot, character development, and story construction. Some offer as an excuse that their films are made for "the people." It is interesting to me that this excuse is offered for shallowness and inconsistencies in plot, character, and story *and* for films that are so densely convoluted they require explication by professors of semiotics to be comprehended.

Thus on the one hand filmmakers say that because their films are about some contemporary problem (drugs, teen pregnancies, abused children), and aimed at those who suffer from that problem, they must make accessible films and need not follow the simplest narrative guidelines. On the other, they resort to some incomprehensible academic mumbo-jumbo to explain why no one can understand their films.

This is, at worst, blaming the audience. At best, it is condescension, for it fails to take into account the visual sophistication of most Black Americans. Most of us, in urban centers at least, grew up watching movies. Most of us, no matter how poor, grew up with television. We are, then, for better or worse, visually sophisticated, though we may not be visually literate. How, then, can we expect to attract an audience with poorly constructed (and poorly made) films, when the audience is used to a product that is, in every way, except perhaps content, superior?

The answer is, of course, that we cannot. In Third World countries where people are not filmgoers and television watchers, certain techniques of filmmaking and distribution meet the needs of particular historical, social, and cultural situations. It is ridiculous to expect that those techniques will work here, ridiculous to expect, for example, that an American audience will sit still for a film that is three hours long, that uses a static camera, a film where the motivations of the characters are poorly shown or not shown at all, where events are not foreshadowed, where characters do not exist as we do; as thinking, feeling human beings facing a bewildering multiplicity of choices. And I have seen that point proven again and again, where audiences have voted with their feet, walking out of films that filmmakers have created with the best (and condescending) intentions of "educating" their audiences.

Here, I want to apologize to the filmmakers whose work I am about to dissect, and express the hope that my criticism will not be taken personally. It is not meant to attack, but to illustrate. The sad thing is that for every example I cite here, there are many others I could also use.

Edward Tim Lewis' *Serving Two Masters*, is a film about two men in Washington, D.C., one a failed Episcopal priest, now a drunk living homeless on the streets, the other a successful businessman in conflict because his firm wants him to sell computers to South Africa.

There is some wonderful camerawork in the film, some wonderful editing. Yet in the end the film fails. Lewis, at a conference recently in Tallahassee where we both spoke, admitted that it was not until he was in the editing room that he began to feel some sympathy for the executive, and that he tried to cut the rest of the film to reflect that sympathy. By then, of course, it was too late; if you didn't get it in the camera, you aren't likely to produce it in the editing room.

The film is also too heavy-handed in too many ways. Character is revealed through dialogue that is actually diatribe, not through action or interaction. In one scene, the executive, Cliff, is playing golf with his superiors. They are condescending; he is compliant. The point is obvious: Cliff is a "Tom," working on the plantation.

But how much richer a scene it would have been had Lewis shown us the humanity of all involved! Is it so unreasonable to expect that in that situation these men might not have liked each other? Cliff was a college football star; in that kind of macho corporate environment where psychological aggression replaces physical violence, would not his peers and his supervisors respect and be envious of a man who has proved his courage in ways that they cannot?

Julie Dash's new film, *Daughters of the Dust*, looks sumptuous, if the trailer is any guide. Yet it seems to spring full-blown from the Alice Walker-Gloria Naylor-Toni Morrison school of writing. Two women on a South Carolina Sea Island, one returning home, the other pregnant after she has been raped. The pregnant woman's husband is grieving for his defiled wife.

The trailer shows brilliant skies, expansive seas, characters in period costume walking in verdant woods, hands opening a book in Arabic I take to be the Koran. These characters are all Black. And what's on the soundtrack? Swelling strings, European quasi-John Williams. That contradiction, together with the languorous pace, a taking-itself-seriously-in-the-worst-kind-of-way quality of the trailer make it hard for me to look forward to the completed film.

Spike Lee's two most recent efforts seem to me mixed. *She's Gotta Have It* richly deserved its success. But *School Daze* subordinated plot and character to idea and the musical numbers seemed tacked-on. Where the sex scenes in *She's Gotta Have It* (arguably exploitative) gave us the sexuality of Black men and women in many forms: passionate, funny, sad, similar scenes in *School Daze* seemed, sadly, obligatory, as if Lee knew what his audience expected and decided to give it to them.

What if the situation in the film had been more complicated, what if the light-skinned woman had found herself falling in love with the dark-skinned man, and vice-versa, and they had all found themselves unable to transcend the taboos of color and caste? Then, perhaps, the film would have earned its "Wake up! Wake up!" ending.

There are many other examples, films where politics, a knee-jerk, infantile Marxism replaces real characters in human situations, facing human dilemmas, characters we can come to know and care about, even if we don't like them. Films where mindless "Afrotarianism," a pointless fascination with the romantic thrill of Africa and things African, replaces any moral questioning or discussion of issues.

The point is that only the worst made-for-television movie fails to observe the elementary rules governing the development of character, the construction of plot and story. And if Black filmmakers want their productions to be viable in an increasingly competitive market, their films must be far better than the worst Hollywood has to offer.

What about films I liked? I've seen two by Reginald Hudlin: *House Party* and *The Kold Waves*. Both shorts, the former is about a boy who sneaks out of his house to party with his friends (and is caught by his father). The latter is about the boys in a Black rock band in search of a drummer. The drummer they find is not only white; he can't play the drums at all.

Both of these films are humorous and deeply rooted in the African-American vernacular. The dialogue is authentic, and each is a highly moral film. In *House Party*, the boy pursues an attractive girl, but honors her by not pressing her for sex as his friends would have him do. He breaks the rules by sneaking out and pays the price for it. In *The Kold Waves*, Black meets white in an urban context and evolves a way to get along that works for both.

If there is any single point that I want to make here it is that the world is so complex, so fraught with ambiguity and contradiction that it imposes a special responsibility on the narrative filmmaker and the novelist or short story writer. We are craftsmen and craftswomen, and it is dishonest and does us no good to retreat to an imagined, idealized past. It does us no good to avoid the world by retreating to fundamentalism, be it Christian, Islamic, or some amalgam of traditional African practices. It does us no good to flatten the contradictions and ambiguities that so threaten the Black community as the 1990s approach by depicting them as easily solved.

I don't pretend to have all the answers to the malaise that threatens the Black independent film movement, but I do know that filmmakers and critics must address them if the movement is to grow and thrive. I do know that the films must be better written, better constructed, and that the filmmakers must stop using budget shortfalls as an excuse for their technical deficiencies. I do know that some of us must have the courage to go against the grain, to look differently at the issues confronting us as a community, instead of repeating the popular panaceas that do absolutely nothing to make us look at the world and ourselves in different ways.

We Black Americans have a long tradition of creation and of criticism in other art forms: Music, literature, dance. That tradition is evolving in cinema, and perhaps it would prove profitable for the new Black filmmakers to consider the example of jazz as the world of Black film evolves.

Surely filmmakers could do worse than to emulate the music's subtleties, its shading, and its nuances and, above all, its absolute fidelity to the human condition. Then, too, jazz musicians have always honed their skills in the most formidable workshop imaginable: the jam session. There the incompetent and the half-hearted are quickly shown for what they are. Until they are prepared, then, the wise woodshed, practicing alone before they go public. And while jazz has always strayed from its roots in blues and gospel, its roots are almost always audible, and from time to time the music as a whole returns whole-heartedly to embrace those roots.

As we go to press, I note with wonderment that the Black Filmmaker Foundation has named Stanley Nelson's *Two Dollars and a Dream* and Julie Dash's *Illusions* as the best films of the preceding 10 years.

How can this be!!?? It is criminal to ignore the work of a great documentary filmmaker like St. Clair Bourne, and *Illusions* is an execrable film, poorly lit with garbled sound.

Here is further evidence of the parochialism that will cripple the Black film movement as it attempts to secure its foothold in the 1990s and beyond. It is a parochialism based on politics (nationalist versus integrationist; Marxist versus capitalist), on sexuality (homophobia versus homophilia), and class.

It is clear that Black filmmakers, must do something similar to the jazzman; it is equally clear that critics must become more like the jazzman's audience. The quality of Black filmmakers' work must improve on all levels, and the work must reflect something uniquely African-American. Anything less will, in too short a time, make the movement and its practitioners only curiosities. And, like Samuel Johnson pondering a dog walking on its hind legs, we will wonder not that it is done well, but that it is done at all.

The Future of Black Film

The Debate Continues

Zeinabu Irene Davis

> I'm tired, Charlie, I'm Tired!
>
> *Line from Billy Woodberry's Black independent classic,* Bless Their Little Hearts

This statement sums up my responses as a filmmaker to recent articles in *Black Film Review* by David Nicholson (Vol. 5, No. 2) and Clyde Taylor (Vol. 4, No. 4).

Although these critics of Black independent film make provocative statements that should move all of us as filmmakers to hone and push the limits of our craft, I can't help thinking how off-base some of their thoughts are and how insidiously they seem to promote "blaming the victim."

Yes, it's true that independent Black feature film production in this country has indeed declined since the early 80s. Many of those from Taylor's "L.A. Rebellion" left institutions and found they had no support to continue making films. Faced with no finances and no equipment access, it has taken filmmakers awhile to save money and begin new films, let alone features.

However, there is no "dearth of production," as David Nicholson would like us to believe. The fact is that in the last eight years Black

Black Film Review, vol. 5, no. 4, 1989, pp. 6–9, 26–28. Reprinted by permission of the publisher.

independents have created a significant number of new short films, videos and experimental works, but they generally have been ignored. Taylor and Nicholson lead us to believe the works have been ignored because of their own shortcomings such as stylistic inaccessibility and structural inadequacy.

However, it is African American critics such as Taylor and Nicholson who are *themselves* stuck in a mode that has them chanting the same names constantly. Perhaps Nicholson should take a closer look at some of the last few issues of *Black Film Review* that were done under his editorial flag. It seems as if African Americans can't get into the *Review* unless they are dead or their last names happen to be Townsend or Lee. There is no room in the stagnant critic's vocabulary for new African American voices like Iverson White, Linda Gibson, S. Torriano Berry, O. Funmilayo Makarah, and Debra Robinson, among others, or even old pros like Jessie Maple.

The fact of the matter is that new Black independent film is alive and well, thank you. Even as compared with 15 years ago, the new work is vibrant, fresh, and more diverse, as more people are working in video, or are fusing film and video technologies. (For a fusion of film and video technologies, see *Zajota and the Boogie Spirit* by Ayoka Chenzira and my *Trumpetistically Clora Bryant*).

Yet, Black American independent filmmakers—the former "darlings" of the late 70s and early 80s—have been put on the shelf and Black British filmmakers have filled the void for those in the academic and art house world.

It is true that the technical quality of Black independent film is a large part of the battle. Nicholson implies that we should try to better duplicate the Hollywood standards to which Black audiences are accustomed. But why should we ape Hollywood, when it already pimps whatever aesthetics we may develop in our films? (Case in point: Melvin Van Peeble's *Sweet Sweetback's Badassss Song* helped to launch Hollywood's Blaxploitation Era).

What independent Black filmmakers are creating is a new cinematic language in our films. We are asking the audience to see things anew. If one really looks at the recent development in Hollywood films, one can see that there are no new explorations of narrative style or any real innovations in terms of content. Studio films are feeding us a pablum of multi-million dollar visual effects that only momentarily dazzle the eye. When was the last time you felt impassioned and moved to discussion after seeing a Hollywood film?

Recent Black independent film takes on new subject matter and often brings interesting perspectives to old themes that we as a community don't care to discuss: sexuality, sexism, slavery, migration, religious fanaticism, the color line, science fiction, and campus politics.

Nicholson would also like us to believe that in Third World countries people are not film-goers and television watchers, so therefore they will accept a three-hour film with a different sense of structure, whereas a Black American audience will not. Contrary to his belief, people in Third World countries watch a lot of film and television, but are exposed to such a variety of film styles—in addition to the Hollywood films that are imposed on their screens—that they can accept and identify with a film that has characters who reflect their life experience.

As a filmmaker, I have found that Black American audiences are also hungry to see their own image or the image of any Black people on the screen, that they can make that jump, that suspension of disbelief, and accept a film that has a style radically different than what they are used to.

Although Nicholson acknowledges that "the filmmaker who wants to make a feature film has a hard way to go," there is no real understanding of what making a feature film means. It takes years to mount a feature. "A hard way to go" is too pat an explanation for the years of sacrifice, bankruptcy and relationship crises that too many filmmakers undergo. Rather than do important case studies of new works that could bring an audience into a deeper understanding of what it means to make a new film—regardless of whether it's a feature or short—the critics do not do their own homework or push the limits of their craft as critics and advocates of the Black film movement.

To some degree, shouldn't the role of a Black film critic be to encourage and promote the completion of new work? After all, if we as filmmakers do not finish our films, what do critics have to write about? Critics should be writing and talking to the people at the Corporation for Public Broadcasting, the National Endowment for the Arts, the Hollywood studios, or the Black middle class who won't finance us or who may finance us with so many strings attached that our creative freedom is stifled.

Though Taylor may argue that the "intellectualist" tradition has "surrendered to Hollywood without a fight the right of Black people to enjoy conflict, drama, humor, poetry, while being challenged or edified," the battle for a vision that is uniquely African American has to be waged by critics along with filmmakers. We can only succeed with pressure on various cultural fronts. To blame the Black filmmaker is to fall victim to the very same syndrome Taylor is attacking.

Taylor also wrote that Black filmmakers should be working with established Black writers to break out of the auteur mode. On the surface, this is a good suggestion. However, the reality is that Black writers, especially recognized ones, need paychecks just like everybody else. Most independent filmmakers don't have the money to finance the production of the film, let alone the writing of a script.

By the same token, just because one can write a novel does not mean that one can also write a screenplay. Alice Walker's attempt to write the screenplay for *The Color Purple* is a case in point. As recent competitions sponsored by the Black Filmmaker Hall of Fame in Oakland have shown us, there are quite a number of Black freelance and independent screenwriters. A system needs to be developed where the filmmakers can more fully utilize the talents of these writers. The quality of writing in Black film is certainly much more complex than Taylor allows us to believe.

Nicholson calls for filmmakers to be like "jazzmen" and to "connect with absolute fidelity to the human condition." Yet, at the same time, he wants the filmmakers to adhere only to a simple narrative structure. History shows us that jazzmen and women broke with form and tradition that they themselves had established in order to take music to a new level. Why won't Nicholson let Black filmmakers emulate this creative process as well? Has Nicholson seen the works of Arthur Rogbodiyan (aka A. J. Fielder) and Philip Mallory Jones? They are artists who make us question, who probe our responses to visuals and music, who move the media of film and video to a higher aesthetic level. Yet Nicholson and Taylor don't even acknowledge that these artists exist.

Nicholson also mentions how jazzmen honed their skills in a workshop or jam session. It would be wonderful if Black filmmakers as a group could have access to such artistic residencies. Unfortunately, filmmakers need the financial backing and support of someone other than filmmakers to make a workshop happen. Like the jazzman with his musical instrument, the camera is our instrument, but, unfortunately, we need money to make it produce a note, let alone keep it out of hock. Simplistic comparisons with other art forms belittle the process of filmmaking and illustrate the critic's ignorance of the field.

A Black film critic would do well if he or she could attach him/herself to a film, from its conception on the paper to its birth on the screen. Instead, critics are being paid to attend international conferences and festivals, while the filmmakers are at home living on credit cards and still hustling to get the word out on their films.

Nicholson laments that Black independent films are "so densely convoluted they require explication by professors of semiotics to be comprehended." Why might some Black independent films be "so densely convoluted?"

One answer might be fear. Fear that you will never be able to make another film again and that you have to say everything in one ten-minute film. You know all the bull that you had to go through to do this film and you're not sure if you have the personal courage or morale to do it all again. Why haven't any of the critics addressed themselves to the many films that never get finished, all the many films that have never gotten out of post-production (probably many stay unfinished), that are still sitting in the filmmaker's garage?

Nicholson is right in claiming that we are too parochial as filmmakers and that we tend to break into narrow political camps. I see us outgrowing this stage and possibly developing some new aesthetic forms, especially if a confluence of ideas and co-production is fostered between filmmakers from the Americas, Europe and Africa.

But what does Nicholson want? I'm not sure he even knows. He criticizes the films for being poorly constructed with inconsistencies in the plot. In particular, I take issue with his criticism of *Illusions,* a film which I have seen time and time again only to be renewed and amazed at all the different levels of meaning and relevance that Julie Dash was able to structure in a 28-minute piece. Nicholson points to a film like *Illusions* for its seeming lack of technical virtuosity. Nicholson would do well to check on the quality of the exhibited print. The quality of lighting in *Illusions* is unquestionable and the poor sound quality was probably due to an electroprint soundtrack, a faulty projector or lack of maintenance by the distribution company.

And where is the unparalleled access to technology for African American filmmakers that Nicholson claims? Without putting too much emphasis on our alma mater, most frequently UCLA, we as filmmakers have found out, all too painfully, that even though we make an award-winning thesis film, and studios may court for a hot moment, the reality is that after film school there is not much out there for us, except exorbitant student loans, the bitter opportunity to work for someone for peanuts, or teaching.

It is, in fact, a myth of the armchair critic which allows people to believe that the entry of a solitary Black figure or two onto the studio scene indicates an opening of the technology to all. Such myths are indeed consistent with Hollywood's own delusions—letting in one or two to convince itself mistakenly of its open-mindedness and plurality, smugly content to keep the door closed to the scores of others who are as equally talented or even more talented and creative.

The fact of the matter is that most filmmakers are not filmmakers full time. They are occupied by other daily activities: doing temp work to pay the bills, teaching on the elementary or university level or working as a craftsperson on some asinine studio production in the hope of "liberating" the so-called unusable expendables such as gels for lighting and short ends (left-overs) of film.

Filmmakers are certifiably crazy people who live and die for putting an image on the screen. We are workaholics who abuse our health and stay holed up in dark rooms at night and on weekends. We desperately need the entire body of Black people in the audience, encouraging us psychologically and offering us financial backing when possible. We as filmmakers are taking the bull by the horns, learning the technology and circumventing its basic premises to create new visions that give rise to hidden voices. Clyde Taylor and David Nicholson, we need your criticism and your thoughts, but, most of all, we need your support.

There seems to be some unspeakable wall that critics and filmmakers cannot cross. There is no communication going on. The critics are formulating theories that are naive, mistaken and have no basis or relevance to Black independent filmmaking. At the same time, filmmakers are getting angry reading these articles but are not responding and challenging these writers.

We can not survive in the 90s unless some deeper understanding of the filmmaking process begins to occur. Black critics seem to take the question of style and form in Black independent filmmaking as a simple matter of aesthetic choice. It is not. It is a matter of extreme hardship—political and economic factors, lack of opportunity and equipment and more that influence and form aesthetic decisions.

Personally, I am not interested in making feature films at this point in my career. I want to put out a new work every year or so, not every ten years. Therefore, I make short film and video works that tend to use an experimental style. In my latest film, *Cycles,* I made a decision to use black and white film and to do without sync sound because it was cheaper to produce. This was an aesthetic choice bound by the limitations of economics and equipment access. Does this mean I am not valid as a filmmaker because I choose not to work strictly under narrative conventions? Are we saying that the Black audience is not sophisticated enough to want to view anything other than strictly fictional work?

The Black independent film and video movement *is* adapting to the current political, economic and social milieu in which it finds itself engaged and by which it is challenged and often repressed. But like music or any other art form that Blacks actively pursue, Black film seemingly submerges and breaks out with new style that challenges the status quo.

The theme for the 90s must be collaboration—open communication lines between filmmakers, critics and those interested in the movement that will allow Black independent media to prosper, flourish and nourish our visual souls always so hungry for images of ourselves.

The Future of Black Film

The Debate Continues

Clyde Taylor

The debt we all owe to David Nicholson cannot be repaid by simple thanks. His vision in starting *Black Film Review* and persistence in bringing it to its present level are among the bright moments of the 80s. David's farewell assessment of Black independents (BFR, Vol. 5, No. 2) comes after one I recently made ("The Paradox of Black Independent Cinema," BFR, Vol. 4, No. 4). Together they suggest that the movement is at a critical juncture. I now see more clearly that this sense of crossroads had been partly provoked by the rise of one filmmaker out of the ranks to popular and critical success. This success has, perhaps dangerously, offered us a subliminal yardstick for ranking independent film efforts across the board.

One danger (and opportunity) here is that this singular triumph raises questions about the direction and cohesiveness of Black filmmaking as a "movement." Should we look at the corps of Black independents as apprentices readying themselves for big studio "breaks?" How was the goal of building a Black film culture beyond Hollywood control and supported by the Black community been affected?

Now that commercial success has been added to the picture, it becomes clearer that Black independents have been pursuing a variety of creative goals all along. Does this welcomed diversity threaten the potential for a vital, unified movement toward cultural democracy?

But a crossroads for filmmaking also implies a crossroads for criticism. One demand of the crossroads is that critics engage each other's ideas

Black Film Review, vol. 5, no. 4, 1989, pp. 7, 9, 27–28. Reprinted by permission of the publisher.

so that they, and others, get a sharper read on what we are all talking about. The most needed criticism for Black independence will come out of engaged dialogue among the widest circle of the interested and the committed. From this spirit of dialogue, I want to air some of my differences—along with some agreements—with David's editorial.

My differences hinge mainly on questions of "craft," ideology and expectations. I confess to a nervous tic every time I hear the wolf cry of "craft" raised against Black independent films—as in, "They've got to learn their craft!" The speaker seems to hold some precise idea that everyone should share, but ignores a range of meanings and ideological implications left floating.

We all love excellence—let's get that out of the way. But the exhortation of *craft* echoes in my ear like a Booker T. Washingtonism. "Drop your intellectualism and political posturing and study the fundamentals of the White man's success."

Maybe that's not what is intended. But the separation of craft from other layers in creative work leads to misperceptions. The best craft is whatever affordable instruments help a film achieve its goals with its intended audience. It may be best for some practitioners to learn all the traditional ABC's before setting any aside. I am still impressed, nevertheless, by the number of important works that break the rules on their way to marvelous realization.

We seem to be repeating old discussions about craft in jazz. David invokes jazz as support for his arguments. But jazz history can be read many ways. It can surely suggest the diversity of ways this question of craft has been successfully handled by different musicians.

Our posture towards the rules, traditions and conventions are rooted in the politics of our personalities and in how we decide to address the world. It is not easily debated. I for one appreciate the "ugliness" of Richard Wright's novel *Native Son* that helped make it an awesome statement. Yes, the book could have been better done, but better does not in this case mean smoother. (The recent movie from the novel was smoother and emptier.)

Something like this same issue was involved when Brent Staples faulted *Do the Right Thing* for not offering rounded characters, as compared to *She's Gotta Have It.* But the point is that the considerable ingenuity in each is directed to different ends, even different audiences.

What Spike was doing included stirring public opinion in New York to help Ed Koch become an ex-mayor. Do Black people need rounded characters or do they need peace and justice? Nicholson's objection to *School Daze* likewise seems to bring preset ideas of craft to a different intention. That *School Daze* "subordinated plot and character to idea," and successfully, though perhaps not perfectly, is what gives this movie distinction, for me.

Right here, I can slip in a snide word about that bete blanc, the aesthetic. In looking at books and movies, I am not interested in how they stack up as "art" or entertainment (though in a private corner of my mind I can get pleasure here). First of all, I'm interested in the politics of representation, then in how they impact on the experience in history of African Americans and other people.

The misapplication of craft makes me wish everybody interested in cinema independence would read, if only one, a piece of theory, Julio García Espinosa's essay, "Toward an Imperfect Cinema." Written from a Cuban context, much of it applies north of Miami. To butcher his point in paraphrase, the triumphs of independent cinema must be appreciated within their "imperfections," even because of them, as they stand opposed to the "perfections" of Hollywood: smoothly crafted, expensively-mounted spectacles where the film does all the work for the viewers and none of that work involves thinking about their own reality.

Once I was chatting about the rightness in many African films' avoidance of glossy spectacle that might be within their reach. Burkinabe filmmaker Gaston Kaboré overheard me and put my casual drift in a well-thought out place. "Yes," he said, "they make use of appropriate means." That's the key: Appropriate Means.

For an African American interpretive community, appropriate means might include livelier expression than in some Third World habitats. But the value of this key is lost if we evaluate these means by comparison with Hollywood's products.

An appropriate understanding of a Black independent film includes the realization that nobody, including the Black community, is making more than a few thousand available for these films. When we talk about subtlety, we should understand the budget connection: so many points of storytelling equals X thousands of dollars.

Somewhere around these issues lie my differences with David Nicholson on interpreting specific films.

He calls *Illusions* "execrable." I recently described the film as "the miniature gem of the Black film movement." Its celebrated recording studio scene I called, "One of the great moments of American filmmaking." Nicholson deplores its bad lighting and poor sound quality. I concede these points in part. The sound track is indecipherable at points unless you use a perfect projector. The lighting in some scenes is uneven with the rest and, at a few points, the print looks underexposed.

The ability to see the brilliant conception of the film, its largely successful execution, its balanced synthesis of idea, character and narrative—through the distractions of goofed lighting and sound—is a part of not confusing our history with somebody else's.

I differ from David's readings of films where he seems to hold fixed ideas of what a film should be without considering the question of appropriate means—artistic, technological and socio-historic. Without this flexibility, his barbs on the trailer for *Daughters of the Dust* fall into cheap shot territory.

The music on the trailer *is* heavy Hollywood (in fact, from *Tess*). But a little slack is in order. This music is righteously appropriate for its purpose: to raise money to complete the film, which the trailer did. I'd be surprised if the music is anything similar on the finished film.

And what's wrong with springing out of the "Alice Walker-Gloria Naylor-Toni Morrison school of writing?" I've been arguing for more attention by filmmakers to good writing, as David has, too, so where's the problem? These aren't good writers? How long has it been a put-down for a film to remind you of, say, Tolstoy? As for taking itself seriously, considering our history, what impresses me is that this film, along with just a couple others from Black women independent filmmakers, is one of the few where Black women do take themselves seriously, like in real life. (There's a subtext under this last discussion that deserves a separate dialogue.)

The issue of "craft" links to the issue of film language. One school of thought argues that there is only one film language and Black filmmakers better stop fiddling around and learn it. This line parallels David Nicholson's repeated prescription of "the elementary rules governing the development of character, the construction of plot and story . . . the simplest narrative guidelines."

For me, the idea of only one all-encompassing film language is incomprehensible. It suggests the speaker knows nothing about language and only those examples of cinema (Hollywood?) that fit the argument. If there were only one "correct" language for cinema and storytelling, they would never change. Language and narrative are culturally and historically determined, often by the *golden* rule: "Them with the gold make the rules." We all talk about Black music as the paragon of creativity. But the spirituals, the blues, jazz, gospel, were all *mistakes* when they appeared, according to the "elementary" rules governing musical expression, and were viciously attacked as such.

One bright sign of hope carried by Black independent film unfurls partly around its search for a more appropriate film language for the expression of African American perspectives and sense of history than the self-betraying classic formulas. But that sign has just appeared on a new horizon. The movement is young; that is not an excuse but a historical perception. Not young in years, but in developmental opportunities, both for practice and reception.

This youth should influence our expectations. I learned a lot from Kathy Collins when she declared Black independent films hardly ready for criticism. "We're just baby filmmakers," she'd say. "I've just begun to learn. I've only made two films." And her shooting partner, Ronald Gray, chimed in with a telling analogy: "Would you like to be judged on the basis of the first thing you wrote? Not published, but wrote?"

Then why show the films? First, because exhibition is part of the learning process. And then because there is such a hunger among many Black people for even a provisional glimpse at what an independent Black voice in cinema might enunciate. Even Toni Morrison has said her writing is going to be dwarfed by the Black writing soon to come.

Then how do I justify the criticisms of the movement that I made in the "Paradox" essay?

David and I agree that the concepts and storytelling strategies of many Black independent narrative films are weaker than they need be, but we probably have different ideas in mind when we make this criticism.

The question I bring to these films is: discounting production values, do the imaginative means of telling the story stand up on their own, or could they have been made more compelling for a wider community through interactive exchanges with other filmmakers, critics, programmers, writers and ordinary viewers *before* the script was filmed?

The questions of craft plus ideology are also complex. I agree with David that some of the political stances in some films are approached simple-mindedly. It's too easy to make characters the vehicles of only one position each, as in a medieval morality play where one is labeled Pride and another Virtue, instead of seeing conflict take place within characters. That's poster cinema. Such storytelling can hardly hope to *advance* the discussion of African American destiny.

But the ideologies themselves call for separate discussion, leading off with the entitlement of filmmakers to pursue whichever line they choose and subject to the judgment of audience response, criticism and history. I frequently suspect some films harbor a self-indulgence of personal "artist" fantasies and personal ideological positions. But a larger audience, once it is developed, may bring a weightier sense of dialogical responsibility.

It's a bit much, though, to ask filmmakers to bring us a more cogent vision of history, the present and the future than they are presently doing, much as one would like to see such films. Who is doing as much outside of films? And who is doing it in a medium where it takes maybe eight to ten years to get one's vision packaged and before an audience?

Further, I am heartened by the sense of commitment and fidelity to a sense of African American destiny—including development, independence of mind and spirit, and the imperishable demand for justice—that I find in these films over all others produced in this country. If I were not, I wouldn't bother with them, except to denounce them.

Black cinema is barely into the Scott Joplin stage, getting ready for the profound innovations of a Louis Armstrong. It is only because our traditions are so rich that we can dream at the level of Monk and Coltrane. But because our traditions are rich, we may just move through these stages more quickly than anyone now anticipates.

David, hang tough! And thanks for the rise you've given us, as publisher, editor and critic.

III
Documents

Resolutions of the Third World Film-makers Meeting Algiers, December 5–14, 1973

Cineaste (Pamphlet No. 1)

The Third World Film-makers Meeting, sponsored by the National Office for Cinematographic Commerce and Industry (O.N.C.I.C.) and the cultural information center, was held in Algiers from December 5 to 14, 1973. The meeting brought together film-makers from all areas of the Third World for the purpose of discussing common problems and goals and to lay the groundwork for an organization of Third World film-makers.

The film-makers attending the conference organized themselves into separate committees to discuss the specific areas of production and distribution as well as how the film-maker fits into the political struggle of the Third World.

The resolutions of the various committees are published here as they were released in Algiers, with only slight modifications in grammar and spelling.

Committee 1: Peoples Cinema

The Committee on Peoples Cinema—the role of cinema and film-makers in the Third World against imperialism and neo-colonialism—consisted of the following film-makers and observers: Fernando Birri (Argentina); Humberto Rios (Bolivia); Manuel Perez (Cuba); Jorge Silva (Colombia); Jorge Cedron (Argentina); Moussa Diakite (Republic of Guinea); Flora Gomez (Guinea-Bissau); Mohamed Abdelwahad (Morocco); El Hachmi Cherif (Algeria); Lamine Merbah (Algeria); Mache Khaled (Algeria); Fettar Sid Ali (Algeria); Bensalah Mohamed (Algeria); Meziani

This document was first published by Cineaste Publishers, Inc.

Abdelhakim (Algeria). Observers: Jan Lindquist (Sweden); Josephine (Guinea-Bissau) and Salvatore Piscicelli (Italy).

The Committee met on December 11, 12 and 13, 1973, in Algiers, under the chairmanship of Lamine Merbah. At the close of its deliberations, the Committee adopted the following analysis.

So-called "underdevelopment" is first of all an economic phenomenon which has direct repercussions on the social and cultural sectors. To analyze such a phenomenon we must refer to the dialectics of the development of capitalism on a world scale.

At a historically determined moment in its development, capitalism extended itself beyond the framework of the national European boundaries and spread—a necessary condition for its growth—to other regions of the world in which the forces of production, being only slightly developed, provided favorable ground for the expansion of capitalism through the existence of immense and virgin material resources, and available and cheap manpower reserves which constituted a new, potential market for the products of capitalist industry.

This expansion manifested itself in different regions, given the power relationships, and in different ways:

a) Through direct and total colonization implying violent invasion and the setting up of an economic and social infrastructure which does not correspond to the real needs of the people but serves more, or exclusively, the interests of the metropolitan countries;

b) In a more or less disguised manner leaving to the countries in question a pretense of autonomy;

c) Finally, through a system of domination of a new type—neo-colonialism.

The result has been that these countries undergo, on the one hand, varying degrees of development and, on the other hand, extremely varied levels of dependency with respect to imperialism: domination, influence and pressures.

The different forms of exploitation and systematic plundering of the natural resources have had grave consequences on the economic, social and cultural levels for the so-called "underdeveloped" countries, resulting in the fact that even though these countries are undergoing extremely diversified degrees of development, they face in their struggle for independence and social progress a common enemy: imperialism which stands in their way as the principal obstacle to their development.

Its consequences can be seen in:

a) The articulation of the economic sectors: imbalance of development on the national level with the creation of poles of economic attraction incompatible with the development of a proportionally planned national economy and with the interests of the popular masses, thereby giving rise to zones of artificial prosperity.

b) The imbalance on the regional and continental levels, thereby revealing the determination of imperialism to create zones of attraction favorable for its own expansion and which are presented as models of development in order to retard the peoples' struggle for real political and economic independence.

The repercussions on the social plane are as serious as they are numerous: they lead to characteristic impoverishment of the majority for the benefit in the first instance of the dominating forces and the national bourgeoisie of which one sector is objectively interested in independent national development, while another sector is parasitic and comprador, the interests of which are bound to those of the dominating forces.

The differentiations and social inequities have seriously affected the living standard of the people, mainly in the rural areas where the expropriated or impoverished peasants find it impossible to reinvest on the spot in order to subsist. Reduced in their majority to self-consumption, unemployment and rural exodus, these factors lead to an intenstification of unemployment and increase under-employment in the urban centers.

In order to legitimize and strengthen its hold over the economies of the colonized and neo-colonized countries, imperialism has recourse to a systematic enterprise of deculturation and acculturation of the people of the Third World.

That deculturation consists of depersonalizing their peoples, of discrediting their culture by presenting it as inferior and inoperative, of blocking their specific development, and of disfiguring their history . . . In other words, creating an actual cultural vacuum favorable to a simultaneous process of acculturation through which the dominator endeavors to make his domination legitimate by introducing his own moral values, his life and thought patterns, his explanation of history: in a word, his culture.

Imperialism, being obliged to take into account the fact that colonized or dominated peoples have their own culture and defend it, infiltrates the culture of the colonized, entertains relationships with it and takes over those elements which it believes can turn it to its favor. This is done by using the social forces which they make their own, the retrograde elements of this culture. In this way, the language of the colonized, which is the carrier of culture, becomes inferior or foreign; it is used only in the family circle or in restricted social circles. It is no longer, therefore, a vehicle for education, culture and science, because in the schools the language of the colonizer is taught, it being indispensable to know it in order to work, to subsist and to assert oneself. Gradually, it infiltrates the social and even the family relationships of the colonized. Language itself becomes a means of alienation, in that the colonized has a tendency to practice the language of the colonizer, while his own language, as well as his personality, his culture and his moral values, become foreign to him.

In the same line of thought, the social sciences, such as sociology, archaeology and ethnology, are for the most part in the service of the colonizer and the dominant class so as to perfect the work of alienation of the people through a pseudo-scientific process which has in fact simply consisted of a retrospective justification for the presence of the colonizer and therefore of the new established order.

This is how sociological studies have attempted to explain social phenomena by fatalistic determinism, foreign to the conscience and the will of man. In the ethnological field, the enterprise has consisted of rooting in the minds of the colonized prejudices of racial and original inferiority and complexes of inadequacy for the mastering of the various acquisitions of knowledge and man's production. Among the colonized people, imperialism has endeavored to play on the pseudo-racial and community differences, giving privilege to one or another ethnic grouping.

As for archaeology, its role in cultural alienation has contributed to distorting history by putting emphasis on the interests and efforts of research and the excavations of historical vestiges which justify the definite paternity of European civilization sublimated and presented as being eternally superior to other civilizations whose slightest traces have been buried.

Whereas, in certain countries, the national culture has continued to develop while at the same time being retarded by the dominant forces, in other countries, given the long period of direct domination, it has been marked by discontinuity which has blocked it in its specific development, so that all that remains are traces of it which are scarcely capable of serving as a basis for a real cultural renaissance, unless it is raised to the present level of development of national and international productive forces.

It should be stated, however, that the culture of the colonizer, while alienating the colonized peoples, does the same to the peoples of the colonizing countries who are themselves exploited by the capitalist system. Cultural alienation presents, therefore, a dual character—national against the totality of the colonized peoples, and social against the working classes in the colonizing countries as well as in the colonized countries.

Imperialist economic, political and social domination, in order to subsist and to reinforce itself, takes root in an ideological system articulated through various channels and mainly through cinema which is in a position to influence the majority of the popular masses because its essential importance is at one and the same time artistic, esthetic, economic and sociological, affecting to a major degree the training of the mind. Cinema, also being an industry, is subjected to the same development as material production within the capitalist system and through the very fact that the North American economy is preponderant with respect to world capitalist production, its cinema becomes preponderant as well and succeeds in invading the screens of the capitalist world and consequently those of the

Third World where it contributes to hiding inequalities, referring them to that ideology which governs the world imperialist system dominated by the United States of America.

With the birth of the national liberation movement, the struggle for independence takes on a certain depth implying, on one hand, the revalorization of national cultural heritage in marking it with a dynamism made necessary by the development of contradictions. On the other hand, the contribution of progressive cultural factors borrowed from the field of universal culture.

The Role of Cinema

The role of cinema in this process consists of manufacturing films reflecting the objective conditions in which the struggling peoples are developing, i.e., films which bring about disalienation of the colonized peoples at the same time as they contribute sound and objective information for the peoples of the entire world, including the oppressed classes of the colonizing countries, and place the struggle of their peoples back in the general context of the struggle of the countries and peoples of the Third World. This requires from the militant film-maker a dialectical analysis of the socio-historic phenomenon of colonization.

Reciprocally, cinema in the already liberated countries and in the progressive countries must accomplish, as their own national tasks, active solidarity with the peoples and film-makers of countries still under colonial and neo-colonial domination and which are struggling for their genuine national sovereignty. The countries enjoying political independence and struggling for varied development are aware of the fact that the struggle against imperialism on the political, economic and social levels is inseparable from its ideological content and that, consequently, action must be taken to seize from imperialism the means to influence ideologically, and forge new methods adapted in content and form to the interests of the struggle of their peoples. This implies control by the people's state of all cultural activities and, in respect to cinema, nationalization in the interest of the masses of people: production, distribution and commercialization. So as to make such a policy operative, it has been seen that the best path requires quantitative and qualitative development of national production capable, with the acquisition of films from the Third World countries and the progressive countries, of swinging the balance of the power relationship in favor of using cinema in the interest of the masses. While influencing the general environment, conditions must be created for a greater awareness on the part of the masses, for the development of their critical senses and varied participation in the cultural life of their countries.

A firm policy based on principle must be introduced in this field so as to eliminate once and for all the films which the foreign monopolies continue to impose upon us either directly or indirectly and which generate reactionary culture and, as a result, thought patterns in contradiction with the basic choices of our people.

The question, however, is not one of separating cinema from the overall cultural context which prevails in our countries, for we must consider that, on the one hand, the action of cinema is accompanied by that of other informational and cultural media, and, on the other hand, cinema operates with materials which are drawn from reality and already existing cultural forms of expression in order to function and operate. It is also necessary to be vigilant and eliminate nefarious action which the information media can have and to purify the forms of popular expression (folklore, music, theatre, etc.) and to modernize them.

The cinema language being thereby linked to other cultural forms, the development of cinema, while demanding the raising of the general cultural level, contributes to this task in an efficient way and can even become an excellent means for the polarization of the various action fields as well as cultural radiation.

Films being a social act within a historical reality, it follows that the task of the Third World film-maker is no longer limited to the making of films but is extended to other fields of action such as: articulating, fostering and making the new films understandable to the masses of people by associating himself with the promoters of people's cinemas, clubs and itinerant film groups in their dynamic action aimed at disalienation and sensitization in favor of a cinema which satisfies the interests of the masses, for at the same time that the struggle against imperialism and for progress develops on the economic, social and political levels, a greater and greater awareness of the masses develops, associating cinema in a more concrete way in this struggle.

In other words, the question of knowing how cinema will develop is linked in a decisive way to the solutions which must be provided to all the problems with which our peoples are confronted and which cinema must face and contribute to resolving. The task of the Third World film-maker thereby becomes even more important and implies that the struggle waged by cinema for independence, freedom and progress must go, and already goes, hand in hand with the struggle within and without the field of cinema, but always in alliance with the popular masses for the triumph of the ideas of freedom and progress.

In these conditions, it becomes obvious that the freedom of expression and movement, the right to practice cinema and research are essential demands of the film-makers of the Third World—freedoms and rights which they have already committed to invest in the service of the working masses against imperialism, colonialism and neo-colonialism for the general emancipation of their peoples.

United and in solidarity against American imperialism, at the head of world imperialism, and direct or indirect aggressor in Vietnam, Cambodia, Laos, Palestine, in Africa through the intermediary of NATO, SEATO and CENTO, and in Latin America, hiding itself behind the fascist coup d'état of the Chilean military junta and the other oligarchies in power, the film-makers present here in Algiers, certain that they express the opinion of their film-maker comrades of the Third World, condemn the interventions, aggressions and pressures of imperialism, condemn the persecutions to which the film-makers of certain Third World countries are subjected and demand the immediate liberation of the film-makers detained and imprisoned and the cessation of measures restricting their freedom.

Committee 2: Production/Co-Production

The Committee on Production/Co-Production, appointed by the General Assembly of the Third World Film-Makers Meeting in Algeria, met on December 11, 12, and 13, 1973, under the chairmanship of Ousmane Sembene. The Committee, which devoted itself to the problems of film production and co-production in the Third World countries, included the following film-makers and observers: Ousmane Sembene (Senegal); Sergio Castilla (Chile); Santiago Alvarez (Cuba); Sebastien Kainba (Congo); Mamadou Sidibe (Mali); Benamar Bakhti (Algeria); Nourredine Touazi (Algeria); Hedi Ben Khelifa (Tunisia); Mostefa Bouali (Palestine); Med Hondo (Mauritania). Observers: Simon Hartog (Great Britain), representing the British film-makers' union, and Theo Robichet (France). Humberto Rios (Argentina) presented an information report to the Committee.

The delegates present, after reporting on the natural production and co-production conditions and the organization of the cinema industries in their countries, noted that the role of cinema in the Third World is to promote culture through films, which are a weapon as well as a means of expression for the development of the awareness of the people, and that the cinema falls within the framework of the class struggle.

Considering:

—that the problems of cinema production in the countries of the Third World are closely linked to the economic, political and social realities of each of them;

—that, consequently, cinema activity does not develop in a similar fashion:

a) in those countries which are waging a liberation struggle,

b) in those countries which have conquered their political independence and which have founded States,

c) in those countries which, while being sovereign, are struggling to seize their economic and cultural independence;

—that those countries which are waging wars of liberation lack a film infrastructure and specialized cadres and, as a result, their production is limited, achieved in difficult circumstances and very often is supported by or is dependent upon sporadic initiatives;

—that in those countries struggling for their economic and cultural independence, the principal characteristic is a private infrastructure which enables them to realize only a portion of their production within the national territory, the remainder being handled in the capitalist countries;

This leads to an appreciable loss of foreign currency and considerable delays which impede the development of an authentic national production.

—that in those countries in which the state assumes the responsibility for production and incorporates it in its cultural activity, there is, nevertheless, in a majority of cases, a lack of technical and industrial development in the cinema field and, as a consequence, production remains limited and does not manage to cover the needs for films in those countries. The national screens, therefore, are submerged with foreign productions coming, for the most part, from the capitalist countries.

—that, if we add as well the fact that world production is economically and ideologically controlled by these countries and, in addition, is of very mediocre quality, our screens bring in an ideological product which serves the interests of the colonizers, creating moreover the habit of seeing films in which lies and social prejudice are the choice subjects and in which these manufacturers of individualistic ideology constantly encourage the habits of an arbitrary and wasteful consumer society;

—that co-productions must, first and foremost, be for the countries of the Third World, a manifestation of anti-imperialist solidarity, although their characteristics many vary and cover different aspects. We do not believe in co-productions in which an imperialist country participates, given the following risks:

1) the imperialist country can shed influence through production methods which are foreign to the realities of our countries,

2) the examples of co-productions have given rise to cases of profit and the cultural and economic exploitation of our countries.

The participants in the Committee therefore concluded that it is necessary to seek jointly concrete means to foster the production and co-production of national films within the Third World countries.

In line with this, a certain number of recommendations were unanimously adopted:

—to provide the revolutionary film-makers of the third world with national cinema infrastructures;

—to put aside the conceptions and film production means of the capitalist countries and to seek new forms, taking into account the authenticity and the realities of the economic means and possibilities of the Third World countries;

—to develop national cinema and television agreements for the benefit of the production and distribution of Third World films and to seek such agreements where they do not exist and to exchange regular programs;

—to organize and develop the teaching of film techniques, to welcome the nationals of countries in which the training is not ensured;

—to use all the audio-visual means available for the political, economic and cultural development of the countries of the Third World;

—to promote co-productions with independent, revolutionary film-makers, while leaving to each country the task of determining the characteristics of these productions;

—to include in the governmental agreements between countries of the Third World those measures likely to facilitate co-productions and film exchanges;

—to influence the establishment of co-productions between national organizations of the Third World in endeavoring to have them accepted by the governmental and professional institutions of their respective countries (through the influence, in particular, of the acting president of the non-aligned countries, Mr. Houari Boumediene);

—to propose the need for the creation of an organization of Third World film-makers, the permanent secretariat of which should be set up in Cuba. While awaiting the creation of this organization, the UAAV (Union of Audio-Visual Arts of Algeria) will provide a temporary secretariat.

The film-makers will henceforth keep each other informed of their respective approaches undertaken within the framework of the FEPACI (Pan-African Federation of Cineastes).

Committee 3: Distribution

The Committee in charge of the distribution of Third World films, after consideration of the different remarks of the members present, proposes: the creation of an office to be called the Third World Cinema Office.

It will be composed of four members including a resident coordinator and one representative per continent. The Committee, in reply to the offer made by Algeria, proposes that the permanent headquarters of the office be established in Algiers.

The goals of the office will be:

1) To coordinate efforts for the production and distribution of Third World films,

2) To establish and strengthen existing relations between Third World film-makers and cinema industries by:

a) the editing of a permanent information bulletin (filmography, technical data sheets, etc.) in four languages: Arabic, English, French and Spanish,

b) making a census of existing documentation on Third World cinema for the elaboration and distribution of a catalogue on the cinema production of the countries of the Third World,

c) fostering other festivals, film markets and film days on the Third World level, alongside the other existing events,

d) the editing of a general compilation of official cinema legislation in the Third World countries (problems of censorship, distribution of film copies, copyright, customs, etc.).

3) To take those measures required for the creation of regional and continental organization leading to the creation of a tricontinental organization for film distribution,

4) To prospect the foreign markets in order to secure other outlets for the productions of the Third World countries (commercial and non-commercial rights, TV and cassettes).

The office will approach the authorities of the OAU, the Arab League and UNESCO in order to obtain from these organizations financial assistance for its functioning. It will also approach the authorities of those countries having effective control of their cinema industries, i.e.: Algeria, Guinea, Upper Volta, Mali, Uganda, Syria and Cuba, as well as other countries which manifest a real desire to struggle against the imperialist monopoly. In addition to the above-mentioned assistance, the operating budget of the office will be composed of donations, grants and commissions on all transactions of Third World films entrusted to the office.

Seminar on "The Role of the African Film-Maker in Rousing an Awareness of Black Civilization" Ouagadougou, April 8–13, 1974

Society of African Culture

Summary

The object of this Seminar* is to discuss the African cinema—not from the purely technical point of view but essentially to see it in perspective and to appreciate its role, content and prospects with respect to what has already been achieved.

As a means of mass education, its place should be reappraised within the context of the colloquium "Black Civilization and Education" which is central to the Second World Black and African Festival of Arts and Culture to be held in Lagos in 1975.

Indeed, the African cinema promises to play a decisive role in the education of the African masses as a consequence of its extension and present development. The Ouagadougou meeting will answer the urgent need of many people who wish to see the African film-maker face to face with his own experience and to give him a clearer responsibility in rousing an awareness of black civilization.

In this connection, we feel he should more than ever draw his strength, authenticity and the beauty of his art from the daily life of the African communities which are the living sources of our civilization values. Therefore we have to promote, or aid the promotion, of a popular cinema—not in the condescending and devalued meaning of the term

* Seminar organized by the Society of African Culture in collaboration with the Cultural and Technical Co-operation Agency under the patronage of the Government of Upper Volta (Ouagadougou: April 8–13, 1974).

First published in *Présence Africaine*, No. 90, 1974, pp. 6–7, 13–20, 59–61, 96–98, 140–141, 166–71, 185–93, 202–203. Reprinted by permission of the publisher.

but in the meaning where this cinema would place itself at the service of the people without, however, curbing its own freedom and creativity inspired by the latter's experiences, the course of their history and the profundity of their destiny. Moreover, the film-maker will not risk losing the quality of his language, individuality and style.

Thus he must resolutely become aware of his responsibilities while creatively developing the form of his language and the content of his messages within the framework of African cultural development and, more particularly, the blossoming of black civilization.

It will be at this price and in such conditions that the African cinema and film-makers will avoid the temptation of producing vulgarities, which, alas, are too profitable, or sophisticated pieces reserved for the delectation of an élite of "connaisseurs."

During this seminar, participants will be invited to examine the following points:

a) how the cinema can inspire a feeling of cultural and historical solidarity between black communities and *rouse an awareness of a common civilization;*
b) how the cinema can present the human, social and cultural realities of Africa and make Africans think about their destiny;
c) how the cinema can work for African independence and cultural authority;
d) how and why the African cinema should use African languages;
e) the problem of an original cinematographic language freed from certain Western models, criteria and myths (e.g. the individualistic hero myth, the all-pervading power of money, unbridled sex, violence, false luxury, etc.).

Basic Text

I. Introduction

The title of this Seminar cannot be more eloquent. It is an open invitation to the African cinema to take part in a global project: to help the African peoples to become aware of their own world, values and fundamental attitudes *vis-à-vis* their existence—in a word, their civilization.

The relations to be established between the action of the cinema and this civilization would gain by being specified and its articulations better defined. This would justify a gathering of film-makers, critics, sociologists, men of letters, etc., in order to avoid a closed and specialized discussion.

The object is to confront the African film-maker with his own experience and that of our peoples in order to make him unbend and speak, through images, a language that reflects the state of our civilization in all its dimensions. From this point, there arises a host of questions which require urgent replies.

II. The African Film-maker as a Responsible Agent

The African film-maker can only act efficiently in the right direction if he becomes fully aware of his responsibilities because he holds *power* in his hands. As a machine which makes people dream, laugh or cry, the cinema does not play an innocent role in our society: it imposes heroes and life-styles, inspires behaviour, serves ideologies or, quite simply, informs and illustrates the diversity of men and cultures.

The problem is to know how the film-maker can inspire a feeling of cultural and historic solidarity between the various black communities and rouse an awareness of black civilization. These communities live artificially in a world that is culturally partitioned and crumbling and it is urgent to show them the true dimensions of their civilization—i.e. *to give them a unified awareness of their identity.* Being essentially a mass medium and through its influence on the masses, the cinema can greatly help to bring this about.

However, the film-maker must engage himself in this project and to make his engagement conclusive he must circumscribe and define his object (problems concerning content, themes, genres, etc.).

III. Films and Problems of Content

To serve this "civilization project," the cinema must avoid at all costs the reproduction of outside cultural models which, alas, are predominant. A true African cinema can only be built by breaking away from the Western cinema which is extroverted with respect to the essence of black civilization. The latter must be the source from which our cinema draws the fecund and rich material of its art.

a) African artistic inspiration has been happily served by historical tradition. Chaka, Soundjata, Lat-Dior, Samory and other leaders are celebrated and their feats have been translated into different languages by our *griots,* story-tellers, poets and novelists. The same applies to the great deeds which have weaved the history of our peoples.

The African film-maker must resuscitate the historical messages hidden in the robes of our past and put them into his work. The first efforts to do this are now taking place (cf. "Emitai" by Ousmane Sembène and "Lat-Dior" which is still being shot). The stakes are high because this action means *the awakening and development of historical awareness among Blacks.* This awareness must not stop at the gates of the present because our history is being made from day to day and is experienced by human beings and the film-maker who is responsible for forging its image.

Independence struggles, armed struggles, *apartheid,* tribalism, famines, etc., are so many realities of our present historical drama. They should be translated into film and this gesture of transmuting life into a speaking image will reconcile the film-maker in his own logic with the essence of his art while reconciling him with his world.

b) Co-operation between black writers and film-makers should find its living expression in producing films directly inspired by the works in our already rich and varied literary heritage. This would prevent the cinema from divorcing itself from literary creativity and thus from shutting itself in isolation.

c) Social realities should also attract the interest of the film-maker and the cinema would find its predilection for describing social contradictions. Once the ethnological stage has been passed, the cinema should vigorously exploit various themes such as the problems posed by the condition of the African woman, youth, racism, migrations, emigration, etc.

Films of social content should evoke current changes such as the duality between town and country, industrial development, the living conditions of different categories of people, etc. Such an orientation which takes root in social life deserves to be broadened as it will allow our peoples to reinforce the image they have of themselves and, consequently, to deepen their awareness.

Other sources of the inspiration should be exploited in order to produce didactic films on traditional and modern techniques, films illustrating Africa and the Black World in their physical and human diversity, but also in their unity (productions dealing with the dance, architecture, sculpture, painting, handicrafts and all forms of rural activity will contribute to rousing cultural awareness).

In our attempt to catalogue our heritage, our main worry is not to forget any new form of artistic creativity. Even entertainment films should not be neglected because, in this important domain which is favoured by the public, there is need to exploit all the resources of African wit, finesse, humour and laughter (e.g. satires, village comedies, love-affairs, stories, etc.).

Generally speaking, the film-maker should listen to the people and exploit the riches of oral tradition and the life of black communities. By contacting the latter and attending their "school" he will be able to find new contents and forge a new language and style for the African cinema.

He must not consider the people from an ethnological point of view or fall into the trap of demagogy. The surest means to make this project fully effective would consist in using national languages without which our cinema would lack a part of its true personality. This factor poses several problems but original solutions must be found. If we wish to rouse the people's interest, we must speak to them in the language of their culture. This question is not merely formal because the incidence of language on the content, extent, character and audience of the cinema is intimately linked to it and the Seminar should pay much attention to it.

In short, our cinema can only be the exclusive vehicle of the images and messages of the African world and its peoples in their contradictions and in the debates in which they are engaged—i.e. a mirror of their proper identity. It will also be the dynamic instrument of their liberation if it provides scenarios whose subjects are African realities.

It is not necessary for the African film to be produced on the Western scale (large budgets, the star system and the use of sophisticated techniques) as it would be better to make several medium-length films than one super-production.

Television techniques (if not its materials), which are more direct than those of the cinema, seem more adequate to give veracity to our reality. They also provide a better contact with human beings and even allow their active participation.

In order to rouse an awareness of the diversity and solidarity of black civilization at the institutional level, governments could be approached with the idea of creating an inter-State organization for facilitating the exchange of films. Such a project would imply the setting up of an inter-African "filmothèque."

IV. Conclusion

a) *Who are we?*
What are the present characteristics of our society?
What do we know about our past?
What future do we wish to build?

b) *How do we live?*
Human relationships: family, religious and political structures.
Habitation.
Work and techniques.
Creations.
Games and sports.
Wisdom: birth, life, death, God, love.
The imaginary world.

c) *Where are we?*
Who are our neighbours?
The place of the community in the Nation.
The place of the Nation in Africa.
The place of Africa in the world.
The Black World.

NOTE: *As all papers presented to the Seminar were in French, we have summarized the main points in English for the benefit of non-bilingual readers. We have also translated a brief report, Workshop Resolutions and the closing speech of the Minister of Information which can be found after the relevant French texts.*

Point 1: How the Cinema Can Inspire a Feeling of Historical and Cultural Solidarity Between the Communities of the Black World

Djibril Tamsir Niane

The cinema and TV are doubtlessly the best instruments for expressing the African personality. Thus the former should primarily play an educative role by transmitting the oral tradition of our griots or the gestures of our artisans. It should also be a popular instrument for rousing national awareness by rehabilitating black civilization and an African culture common to the whole continent.

The first duty of the film-maker is to make Africa known in her generality and particularity and this is why we should fight against foreign films which depict social problems that do not concern us and provide a factor of alienation.

The cinema can inspire a feeling of cultural and historical solidarity between the communities of the Black World and rouse an awareness of our civilization by showing the cultural unity that exists between the different regions of Africa, despite linguistic barriers, and our historical solidarity, despite the colonial conquest and European domination.

Thus it is necessary to produce historical films and documentaries covering each region in an effort to educate. This demands complete freedom of expression and, often, the need to use African languages. The problem is therefore political, financial and technical at one and the same time and it is indispensable for our governments to give financial support to our film-makers and have confidence in them.

Lastly, audio-visual techniques are the best means of revealing a black civilization which is aware of what it should contribute to universal civilization.

Richard B. de Medeiros

Almost fifteen years after independence, the African cinema is still enslaved by foreign capital and dominated by foreign groups in the purest colonial tradition. The problems of this profession are mainly material and solidarity between States would be particularly necessary to solve them.

African film-makers should first of all deal with the innumerable historical subjects that illustrate the profound solidarity existing between black peoples, but the difficulty resides in finding African script-writers.

By choosing these historical subjects, our film-makers will be able to show that Africa is one nation, and by producing authentically African films we shall be able to fight against the intellectual colonialism wished upon us by the foreign cinema.

However, our cinema should not deal uniquely with History. It should tackle all subjects in order to show our society as it actually is and to reconcile us with the realities of our daily lives and with ourselves.

It should also be an inventory of all our riches and all aspects of our civilization in order to play a role of popular educator.

Therefore the cinema has a unifying function which must be developed and the Pan-African Federation of Film-Makers, founded in 1970, is a symbol of this solidarity.

Black Americans are showing renewed interest in African culture which is a veritable return to the sources. Another expression of the awareness of a growing solidarity is the interest in films depicting the difficulties of African immigrants in Europe.

Finally, African civilization appears more and more as a pole of attraction for all the black communities scattered throughout the world.

Lucien Mailli

The mass media have brought about a complete change in the behaviour of the individual because the image penetrates his very subconscious.

However, this aspect should not make us forget the positive role of the media in popularizing science, making man discover himself and the Universe, educating the masses, etc.

Therefore the film-maker has great power and heavy responsibility because he can influence certain behaviour-patterns. His mission is an educative one in order to rouse a feeling of cultural and historical solidarity.

The "colonial" cinema was a dangerous weapon because it gave the world an image of our continent that was false, paternalistic or "common" (exotic and folklorish).

The problems of our cinema are essentially financial. Each State should consider that information is a priority investment for the collective transformation of attitudes brought about by entirely African films.

To achieve this, more cinemas should be built and more mobile-cinemas should go to the furthest villages with the aim of decentralizing information and giving priority to small localities.

The spectator should also be taught how to "see" a film by means of radio, TV talks and press articles so that culture may be accessible to all.

The role of our cinema is to enhance black civilization and culture—particularly in the eyes of those who are aware of Western civilization. For the latter, the recognition of our traditional and sacred values could make them at one with themselves and recreate a spiritual communion and a feeling of solidarity. One would also see the disappearance of certain tribal barriers.

In order to achieve this return to the sources, it seems indispensable to express our traditions by the use of African languages.

Finally, as a means of expression and revolutionary education, our cinema must accentuate the human values and the will to understand between our various peoples.

Point 2: How the Cinema Can Present the Human, Social and Cultural Realities of Africa

Sékou Tall

It is a fact that African people are becoming more and more aware of their past and their future, their civilization and their culture.

Whenever people become aware of their civilization, they always make an effort to master themselves and to bring about their unity.

With this aim in mind, the African film-maker must inform, entertain, teach and educate. He should not look for the exotic but for art and express his artistic feelings by symbols that represent African realities.

Firstly, he must express the mystical side of the African universe—what is called magic, sorcery, fates, rites and mystique—through the legends and symbolic divinities particular to our different peoples and which only they can understand.

He must refuse the negative aspect of colonialism which influences African art through publicity and the cinema by imposing its consumer society and creating larger differences between men.

He must destroy the myths given to our continent by the Western cinema and find new faith in enriching our spiritual, economic, social and cultural experience.

Knowing how to interpret signs and symbols is to know Africa's deepest aspirations. An authentic African cinema is that where each spectator can recognize himself through the attitudes and behaviour of the actors who express the struggles of the individual with himself or with Nature, and in a language that he can understand.

It is up to our governments to motivate, promote and support unconditionally, morally and financially all those artists who contribute to the awakening of an awareness of black civilization.

Solo Randrasana

We can ascertain that, very often, our artistic heritage has been recognized by "Western judgement" which, at the beginning of the century, gave it value, as witness the birth of Cubism which was inspired by African sculpture.

But what place do we ourselves give to our art within the African family? This sets the problem of our awareness of civilization.

Our greatest poets are "French-speaking" and not accessible to the ordinary man. We must become Africans again in order to communicate. The same malaise affects the cinema which gives priority to "profitability" rather than to authenticity, in accordance with Western standards.

Who is the African film-maker? More than any other artist, he has been cut off from the cultural realities of his country and milieu since his early years. He has attended schools in order to assimilate an "imported" culture under the guidance of a "Greco-Roman" civilization. Once he has acquired his foreign diplomas he can enter the film-studios of London, Paris or Moscow . . .

In contemporary African society, he belongs to a cultural élite and is far removed from the human and social realities which he should depict. Our cinema should have social and cultural objectives, particularly in the fight against under-development and express the class-struggle because art must be influenced by the social, political and economic problems of the age.

Finally, the cinema should awaken, motivate and serve the masses—i.e. be a militant art and not subjected to the demands of profit.

Alkaly Kaba

An African cinema must be born and to achieve this we must question two conceptions which dangerously divide our peoples: the first is a fundamental modification of everything that reminds us of colonialism; the second is an orientation and utilization of the acquired knowledge contributed by colonialism.

The African film-maker must choose the first conception—i.e. he must completely replace imitated colonial behaviour and habits by the behaviour and habits of the black man which he himself has discovered and adapted to his own ambitions, aspirations and means.

The cinema is an instrument which can contribute to getting rid of the inertia and ignorance of the African in all media and help him face the problems of development in order to get out of the rut of a subsistence economy.

Our peoples have been more uprooted by the Western cinema than by the "white school" and it is to be noted that the atmosphere of mistrust and fear existing between governments and creators and artists paralyses artistic creativity.

The African film-maker must avoid resembling his Western colleague. He must take account of the gains and losses of our societies, express himself in African languages and see African and foreign situations through African eyes.

In conclusion, the production, distribution and exploitation circuits of the film industry on our continent still remain in the hands of foreign monopolies.

Point 3: How the Cinema Can Work for African Independence and Cultural Authority

Ntite Mukendi

In a world made smaller by modern means of transport and communication, we must accept the exigencies of the new means of information which are called the mass media.

When information circulates at group level, the individual is in a better position to integrate himself into it in order to live a better life. Information plays a capital role in making people aware of their belonging to a group and hence maintaining group cohesion.

The spoken word cannot resist time or space because human memory is not trustworthy. Modern communication techniques allow a retransmission of messages that is all the more accurate when accompanied by an image.

The written word provides only a secondary form of communication and demands an effort of abstraction which cannot always be made by all readers.

Audio-visual techniques are quite different to the written and spoken word: the number of speakers is very limited and the mass of listeners and spectators is amorphous. Thus these techniques can be powerful instruments of domination because there is no dialogue and the image provides a strong force of persuasion. Moreover, these techniques take the spectator into a world where it is difficult to separate reality and true experience from the dream. This can disturb behaviour-patterns and create a certain disequilibrium in our societies.

Therefore it is necessary for our cinema to express African realities and a certain censorship should be exercised against foreign films. On the other hand, the cinema should be helped by the State but in a spirit of great tolerance and freedom so that it does not become an instrument of propaganda.

For Africa, the cinema should be a support of national culture which will allow her to have a better place in the world. However, the film-maker must express his art in his mother-tongue in order to break the spectator's indifference, make him feel more "concerned" and thus awaken his critical spirit.

The content of our films should express our culture, traditional values and personality and make a plea to the world for it to give validity to our vision of the Universe so that our fundamental unity be better understood by ourselves and foreigners.

Férid Boughedir

The principle merit of the cinema is its educative aspect—the effect it has on the public. Only two types of cinema exist: that which "awakens" the spectator by taking him along the path of progress, and that which "lulls him to sleep" by making him sluggish and escape from reality by merely entertaining him.

The present-day African cinema should be "useful"—i.e. make people aware of realities, which seems more important than the notions of "art" and "entertainment."

The colonial cinema—especially that of the USA—has contributed myths, false values, misrepresentation of reality, a concept of world violence and frenzied individualism. Its films depict society and the masses as mediocre, stupid and hostile to the "heroic individual." The White is always superior and the African or the Indian must endeavour to be like him. It should also be noted that what differentiates the white hero from the African hero is that the latter is deeply attached to his community.

Our cinema should be one of "decolonization" and restore the value of African culture. It must be culturally independent to be economically and politically independent and must also be wary of the "folklore approach."

It should give maximum priority to screening national films on TV. As a first step, every film produced by an African in a national language should be considered to be "good" whatever its real value. Next, preference should be given to films which "awaken" the people, make them "live" on the screen and give them a better knowledge of our many regions. This will rouse an awareness of black civilization through a cultural cinema. As a last step, we could demand an "art" cinema but it would be necessary to make this "useful" instead of "luxurious."

As far as language is concerned, the first necessity is for it to be "readable." It must be simply expressed to be understood by all spectators and classical dramatic methods must be used to sustain interest.

As for economics, we must, above all, avoid the concept of "profitability at all costs" which benefits the commercial and "lulling" cinema. We must organize a new system of distribution particular to Africa by taking mobile-cinemas to the countryside to give more or less free shows and also use TV to give films a wider coverage. This popular "awakening" cinema can only function with State aid.

Point 4: The Problem of an Original Cinematographic Language

Bassori Timite

The language of the cinema is the expression of a country's proper cultural ambience and a reproduction of life.

In present-day Africa, there is an ideological partitioning between States which is harmful to the building of a common culture and to cultural exchanges between the former.

Our cinema needs to be industrially organized and can only blossom when political, administrative, financial and intellectual conditions permit it to produce films regularly and when films can freely circulate between countries. Only a continental market will guarantee its development at the production, distribution and exploitation levels. The co-ordination of these three sectors could be assisted by an institution created at the level of a group of States. Its task would also include refusing a plethora of foreign productions and encouraging local ones.

In these conditions, the cinema could look for an original language to explain our cultural unity through pre-colonial history and population movements. Through traditions and customs, beliefs and cultures we could seize African thought before these societies disappear.

Starting from our experiences of the past, we can evaluate what we are and what we can be and create an original aesthetic adapted to our modern world.

The value of the cinema can only be asserted in a general cultural ambience—i.e. with the spreading of literature, music, painting and all the arts that impregnate our daily lives.

The cinema must help to make us aware of our realities and personality and to better understand our problems. Therefore it will not be a cinema of leisure but one of reflection in complete freedom.

In conclusion, we may say that our cinema has made a good start and will be able to avoid Western models, criteria and myths.

Jean-Claude Rahaga

It would be very tempting for the African film-maker to keep the public in the land of dreams and fantasies provided by the Western commercial cinema but our cinema wishes to be an instrument in rousing an awareness of civilization, a public spectacle that is not reserved for intellectuals. The difficulty resides in confronting a public accustomed to the American cinema and in harmonizing the economic side of the profession with the artistic side.

The African film-maker finds himself in a state of revolt against Europe which makes him produce films of social and racial conflict and those dealing with problems of inadaptibility. Thus our cinema is social, critical and combative and not always an instrument for affirming Negro-African values but for demanding greater social justice.

We must stop this obsessive reference to the Western world and deal with the African man in his true world of values: his attachment to the surreal, his faculty of emotion, his ties with the ancestors and family codes. These four points form the main bases of African culture.

The fear of exoticism and folklore has turned the African intellectual towards a more modern cinema which is not well understood by the public at large. No popular cinematographic education exists although the African is traditionally sensitive to what can be seen and heard. The creation of "cinemathèques" would be a step in the right direction to create this form of education.

It is to be feared that full-length films will run counter to a truly African expression which should firstly be satisfied by documentaries and educative films, especially those dealing with Negro art.

In conclusion, not being divorced from the people presumes great modesty and lack of personal ambition on the part of the film-maker as well as an attempt to approach his work in a "marvellous" way such as practised by our griots and storytellers.

Brief Report on the Seminar*

Papers presented to the Seminar dealt essentially with the following fundamental points:

(a) how the cinema can inspire a feeling of historical and cultural solidarity among black communities and rouse in them an awareness of civilization;

(b) how the cinema can portray African human, social and cultural realities and fortify the reflexions of our peoples on their proper destiny;

(c) the problem of an original cinematographic language.

The various papers which dealt with these points offered food for deep discussion among the different participants coming from ten African countries and cultural organizations like the F.E.P.A.C.I. (the Pan-African Federation of Film-Makers).

On the first point, it was agreed that the African cinema has at its disposal the necessary materials to forge and maintain the feeling of historical solidarity among black peoples. *"What fantastic epics lie ready for the cinema! The history of our medieval empires abounds with exploits and terrible reversals suitable for constituting full-length films."*

*Sentences in italics are translations of various comments made by participants.

Thus, script-writers and film-makers directly inspired by the sources of our history, oral tradition and legends, can make the cinema play a big part in revealing and amplifying this feeling.

The problem lies in organizing this inspiration and making it flow in an authentically African language rid of dominating Western structures.

As a mass medium, the cinema has a wonderful power of addressing itself to the masses across the barriers of language, education and customs. In such a perspective, it powerfully contributes to bringing closer and unifying the common awareness of our identity, alas! torn apart by colonization. *"Of course, it is necessary for this service to be rendered by men of value to avoid the vulgarities which we see today in the European cinema. For Africa, cinema artists must be educators."*

This does not mean obliging the African film-maker to nail himself to history, at the risk of making his inspiration sterile. His object should simply be to take possession of Africa and restore her to his brothers.

As explorers of the interior, film-makers should take stock of all the riches of Africa and all aspects of her civilization, be they geographic, sociological, ethnological, musical, etc. This task has already begun but we need to follow up the effort and accentuate current trends.

In fact, the unifying action of the cinema in the service of culture has gone beyond the boundaries of Africa, as witness the movements of black revival in the United States with their interest in our clothing, travelling to our continent and the inclusion of African Studies in American university programmes. All this desire to go back to the sources, to "Mother Africa," has increased the curiosity of Black Americans for Africa and her film-makers from both sides of the Atlantic express an awareness of belonging to the same community, despite material distances and the disparities of living conditions. The same exchange is taking place between Blacks of Central and South America and ourselves.

The above points gave rise to lively debate. Among others, we quote the following exchanges between various speakers concerning topics discussed on the first day.

M. Damiba (Upper Volta)

Why are there not enough historical films? Their production must be difficult and costly . . .

The Abbé Bilgho (Upper Volta)

It is a fact that writers existed before film-makers but no attention has been paid to their necessary collaboration.

How can this relationship be envisaged? Don't you think that one has the tendency to evoke legends more than the peaceful relations among African Peoples? (. . .)

I have been touched by examples concerning the diaspora. To my knowledge, it is difficult for African religions not to be subjected to certain Moslem and Christian influences.

Mme. Annette M'Baye (Senegal)

De Medeiros said that every work produced is a step forward. Yes, but there is the problem of commitment. What does he think of this?

R. de Medeiros

Firstly, concerning the collaboration between film-makers and writers, there is need for real contact, there must be an association of writers. On the whole, personal relations are very effective. A good writer can be a good script-writer but there is a regrettable absence of the latter.

Secondly, the vision of history of which you are talking is very modern.

The history of epics and legends is more easily transmissible, but the history of the people, as witnessed by the people, is difficult to be transmitted by images. An example is *Emitai* in which only the action is seen.

As for the question of commitment, there is no standard-type African film-maker as wishing to standardize would mean aiming at depersonalizing the cinema (. . .)

One has to conquer this complex of pseudo-political complacency. The camera has no machine-gun. Commitment is first of all a problem of political awareness.

F. Boughedir (Delegate of the F.E.P.A.C.I.)

I wouldn't like to engage myself in the debate on commitment. There are two forms of function to be distinguished: the cinema as a witness and the social cinema (where the aim is to make the individual react).

The committed cinema can exist with the other. Everything depends on the arrangement of the creator (. . .)

The speech has the merit of widening the scope of historical solidarity (solidarity of destiny, etc.) and I thank the speaker for foreseeing the solidarity of the Third World, but how can the African film-maker react to his government in the case of opposition to his ideas?

D. T. Niane (Delegate of the S.A.C.)

On the problem of commitment, one has to be able to locate two phases: the first is waging war on colonialism; the second begins with independence. Commitment in this phase means taking up one's responsibilities against cultural oppression. This is the difficult task.

How can the film-maker react against an official truth that one would like to make a canon? Our cinema has not the literary substructure that one finds in the French cinema, for example (. . .)

Oral literature has been insufficiently exploited. As for our history, we discovered it in books. Film-makers must collaborate closely with historians and novelists.

M. Some (Director of Sonavoci, Upper Volta)

Are we going to guarantee every film provided it is African? The African cinema must make Africa known to the outside world.

The second day of the Seminar was dominated by the question of the use of African languages as the medium of expression.

Just as it is urgent to promote these languages in written literature, it is equally necessary for the cinema to do something to solve this problem.

First of all, for reasons of communication, the cinema will have to speak the language of its public. This is a necessity which faces a lot of difficulties—the multiplicity of African languages, the problems of translation, films to be produced in many versions, etc.

The personality of our cinema will only be established when it uses the mode of expression which reflects most accurately the cultural genius of our peoples.

The use of foreign languages helps to reinforce cultural alienation and to widen the gap between film-makers and their immediate and natural publics who are denied all means of receiving the full message of the cinema.

Encouraging moves are being made: for example, *Le Mandat* has been filmed in Wolof and *Emitaï* in Diola.

The debates brought out the following points.

Mme. Ki-Zerbo (Upper Volta)

African films in African languages have the advantage of demystifying the French language. Seeing a character and listening to him speak Wolof (or any other African language) is a way of enhancing our languages (. . .).

D. T. Niane

The problem of the multiplicity of languages will be solved from the moment when a film is well produced and harnesses African feeling. Our films can be educative instead of artificial. The psychological factor of demystifying French can be extremely important.

Certain languages have very wide geographical extension and this is an asset. As for real cinematographic language, it is a question of making it respect the forms specific to Africa. The three speakers insisted on this point and stressed the relation between the content of the cinematographic message and the language in which it is expressed. African peoples want to enjoy not only a cinema that speaks their language but also that which adopts a language that can reflect their culture, the way they feel and can explain the world to them.

This is why the Working Group made the following resolution:

> It can be said that a cinematographic language is specific when the sum of its elements belong to a nation, a culture and a given civilization. One can therefore see what an African cinematographic language could be: one which uses African languages and music and is associated with our pictorial, architectural and dramatic traditions.

There lies the road to research for the African cinema because it cannot discover its true identity unless it forges its proper and specific language.

It is therefore necessary for our cinema not to lend itself to the easy temptations offered by imitation, imported Western languages, formalism, etc. These blemishes which characterize much of the profession today have produced bad results known to everybody—the depersonalization of the African and his tendency to run away from the fact that he is a colonized man. In this respect, a cinematographic language is not passive: it has an influence on the message and even determines it. It is therefore urgent for the African cinema to search for a style, a cinematographic expression—in brief, a language properly adapted to Black Civilization.

Workshop Resolutions

First Workshop

Resolution 1: On African History

Since the history of Africa is badly known by Africans, African film-makers will have a major role to play in promoting the knowledge of this history by producing historical films.

They will have to recall and illustrate the great empires of the past, evoke legends and mythical stories through which the profound life of the people is expressed. All these concern ancient history.

As for modern and contemporary history, African film-makers will have to retrace the different episodes of colonization and analyse them in order to learn the lessons that explain what Africa is today. They will also have to fight against the after-effects and relics of systematically organized cultural subjugation.

In all cases, the cinema will have to strive to highlight the unity of Africa, while going beyond the dismantlement imposed by colonization: it will have to portray the cultural unity of Black Civilization, a unity which is evident.

Such an assignment necessitates thorough historical research so that film-makers can avoid clichés or errors despite their good intentions. That is why there must be close collaboration between film-makers, writers, traditionalists (*griots,* poets, storytellers, etc.), historians and researchers.

Resolution 2: On Civilization and African Culture

The Seminar began with a premise unanimously accepted as true: the existence of civilization and a culture that are African, perceived as the sum of values, usages and customs of black peoples which establishes the relations between man and his environment. It will be the job of another Seminar to go deeper into this very general definition and, eventually, discuss it.

On this basis, the Seminar has emphasized the necessity for the African cinema to show all aspects of this civilization and culture, as well as its problems, contradictions and the dynamics of its modern evolution, so that they can serve Africans as a basis of reflexion on their own world.

The cinema will therefore have to

a) react against the evils of colonization and cultural alienation carried and spread for decades by the foreign cinema;

b) take stock of and enhance all domains of our culture and civilization in order to inculcate a deeper knowledge of Africa upon Africans themselves. In this respect, there is no question of painting an imaginative or idyllic Africa but, rather, the realities of the continent as they are, so as to make the spectator reflect on them;

c) produce, in addition to fiction-films, documentaries with a view to describing all aspects of the life of Africans, their techniques and traditional arts.

Here again, close collaboration between film-makers and other specialists is indispensable.

Various organizations and the research services of different States can play a determinant role in acquiring documents and magazines on their research activities and discoveries.

Resolution 3: The Responsibility of Film-makers

From the proceedings of this Seminar, it is clear that the African film-maker has an important mission—that of educating the masses through the power of this tool which is the cinema.

To carry out this heavy task of restoring Africa to the Africans, the film-maker has to

a) educate and cultivate himself in the realities of Africa and her people; by this act of humility, he will overcome the problem of cultural insufficiency and ignorance noticed in a good number of present-day film-makers who have been generally trained in European or American schools;

b) spare no effort in producing documentaries that are easily accessible to the masses for whom they are made; to this end he must be fully involved in the realities which he is presenting.

c) He is therefore advised to be in touch with research specialists, organizations and institutes of all kinds to complete his culture and elaborate his projects.

Resolution 4: The Responsibility of Governments

African governments are urged to recognize the importance of the cinema as a means of education and formation, beyond its short-sighted use as a simple means of information and propaganda.

a) While promulgating charters guaranteeing the rights and obligations of artists, African States are advised to understand the usefulness to individual States and to Africa of according freedom of thought and creation to film-makers, other artists and intellectuals.

b) From the material point of view, the cinema should be considered as a primary investment since it is invaluably efficacious as a factor of economic, social and cultural development. Governments are therefore advised, each according to its means, to establish rapidly small production units by training competent personnel and assuring adequate technical material, with the understanding that the least effort speedily made will be developed progressively so as to give the cinematographic industry all the necessary conditions for growth.

c) These national efforts will be grouped into inter-African units whose greater capacity will be a determining element of effectiveness.

d) On the other hand, the various States will give strength and brilliance to the African cinema by creating an inter-African Organization to carry out documentation, research, etc.

e) In view of the situation inherited from colonialism, African States will have to try to find solutions which favour the birth of an independent cinema at all stages—production, distribution and exploitation.

f) There is a need for an African Cinema School in order to train film-makers deeply rooted in African culture and civilization. African States have a duty to promote rapidly such a school at a Pan-African level or even at regional levels.

The State has a fundamental role to play in the development of the cinema in Africa.

Special Mention no 1: The Financial Problem

Since this problem is of crucial importance, a working group has envisaged certain solutions to be proposed to different authorities.

a) Various forms of State aid in addition to financial help from departments and communities.

b) A sinking-fund to be created from cinema returns.

c) Help from individuals, businessmen, etc.

d) The intervention of Universities, institutes and organizations, as well as the creation of particular development programmes.

e) The idea of offering scholarships and bursaries in the service of cinema development has been raised and merits attentive study.

In any event, it has to be noted that neglect of this problem runs the risk of bringing to naught the efforts made in other sectors.

Special Mention no 2: Francophones and Anglophones

Many delegates deplore the absence of English-speaking participants at this Seminar which is, in a way, preparatory to the Festival of Lagos.

They want to remind Africans that, in order to restore the unity of our civilization and culture, it is important to overcome the barriers created by the colonial masters, especially in the domain of vehicular languages.

Special Mention no 3: Exchanges

The Seminar recommends the development of exchanges in the domain of arts throughout the different countries of Africa.

It invites governments, inter-African and other organizations to favour these exchanges by organizing co-productions, festivals, congresses, study tours, scholarships, etc.

The Seminar lays emphasis on the fact that the free movement of artists throughout African countries is an important motivating factor of unity which African States can powerfully help to achieve.

Second Workshop

African Languages and the African Cinema

The Seminar studies the relationships between African languages, culture and the cinema, and makes the following resolutions.

a) The African cinema cannot afford to be simply a cinema of entertainment; it has to become a means of educating the people, especially if the cinematographic message is expressed in African languages. In fact, this will favour, from all points of view, a more active participation of the population in cinematographic creation.

b) The Seminar is, moreover, conscious of the fact that every language moulds a particular mental structure and determines individual behaviour patterns.

The Seminar is therefore convinced that one cannot liberate cultural exiles politically and economically, and that the use of African languages is an effective weapon against foreign cultural and economic domination. This can be done by demystifying Western languages, often considered as the only valid means of expression, and by enhancing African languages which are the soul of the African personality that they model and whose richness only they can express in the final analysis.

c) The Seminar considers that, far from favouring tribalism or exacerbating it, the use of African languages can highlight the fundamental cultural unity of the Black World.

In view of these facts, the Seminar

i) strongly encourages African film-makers to continue to use African languages with the aid of all the technical resources at their disposal;

ii) invites African States to favour the diffusion of African films in cinemas, on television, by the use of mobile cinemas, etc., so that a vaster public can be involved.

To this end, States would need to establish the necessary infrastructure, not only by installing the equipment necessary for diffusion (projection-rooms, televised circuits, mobile cinemas), but also those necessary for production (studios, etc.).

In this regard, the Seminar recalls the need to apply rapidly the decisions to create the necessary circuits of distribution and production at a continental or regional level.

d) The seminar suggests that, where possible, the best foreign films be commented in African languages.

e) It once more draws the attention of African States to the necessity of intensifying the teaching of African languages; it equally seizes the opportunity to congratulate the efforts of both African States and international organizations like the OUA and UNESCO in this domain.

f) Finally, the Seminar wishes African governments to encourage linguists and teachers in their efforts to make African languages vehicles of scientific, technical and technological knowledge.

Third Workshop

1) What is the Cinema?

The cinema is an instrument of communication and education through image and sound; it is therefore a language. This language is particularly adapted to Africa because it can cut across linguistic barriers through its power of image and can adapt itself essentially to our oral civilizations which are based on spectacle and speech.

2) What is a Cinematographic Language?

One should not confuse cinematographic language with cinematographic technique. Since the invention of the cinema, technique has been practically the same in all countries of the world; the language in itself is different according to whether the film is Japanese, Brazilian, Russian or American. That is to say that the language is a component of many elements which can differ according to countries. These elements are of three kinds: visual, sonant and rhythmic. In the visual element can be classed the actor's countenance, his clothing, the background, movement, gesticulations and the dramatic act: in brief, the signs and symbols of a civilization. In the sonant element can be classed essentially the language used by the actors, sounds, the surroundings and, finally, music. As for the rhythm, it is the manner of arranging all these elements according to cadence, respiration and a dramatic curve characteristic of the sensibility of each people. One will not fail to remark that the cinema embodies all these elements to the point that it has been said that the cinema is the synthesis of all acts. One can therefore say that a cinematographic language is specific when the sum of its elements belongs to a nation, a culture, a given civilization. From this it can be seen what an African cinematographic language could be—that which uses African languages and music and which should be joined to African pictorial, architectural, gesticular and dramatic traditions.

3) The Function of Cinematographic Language

Like the spoken word, cinematographic language can convey true or false information. It can therefore be used either to educate and make the spectator aware of realities or, on the contrary, mask these realities and lull him to sleep. In Africa, it has been used with two principal aims in mind: to depersonalize the African by imposing a language, a culture and a vision of the world that are foreign to him; to make him close his eyes to the reality of his being a colonized man, and thereby perpetuating economic domination. Therefore the role of an African cinematographic language in the post-colonial and socio-historical context which we are now in should be to

a) restore to the African his lost identity by bringing him closer to his civilization so as to guarantee his cultural independence;

b) give him the means of analysing the contradictions of present realities so as to allow him to fight for his economic independence.

4) How to Use This Language in Africa and Why?

The word *language* implies a speaker and a listener; in this precise case, it means the film-maker and the public. In Africa, the former have been formed and the latter deformed in the manner of the West—that is to say, conditioned by a certain type of cinema. The African film-maker should abandon the Western notion of a learned artist cut off from the popular masses and reality; let him accept the need to re-educate himself by contact with historians and African men of culture and involve his art in the daily life of the people; let him take his people into consideration by showing them their reflection without shocking them by too difficult intricacies of form; when he is working on fiction, let him try to win interest by way of presenting events related to African stories. In all cases, he should try to readjust his language with his public in mind.

5) Diffusion

In all the countries of the world, it has been noticed that the traditional commercial structure of the cinema, based on money, has always favoured the triumph of escapist and violent films. The State has a big role to play in diffusion. Non-commercial structures should henceforth be established (mobile cinemas and television networks for collectivities) to bring the cinema to Africans of all social classes at the lowest cost. As for the commercial cinema, it is advisable for admittance to be cheap enough to allow as many people as possible to see films. It is equally advisable that projections be preceded or followed by explanations in local dialects or languages. It is also desirable that part of the receipts realised be used to mount and finance a general programme of education through the cinema.

6) Conclusion

We congratulate the different initiatives taken by African governments to promote their cinema. However, it has been noticed that there is a lack of co-ordination between States on the decisions taken in this domain. Consequently, we are inviting them to organise, at the level of the OUA, a special conference on "The cinema" so as to agree on the policy of distributing African films on our continent and in the rest of the world on the one hand and, on the other hand, that of our consumption of foreign films, while respecting the sovereignty of each State.

Closing Speech by the Minister of Information

Honourable Minister,
Your Excellencies,
Madam Chairman,
Ladies and Gentlemen,

For four days you have worked steadily at your task and the points on the Agenda have been carefully examined.

Faithful to the promise of realism that you made at the beginning of this Seminar, you have made a spirit of just compromise prevail over the temptation to indulge in dogmatic and sterile debate.

Your Seminar has been sincere to itself, frank with respect to others and realistic in the face of events. If your discussions have sometimes been very passionate (as I have already said), it is doubtlessly because each one of you has acted as the defender of a just cause and the higher interest of our heritage and civilization. Each one of you has shown his talents during the various discussions which have been full of lessons and this is proved by the quantity and quality of your motions and recommendations.

Among the themes developed I will retain the first, second and fourth which seem to me to resume the whole of our current preoccupations in the search for an African cinema. If your meeting has stressed the urgent need to create a cinema adapted to our realities, you have not indulged in futile protest. What pleases me to mention is the fact that you have insisted that the cinema is a powerful means of education and persuasion and on the need to train film-makers who are aware of their responsibilities.

The latter must be impregnated with a sense of our realities and have a feeling of humility and the knack of adapting themselves. In order to be operative, their task must have the full understanding of our peoples and, especially, of a certain elite. In a world that is seeking its true self and constructing itself, the cinema cannot merely be one of entertainment and its first task is to rouse in us an awareness of belonging to this same world and to a common civilization, despite an apparent diversity.

It seems that I have dwelt too much on the role of the cinema and too little on that of governments. In the final analysis it is they who will have the last word. Not only must they define a policy of the cinema but also supply the necessary financial and material means to bring it about.

It would have been a pity if only kindness and civility had presided over your work because this would mean that four days of thoughtful discussion will never come to maturity.

For my part, I am convinced that your Seminar has resulted in a true awareness of our situation in this domain and that you all feel concerned by all that has been said and done here. This is why I hope that you will all work together to ensure that the motions and recommendations adopted during this period will not add to the pile of pious wishes . . .

In conclusion, I would like to say again how pleased we have been to accommodate you. Our hospitality, whose insufficiencies we are aware of and which have happily met with your indulgence, is like any other human endeavour.

As pioneers of a new conception of the African cinema, I do not doubt that you will always be the fierce but enlightened defenders of your work. I wish you a safe return to your respective countries and declare closed the Seminar on "The Role of the African Film-Maker in Rousing an Awareness of Black Civilization."

LONG LIVE THE AFRICAN CINEMA!

Niamey Manifesto of African Film Makers: First International Conference on Cinema Production in Africa, Niamey, 1–4 March, 1982

The first international conference on cinema was held in Niamey from March 1st to 4th. The participants were film makers, critics, officials from several African countries and international cinema experts.

The participants recognised the under-development of cinema, including regular film productions in the majority of African countries.

Convinced that African cinema must have a commitment to assert the cultural identity of African peoples; be a means for international understanding; an effective means of education and entertainment; an incentive for development, contributing to national and regional economic policies.

The Conference started by making a serious evaluation of African and international policies on cinema.

The participants then studied proposals for the development of African Cinema, production and the financing of productions and the possibilities of legislation that would promote pan-African strategies for the development of African cinema industry. They examined ways of implementing the proposals.

The conference finally adopted the following resolutions and recommendations:

General Principles

The participants considered and set up the following principles:

—The viability of cinema production is closely tied to the complementary viability of the other four main sectors of cinema, namely the exploitation of cinema theatres, importation of films, distribution of films, technical infrastructure and training.

—There cannot be any viable cinema without the involvement of African States for the organisation, the support, the stabilisation of cinema and the encouragement and protection of private public investment in cinema.

—It is not possible to have a viable cinema industry on a national level in Africa. The development of national cinema should take into consideration regional and pan-African co-operation by integrating cinema to political and economic ties that already exist between states.

—At the present stage of development of audio-visual facilities in the world and particularly in Africa, television should be complementary to cinema.

—It is possible to finance African film productions from the present revenue from the millions who patronise cinemas in Africa. What is required is a strategy that will ensure that part of this revenue legitimately returns to the production of films. Production should not rely solely on patronage.

Recommendations

Cinema Market (Exploitation and projection):

Every State should organise, support safeguard and develop its movie theatre market and encourage and collaborate with neighbouring states to form a regional common market for the importation and exploitation of films.

—*Measures to be taken:*

a. The setting up of national ticket agencies to monitor receipts of cinemas for the benefit of the exchequer, the cinema owners and film producers.

b. The provision of cinemas and other appropriate film projection venues and facilities.

c. To make available funds from cinema taxes to encourage exhibitors to expand their cinema circuits, thus enlarging the market.

d. States to exempt taxation on equipment imported for film projections.

e. States to encourage investment to build cinemas by creating incentives for would be investors.

Importation and Distribution of Films

We have to control and organise the importation and distribution of foreign films to ensure the projection of African films on national, regional and continental levels. We have to limit the dependence on foreign suppliers, and ensure cultural diversification of foreign films thus preventing the domination of films from particular areas. All this must be done with the aim of reconquering and enlarging our cultural and economic space.

—*Measures to be taken:*

a. The setting up of national distribution corporations in countries where they don't already exist, be they state run or in the private sector.

b. The setting up of regional film importation companies that would function as cooperatives, e.g. C.I.D.C. Where possible representative film purchasing companies based in foreign countries should have African status so that taxes related to their activities be paid in Africa. These companies should promote African films abroad and their diffusion.

c. To strengthen existing importing companies like C.I.D.C. by the participation of other States.

d. Enact laws on distribution to favour African films nationally, regionally and continentally. This can be achieved by decreasing the share of revenue to distributors when dealing with African films. This would contribute to the financing of future productions.

Production

Cinema productions, whether national, regional or inter-African, should be financed, not necessarily by state funds, but mainly by revenue from distribution and from various forms of cinema taxation including taxes on earnings by foreign films. Thus cinema will finance cinema.

—*Measures to be taken to finance Productions:*

a. The creation of Film Finance Corporations funded by revenue from cinema.

b. The creation of support funds to be administered by the corporations. The Support Funds help the production of film on the approval of scenarios.

c. The exemption from taxes of imported products and equipment required for the production of films. This would reduce the production costs.

d. Increase of African Producers' shares of box-office receipts.

e. Advance payments to producers by distributors.

f. Governments to legislate that television participate in financing of film production in various ways.

g. To create by legislation incentives for capital investments in film productions. This can be accomplished by offering tax exemptions.

h. To make bank loans at low interest available to Producers by national banks. These loans to be guaranteed by support funds.

i. To have inter-governmental agreements, bilaterally, regionally and continentally, for the free circulation of technicians, equipment and other production facilities, and to reciprocal support funds and to infrastructure.

j. To reinforce and encourage the activities of existing production organisations such as CIPROFILMS (International Centre for Film Production), through participation by States, by paying subscriptions and by contributions from revenues acquired through cinema taxes.

k. To support the production of Short Feature Films through finance from Support Funds. These will give added experience to film makers and be an additional source of labour for technicians. Cinemas should also be compelled to screen these films.

l. Another source of finance for productions can be obtained from theatrical and non-theatrical rights from distributors and television.

Technical Infrastructure

—*Measures to be taken:*

a. The last twenty years' experience having proved that cinema technical infrastructures was impossible to be maintained and made profitable on a national level because of the high costs of maintenance and management. The conference recommends that the future establishments of these structures should be on regional levels after joint studies and agreements between parties involved.

b. To create archives and film libraries on regional and continental levels.

Training

It is preferable that the training of technicians and other disciplines related to cinema be in centers established in regions and within the framework of any cinema activities in Africa. Wherever foreign technicians are employed it should be obligatory that African technicians are attached.

African film makers and technicians working abroad should be encouraged to return to the continent to contribute to the development of African cinema.

—*Measures to be taken:*

a. Vocational training centres should be established to ensure the training of film and television technicians and their absorption in both media.

b. Ensure the training of managerial staff and other non-technical personnel e.g. lawyers, producers, production managers etc.

c. To facilitate efficient distribution, the training in programming, promotion and public relations.

d. Ensure the training of projectionists, cinema managers and other activities related to exhibition of films.

e. The development of film critics through continuous dialogue between film makers and critics.

Legislation

Cinema legislation of any State should take into consideration the joint development of its Cinema industry with that of its neighbouring States and also of the Region.

National Film Corporations

National Film Corporations should be established in every country. The corporations should be autonomous in decision making yet being under a ministry. The role of these corporations should be to centralise all activities and matters relating to cinema in the country. There can be a management committee representing the government and the corporation.

A complementary authority should be established on a regional level to ensure coordination of cinema policies of regions.

Final Recommendation

Any decision made executively or regarding legislation on cinema, nationally or regionally, should be considered by a committee representing the State, film makers, cinema professionals, and investors and cinema owners, to avoid individual or bureaucratic decisions abitrarily taken against the interests of African Cinema. On the other hand film makers should maintain a sense of responsibility and morality in dealing with their governments and others they have dealings with.

Sections of this document were translated by Louise Jefferson.

FeCAViP Manifesto (1990)

We, producers, filmmakers, screenwriters, technicians and actors of the second Images Caraïbes Festival 1990, being aware of the need to further develop the space within the Caribbean, for professional workers in film and video, reflecting our special needs, and after having made a deep analysis of our reality, acknowledging the importance of film, TV, and video, decided to give ourselves the means in order to obtain the conditions necessary for the realization of the expression of the professionals working in film and video.

So together, we have to:

1. CREATE, PRODUCE, DISTRIBUTE, AND BROADCAST THE WORKS OF OUR YOUNG CARIBBEAN ARTISTS
2. CONTRIBUTE TO THE TRAINING OF OUR YOUNG ARTISTS AND TECHNICIANS
3. COLLECT, RECORD, ARCHIVE, AND PRESERVE OUR CULTURAL HERITAGE
4. OVERCOME THE EXISTING LINGUISTIC, LEGAL, TECHNICAL, AND COMMERCIAL BARRIERS
5. PROMOTE CARIBBEAN CINEMA, VIDEO, AND TV PRODUCTIONS
6. DEVELOP THE EXCHANGE OF INFORMATION BETWEEN CARIBBEAN PROFESSIONALS
7. ESTABLISH RELATIONSHIPS BETWEEN ALL THE ASSOCIATIONS AND AUDIOVISUAL EVENTS OF THE CARRIBEAN AND ITS DIASPORA

Distributed by APDCC/Images Caraïbes, Fort-de-France, 1990

8. CREATE NEW CONTACTS WITH COUNTRIES FACING SIMILAR PROBLEMS (in Africa, South America, for example)

In order to achieve our goals, it has been decided a Federation be created which name will be Federation of Caribbean Audiovisual Professionals (FeCAViP). Fort-de-France, June 8, 1990.

Select Bibliography

"African Dossier: Hondo, Gerima, Sembene," *Framework* no. 7/8 (1978): 26–28.

Amin, Samir. *Eurocentrism*. New York: Monthly Review Press, 1989.

———. *Unequal Development*. New York: Monthly Review Press, 1973; 1976.

Armes, Roy. *Third World Film Making and the West*. Berkeley: University of California Press, 1987.

Barclay, William, Krishna Kumar and Ruth P. Simms, ed. *Racial Conflict, Discrimination, & Power: Historical & Contemporary Studies*. New York: AMS Press, 1976.

Barnouw, Erik and S. Krishnaswamy. *Indian Film*. New York: Oxford University Press, 1963; 1980.

Berger, Mark. "The End of the 'Third World'? *Third World Quarterly* 15, no. 2 (1994):257–275.

"Black Film British Cinema." ICA Documents No. 7. London: ICA, 1988.

"Black Film Issue." Smith, Valery, Camille Billops, and Ada Grittin, ed. *Black American Literature Forum,* 25, no. 2 (1991).

Blaut, J. M. "Colonialism and the Rise of Capitalism." *Science & Society* 53, no. 3 (1990): 260–296.

Bowser, Pearl and Renee Tajima. *Journey Across Three Continents*. New York: Third World Newsreel, 1985.

Cabral, Amilcar. *Return to the Source*. New York: Monthly Review Press, 1973.

Cham, Mbye B. *Ex-Iles: Essays on Caribbean Cinema*. Trenton, N.J.: Africa World Press, 1992.

Cham, Mbye B. and Claire Andrade-Watkins, ed. *Blackframes: Critical Perspectives on Black Independent Cinema*. Mass.: MIT Press, 1988.

Cox, Oliver C. *Cast, Class, & Race: A Study in Social Dynamics*. New York: Doubleday, 1948.

———. *Capitalism as a System*. New York: Monthly Review Press, 1964.

Cripps, Thomas. *Black Film as Genre*. Bloomington: Indiana University Press, 1979.

Crusz, Robert. "Black Cinemas, Film Theory and Dependent Knowledge." *Screen* 26 (1985): 152–156.

Daniels, Therese and Jane Gerson, ed. *The Colour Black: Black Images in British Television.* London: BFI, 1989.

Debrunner, Hans Werner. *Presence and Prestige: Africans in Europe.* Basel: Basler Afrika Bibliographien, 1979.

Diawara, Manthia. *African Cinema.* Bloomington: Indiana University Press, 1992.

———, ed. *Black American Cinema: Aesthetics and Spectatorship.* New York: Routledge, 1993.

Downing, John D. H., ed. *Film and Politics in the Third World.* New York: Autonomedia, 1987.

Drake, St. Clair. *Black Folk Here and There,* 2 vols. Los Angeles: UCLA Center for Afro-American Studies, 1987; 1990.

DuBois, W. E. B. "The Conservation of the Races," *Negro Social and Political Thought 1850–1920,* ed. Howard Brotz. New York: Basic, 1966.

Fanon, Frantz. *The Wretched of the Earth.* New York: Grove Press, 1964.

Fusco, Coco. *Young British and Black: The Work of Sankofa and Black Audio Film Collective.* Buffalo, N.Y.: Hallwalls, 1988.

Gabriel, Teshome. *Third Cinema in the Third World: The Aesthetics of Liberation.* Ann Arbor, Mich.: UMI Research Press, 1982.

Gauhar, Altaf. *Third World Affairs 1985.* London: Third World Foundation, 1985.

———. *Third World Affairs 1988.* London: Third World Foundation, 1988.

Gilroy, Paul. *The Black Atlantic: Modernity and Double Consciousness.* Cambridge, Mass.: Harvard University Press, 1993.

Gramsci, Antonio. *Prison Notebooks.* Translated by Quintin Hoare and Geoffrey Nowell Smith. New York: International Publishers, 1983.

Guerrero, Ed. *Framing Blackness.* Philadelphia: Temple University Press, 1993.

Hall, Stuart. "Cultural Identity and Cinematic Representation." *Framework* 36 (1989): 68–81.

———. "What is This 'Black' in Black Popular Culture?" *Social Justice* 20, nos. 1–2 (1990/91): 104–114.

Harris, Joseph E. ed. *Global Dimensions of the African Diaspora.* Washington, D.C.: Howard University Press, 1982.

Jameson, Fredric. *The Geopolitical Aesthetic: Cinema and Space in the World System.* Bloomington: Indiana University Press, 1992.

Johnson, Randal and Robert Stam, ed. *Brazilian Cinema.* East Brunswick, N.J.: Associated University Presses, 1982.

Julien, Isaac and Colin MacCabe. *Diary of a Young Soul Rebel.* London: BFI, 1991.

Landau, Jacob M. *Studies in the Arab Theater and Cinema.* Philadelphia: University of Pennsylvania Press, 1958.

Malkmus, Lizbeth and Roy Armes. *Arab and African Filmmaking.* London: Zed, 1991.

Marks, Laura. "A Deleuzian Politics of Hybrid Cinema." *Screen* 35, no. 3 (1994): 244–264.

Mercer, Kobena, ed. *Black Film/British Cinema.* London: Institute of Contemporary Studies, 1988.

Mercer, Kobena and Isaac Julien. "De Margin and De Centre." *Screen* 29, no. 4 (1988): 2–10.

Miles, Robert. *Capitalism and Unfree Labour: Anomaly or Necessity?* London: Tavistock, 1987.

Naficy, Hamid. *The Making of Exile Cultures: Iranian Television in Los Angeles.* Minneapolis: University of Minnesota Press, 1993.

———. "Phobic Spaces and Liminal Panics: Independent Transnational Film Genre." *East-West Film Journal* 8, no. 2 (1994): 1–30.

Pfaff, Françoise. *The Cinema of Ousmane Sembene, A Pioneer of African Film.* Westport, Conn.: Greenwood Press, 1984.

———. *Twenty-Five Black African Filmmakers.* Westport, Conn.: Greenwood Press, 1988.

Pick, Zuzana M. *The New Latin American Cinema: A Continental Project.* Austin: University of Texas Press, 1993.

Pieterse, Jan Nederveen. *White on Black: Images of Africa and Blacks in Western Popular Culture.* New Haven, Conn.: Yale University Press, 1992.

Pines, Jim and Paul Willemen, ed. *Questions of Third Cinema.* London: BFI, 1989.

Resolutions of the Third World Film-makers Meeting, Algiers, Dec. 5–14. New York: Cineaste, 1973.

Said, Edward W. *Culture and Imperialism.* New York: Knopf, 1993.

Salmane, Hala Simon Hartog and David Wilson, ed. *Algerian Cinema.* London: BFI, 1976.

Screen 29, no. 4 (1988). "The Last 'Special Issue' on Race?"

Shaw, Timothy. "The South in the 'New World (Dis) Order': Towards a Political Economy of Third World Foreign Policy in the 1990s." *Third World Quarterly* 15, no. 1 (1994): 17–30.

Shiri, Keith, ed. *Directory of African Film-Makers and Films.* Westport, Conn.: Greenwood Press, 1992.

Shohat, Ella. *Israeli Cinema: East/West and the Politics of Representation.* Austin: University of Texas Press, 1989.

Shohat, Ella and Robert Stam. *Unthinking Eurocentrism: Multiculturalism and the Media.* New York: Routledge, 1994.

Snowden, Frank M. *Before Color Prejudice, An Ancient View of Blacks.* Cambridge, Mass.: Harvard University Press, 1983.

Stam, Robert. "Samba, Condomblé, Quilombo: Black Performance and Brazilian Cinema." *Journal of Ethnic Studies* 13, no. 3 (1985): 55–84.

Stam, Robert and Louise Spence. "Colonialism, Racism, and Representation." *Screen* 24, no. 2 (1983): 2–20.

Stavrianos, L. S. *Global Rift: The Third World Comes of Age.* New York: William Morrow, 1981.

Steven, Peter, ed. *Jump Cut: Hollywood, Politics, and Counter Cinema.* New York: Praeger, 1985.

Thompson, Felix. "Metaphors of Space: Polarization, Dualism and Third World Cinema." *Screen* 34, no. 1 (1993): 38–53.

Tomaselli, Keyan. *The Cinema of Apartheid: Race and Class in South African Film.* New York: Smyrna/Lake View Press, 1988.

Tomaselli, Keyan, Alan Williams, Lynette Steenveld and Ruth Thomaselli. *Myth, Race and Power: South Africans Imaged on Film and TV.* Bellville, South Africa: Anthropos Publishers, 1986.

Wallerstein, Immanuel. *The Modern World-System.* New York: Academic Press, 1974.

Watson, Hilbourne A. "Surplus Labor, Unequal Exchange, and Merchant Capital: Rethinking Caribbean Migration Theory." *Studies of Development and Change in the Modern World,* ed. Michael T. Martin and Terry R. Kandal, 242–266. New York: Oxford University Press, 1989.

Wiley, David S., ed. *Africa on Film and Videotape 1960–1981.* East Lansing: Michigan State University African Studies Center, 1982.

Williams, Raymond. *Marxism and Literature.* New York: Oxford University Press, 1977.

———. *Problems in Materialism and Culture.* London: NLB, 1980.

Wolf-Phillips, Leslie. "Why 'Third World'?: Origin, Definition and Usage." *Third World Quarterly* 9, no. 4 (1987): 1311–1327.

Woll, Allen L. *The Latin Image in American Film.* Los Angeles: UCLA Latin American Center Publications, 1977.

Yearwood, Gladstone L., ed. *Black Cinema Aesthetics: Issues in Independent Black Filmmaking.* Athens: Ohio University Center for Afro-American Studies, 1982.

Contributors

Claire Andrade-Watkins is Associate Professor of mass communication at Emerson College and co-editor (with Mbye Cham) of *Blackframes: Critical Perspectives on Black Independent Cinema* (MIT, 1988).

Roy Armes is a Professor of Film at Middlesex University. He is the author of numerous books including *Third World Film Making and the West* (California, 1987) and *Arab & African Film Making* (Zed, 1991).

José Arroyo is a lecturer in the Joint School of Film and Literature at the University of Warwick.

Férid Boughedir is a film critic from Tunisia.

Mbye Cham is Associate Professor of African Studies at Howard University and co-editor of *Black Frames: Critical Perspectives on Black Independent Cinema* (MIT, 1988) and *Ex-Iles: Essays on Caribbean Cinema* (Africa World Press, 1992).

Thomas Cripps is University Distinguished Professor at Morgan State University and author of numerous scholarly publications. His most recent book is *Making Movies Black* (Oxford, 1993).

Julie Dash is an independent filmmaker and director of the award winning film, *Daughters of the Dust,* and author of *Daughters of the Dust: The Making of An African American Woman's Film* (New Press, 1992).

Zeinabu Irene Davis is an independent filmmaker and Assistant Professor of Film at Antioch College.

Manthia Diawara is Professor of Africana Studies at New York University. His recent publications include *African Cinema* (Indiana, 1992) and *Black American Cinema* (Routledge, 1993).

Coco Fusco is a journalist and writer whose essays are published in *Afterimage, Cineaste* and numerous other publications.

Teshome H. Gabriel is an Associate Professor in the Department of Film and Television, University of California, Los Angeles. He is the author of *Third Cinema in the Third World* (UMI, 1982) among other publications.

Abid Med Hondo is a Mauritanian filmmaker living in exile in Paris.

Nicholas Peter Humy is a visiting professor in the Department of English at Tulane University.

Adam Knee has taught film history and theory at New York University, Penn State University, and the School of Visual Arts. His writing on film has appeared in various journals and in the anthologies, *Screening the Male* (Routledge, 1993) and *Representing Jazz* (Duke, 1995).

Ana M. Lopez is an Associate Professor in the Department of Communication at Tulane University. She is the author of numerous articles on Latin American cinema and co-editor (with John King and Manuel Alvarado) of *Mediating Two Worlds: Cinematic Encounters in the Americas* (BFI, 1993).

Tommy L. Lott is a Professor in the Department of Philosophy at San Jose State University and the author of numerous publications.

Michael T. Martin is a Professor and Chair of the Department of Africana Studies at Wayne State University. He is co-editor of *Studies of Development and Change in the Modern World* (Oxford, 1989), and editor of *New Latin American Cinema*, volumes 1 and 2 (Wayne State University Press, forthcoming). He is also director and co-producer (with Robert Shepard) of the award winning feature documentary, *In the Absence of Peace* (1989).

Charles Musser is Assistant Professor of American Studies and Film Studies at Yale University, where he directs the Film Studies Program. His books include *The Emergence of Cinema: The American Screen to 1907,* which was awarded the Jay Leyda Prize in Cinema Studies among others. He is currently co-curating (with Pearl Bowser and Jane Gaines) *Oscar Micheaux and His Circle: African American Filmmaking and Race Cinema of the Silent Era,* a touring six-part program.

David Nicholson is a writer and the foundling editor of *Black Film Review.*

Françoise Pfaff is a film historian and Professor in the Department of Romance Languages at Howard University. Her books include *The Cinema of Ousmane Sembene, a Pioneer of African Film* (Greenwood, 1984) and *Twenty-Five Black African Filmmakers* (Greenwood, 1988).

Mark A. Reid is Assistant Professor of English at the University of Florida, Gainesville and author of *Redefining Black Film* (California, 1993) among other publications.

James A. Snead was, before his death, Associate Professor of English and Film Studies at the University of Pittsburgh.

Robert Stam is author of two books in Portuguese *(Espetaculo Interrompido* and *Bakhtin)*, and five books in English: *Brazilian Cinema* (with Randal Johnson); *Reflexivity in Film and Literature; Subversive Pleasures; New Vocabularies in Film Semiotics;* and (with Ella Shohat) *Unthinking Eurocentrism: Multiculturalism and the Media.* He is Professor of cinema studies at New York University.

Clyde Taylor is Professor of English at Tufts University and the author of numerous articles about Black cinema. His book, *Breaking the Aesthetic Contract* is forthcoming.

Keyan Tomaselli is a Professor and Director of the Centre for Cultural and Media Studies at the University of Natal, and founder and editor of *Critical Arts.* His numerous publications include *The Cinema of Apartheid* (Smyrna/Lake View, 1988).

Nwachukwu Frank Ukadike teaches in the Department of Communication and in the Center for Afroamerican and African Studies at the University of Michigan, Ann Arbor. He has published articles in *Jump Cut* and *Transition* and recently authored the book, *Black African Cinema* (California, 1994).

Keith Q. Warner is a Professor in the Department of Romance Languages at Howard University.

Index

Books in the Contemporary Film and Television Series

Cinema and History, by Marc Ferro, translated by Naomi Greene, 1988

Germany on Film: Theme and Content in the Cinema of the Federal Republic of Germany, by Hans Gunther Pflaum, translated by Richard C. Helt and Roland Richter, 1990

Canadian Dreams and American Control: The Political Economy of the Canadian Film Industry, by Manjunath Pendakur, 1990

Imitations of Life: A Reader on Film and Television Melodrama, edited by Marcia Landy, 1991

Bertolucci's 1990: A Narrative and Historical Analysis, by Robert Burgoyne, 1991

Hitchcock's Rereleased Films: From Rope to Vertigo, edited by Walter Raubicheck and Walter Srebnick, 1991

Star Texts: Image and Performance in Film and Television, edited by Jeremy G. Butler, 1991

Sex in the Head: Visions of Femininity and Film in D.H. Lawrence, by Linda Ruth Williams, 1993

Dreams of Chaos, Visions of Order: Understanding the American Avant-garde Cinema, by James Peterson, 1994

Full of Secrets: Critical Approaches to Twin Peaks, edited by David Lavery, 1995

The Radical Faces of Godard and Bertolucci, by Yosefa Loshitzky, 1995

The End: Narration and Closure in the Cinema, by Richard Neupert, 1995

German Cinema: Texts in Context, by Marc Silberman, 1995

Cinemas of the Black Diaspora: Diversity, Dependence, and Oppositionality, edited by Michael T. Martin, 1995